These safety symbols are used in laboratory and field investigations in this book to indicate possible hazards. Learn the meaning of each symbol and refer to this page often. *Remember to wash your hands thoroughly after completing lab procedures.*

PROTECTIVE EQUIPMENT

Do not begin any lab without the proper protection equipment.

GOGGLES

Proper eye protection must be worn when performing or observing science activities that involve items or conditions as listed below.

APRON

Wear an approved apron when using substances that could stain, wet, or destroy cloth.

SOAP

Wash hands with soap and water before removing goggles and after all lab activities.

GLOVES

Wear gloves when working with biological materials, chemicals, animals, or materials that can stain or irritate hands.

LABORATORY HAZARDS

Symbols	Potential Hazards	Precaution	Response
DISPOSAL	contamination of classroom or environment due to improper disposal of materials such as chemicals and live specimens	• DO NOT dispose of hazardous materials in the sink or trash can. • Dispose of wastes as directed by your teacher.	• If hazardous materials are disposed of improperly, notify your teacher immediately.
EXTREME TEMPERATURE	skin burns due to extremely hot or cold materials such as hot glass, liquids, or metals; liquid nitrogen; dry ice	• Use proper protective equipment, such as hot mitts and/or tongs, when handling objects with extreme temperatures.	• If injury occurs, notify your teacher immediately.
SHARP OBJECTS	punctures or cuts from sharp objects such as razor blades, pins, scalpels, and broken glass	• Handle glassware carefully to avoid breakage. • Walk with sharp objects pointed downward, away from you and others.	• If broken glass or injury occurs, notify your teacher immediately.
ELECTRICAL	electric shock or skin burn due to improper grounding, short circuits, liquid spills, or exposed wires	• Check condition of wires and apparatus for fraying or uninsulated wires, and broken or cracked equipment. • Use only GFCI-protected outlets	• DO NOT attempt to fix electrical problems. Notify your teacher immediately.
CHEMICAL	skin irritation or burns, breathing difficulty, and/or poisoning due to touching, swallowing, or inhalation of chemicals such as acids, bases, bleach, metal compounds, iodine, poinsettias, pollen, ammonia, acetone, nail polish remover, heated chemicals, mothballs, and any other chemicals labeled or known to be dangerous	• Wear proper protective equipment such as goggles, apron, and gloves when using chemicals. • Ensure proper room ventilation or use a fume hood when using materials that produce fumes. • NEVER smell fumes directly. • NEVER taste or eat any material in the laboratory.	• If contact occurs, immediately flush affected area with water and notify your teacher. • If a spill occurs, leave the area immediately and notify your teacher.
FLAMMABLE	unexpected fire due to liquids or gases that ignite easily such as rubbing alcohol	• Avoid open flames, sparks, or heat when flammable liquids are present.	• If a fire occurs, leave the area immediately and notify your teacher.
OPEN FLAME	burns or fire due to open flame from matches, Bunsen burners, or burning materials	• Tie back loose hair and clothing. • Keep flame away from all materials. • Follow teacher instructions when lighting and extinguishing flames. • Use proper protection, such as hot mitts or tongs, when handling hot objects.	• If a fire occurs, leave the area immediately and notify your teacher.
ANIMAL SAFETY	injury to or from laboratory animals	• Wear proper protective equipment such as gloves, apron, and goggles when working with animals. • Wash hands after handling animals.	• If injury occurs, notify your teacher immediately.
BIOLOGICAL	infection or adverse reaction due to contact with organisms such as bacteria, fungi, and biological materials such as blood, animal or plant materials	• Wear proper protective equipment such as gloves, goggles, and apron when working with biological materials. • Avoid skin contact with an organism or any part of the organism. • Wash hands after handling o	• If contact occurs, wash the affected area and notify your teacher immediately.
FUME	breathing difficulties from inhalation of fumes from substances such as ammonia, acetone, nail polish remover, heated chemicals, and mothballs	• Wear goggles, apron, and glo • Ensure proper room ventilati when using substances that • NEVER smell fumes directly.	e area ier
IRRITANT	irritation of skin, mucous membranes, or respiratory tract due to materials such as acids, bases, bleach, pollen, mothballs, steel wool, and potassium permanganate	• Wear goggles, apron, and glo • Wear a dust mask to protect	...the affected area with water and notify your teacher.
RADIOACTIVE	excessive exposure from alpha, beta, and gamma particles	• Remove gloves and wash hands with soap and water before removing remainder of protective equipment.	• If cracks or holes are found in the container, notify your teacher immediately.

HUMAN BODY SYSTEMS

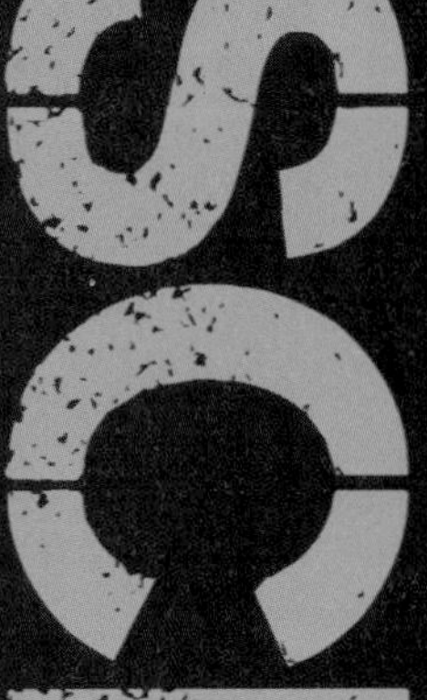

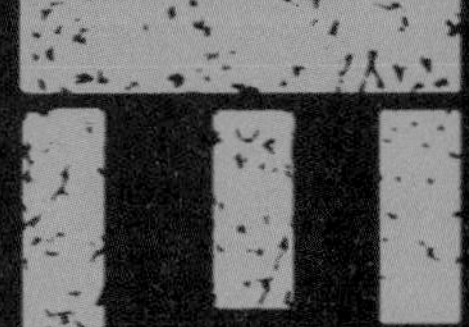

Glencoe

Your online portal to everything you need

connectED.mcgraw-hill.com

Look for these icons to access exciting digital resources

Inquiry

WebQuest

Assessment

Concepts in Motion

Mc Graw Hill Education

Snow Leopard, ***Uncia uncia***

The snow leopard lives in central Asia at altitudes of 3,000 m–5,500 m. Its thick fur and broad, furry feet are two of its adaptations that make it well suited to a snowy environment. Snow leopards cannot roar but can hiss, growl, and make other sounds.

The McGraw·Hill Companies

 Education

Send all inquiries to:
McGraw-Hill Education
8787 Orion Place
Columbus, OH 43240-4027

ISBN: 978-0-07-888017-9
MHID: 0-07-888017-3

Printed in the United States of America.

3 4 5 6 7 8 9 10 11 DOW 15 14 13 12

Authors and Contributors

Authors

American Museum of Natural History
New York, NY

Michelle Anderson, MS
Lecturer
The Ohio State University
Columbus, OH

Juli Berwald, PhD
Science Writer
Austin, TX

John F. Bolzan, PhD
Science Writer
Columbus, OH

Rachel Clark, MS
Science Writer
Moscow, ID

Patricia Craig, MS
Science Writer
Bozeman, MT

Randall Frost, PhD
Science Writer
Pleasanton, CA

Lisa S. Gardiner, PhD
Science Writer
Denver, CO

Jennifer Gonya, PhD
The Ohio State University
Columbus, OH

Mary Ann Grobbel, MD
Science Writer
Grand Rapids, MI

Whitney Crispen Hagins, MA, MAT
Biology Teacher
Lexington High School
Lexington, MA

Carole Holmberg, BS
Planetarium Director
Calusa Nature Center and Planetarium, Inc.
Fort Myers, FL

Tina C. Hopper
Science Writer
Rockwall, TX

Jonathan D. W. Kahl, PhD
Professor of Atmospheric Science
University of Wisconsin-Milwaukee
Milwaukee, WI

Nanette Kalis
Science Writer
Athens, OH

S. Page Keeley, MEd
Maine Mathematics and Science Alliance
Augusta, ME

Cindy Klevickis, PhD
Professor of Integrated Science and Technology
James Madison University
Harrisonburg, VA

Kimberly Fekany Lee, PhD
Science Writer
La Grange, IL

Michael Manga, PhD
Professor
University of California, Berkeley
Berkeley, CA

Devi Ried Mathieu
Science Writer
Sebastopol, CA

Elizabeth A. Nagy-Shadman, PhD
Geology Professor
Pasadena City College
Pasadena, CA

William D. Rogers, DA
Professor of Biology
Ball State University
Muncie, IN

Donna L. Ross, PhD
Associate Professor
San Diego State University
San Diego, CA

Marion B. Sewer, PhD
Assistant Professor
School of Biology
Georgia Institute of Technology
Atlanta, GA

Julia Meyer Sheets, PhD
Lecturer
School of Earth Sciences
The Ohio State University
Columbus, OH

Michael J. Singer, PhD
Professor of Soil Science
Department of Land, Air and Water Resources
University of California
Davis, CA

Karen S. Sottosanti, MA
Science Writer
Pickerington, Ohio

Paul K. Strode, PhD
I.B. Biology Teacher
Fairview High School
Boulder, CO

Jan M. Vermilye, PhD
Research Geologist
Seismo-Tectonic Reservoir Monitoring (STRM)
Boulder, CO

Judith A. Yero, MA
Director
Teacher's Mind Resources
Hamilton, MT

Dinah Zike, MEd
Author, Consultant,
Inventor of Foldables
Dinah Zike Academy;
Dinah-Might Adventures, LP
San Antonio, TX

Margaret Zorn, MS
Science Writer
Yorktown, VA

Consulting Authors

Alton L. Biggs
Biggs Educational Consulting
Commerce, TX

Ralph M. Feather, Jr., PhD
Assistant Professor
Department of Educational Studies and Secondary Education
Bloomsburg University
Bloomsburg, PA

Douglas Fisher, PhD
Professor of Teacher Education
San Diego State University
San Diego, CA

Edward P. Ortleb
Science/Safety Consultant
St. Louis, MO

Series Consultants

Science

Solomon Bililign, PhD
Professor
Department of Physics
North Carolina Agricultural and Technical State University
Greensboro, NC

John Choinski
Professor
Department of Biology
University of Central Arkansas
Conway, AR

Anastasia Chopelas, PhD
Research Professor
Department of Earth and Space Sciences
UCLA
Los Angeles, CA

David T. Crowther, PhD
Professor of Science Education
University of Nevada, Reno
Reno, NV

A. John Gatz
Professor of Zoology
Ohio Wesleyan University
Delaware, OH

Sarah Gille, PhD
Professor
University of California San Diego
La Jolla, CA

David G. Haase, PhD
Professor of Physics
North Carolina State University
Raleigh, NC

Janet S. Herman, PhD
Professor
Department of Environmental Sciences
University of Virginia
Charlottesville, VA

David T. Ho, PhD
Associate Professor
Department of Oceanography
University of Hawaii
Honolulu, HI

Ruth Howes, PhD
Professor of Physics
Marquette University
Milwaukee, WI

Jose Miguel Hurtado, Jr., PhD
Associate Professor
Department of Geological Sciences
University of Texas at El Paso
El Paso, TX

Monika Kress, PhD
Assistant Professor
San Jose State University
San Jose, CA

Mark E. Lee, PhD
Associate Chair & Assistant Professor
Department of Biology
Spelman College
Atlanta, GA

Linda Lundgren
Science writer
Lakewood, CO

Series Consultants, continued

Keith O. Mann, PhD
Ohio Wesleyan University
Delaware, OH

Charles W. McLaughlin, PhD
Adjunct Professor of Chemistry
Montana State University
Bozeman, MT

Katharina Pahnke, PhD
Research Professor
Department of Geology and Geophysics
University of Hawaii
Honolulu, HI

Jesús Pando, PhD
Associate Professor
DePaul University
Chicago, IL

Hay-Oak Park, PhD
Associate Professor
Department of Molecular Genetics
Ohio State University
Columbus, OH

David A. Rubin, PhD
Associate Professor of Physiology
School of Biological Sciences
Illinois State University
Normal, IL

Toni D. Sauncy
Assistant Professor of Physics
Department of Physics
Angelo State University
San Angelo, TX

Malathi Srivatsan, PhD
Associate Professor of Neurobiology
College of Sciences and Mathematics
Arkansas State University
Jonesboro, AR

Cheryl Wistrom, PhD
Associate Professor of Chemistry
Saint Joseph's College
Rensselaer, IN

Reading

ReLeah Cossett Lent
Author/Educational Consultant
Blue Ridge, GA

Math

Vik Hovsepian
Professor of Mathematics
Rio Hondo College
Whittier, CA

Series Reviewers

Thad Boggs
Mandarin High School
Jacksonville, FL

Catherine Butcher
Webster Junior High School
Minden, LA

Erin Darichuk
West Frederick Middle School
Frederick, MD

Joanne Hedrick Davis
Murphy High School
Murphy, NC

Anthony J. DiSipio, Jr.
Octorara Middle School
Atglen, PA

Adrienne Elder
Tulsa Public Schools
Tulsa, OK

Series Reviewers, continued

Carolyn Elliott
Iredell-Statesville Schools
Statesville, NC

Christine M. Jacobs
Ranger Middle School
Murphy, NC

Jason O. L. Johnson
Thurmont Middle School
Thurmont, MD

Felecia Joiner
Stony Point Ninth Grade Center
Round Rock, TX

Joseph L. Kowalski, MS
Lamar Academy
McAllen, TX

Brian McClain
Amos P. Godby High School
Tallahassee, FL

Von W. Mosser
Thurmont Middle School
Thurmont, MD

Ashlea Peterson
Heritage Intermediate Grade Center
Coweta, OK

Nicole Lenihan Rhoades
Walkersville Middle School
Walkersvillle, MD

Maria A. Rozenberg
Indian Ridge Middle School
Davie, FL

Barb Seymour
Westridge Middle School
Overland Park, KS

Ginger Shirley
Our Lady of Providence
Junior-Senior High School
Clarksville, IN

Curtis Smith
Elmwood Middle School
Rogers, AR

Sheila Smith
Jackson Public School
Jackson, MS

Sabra Soileau
Moss Bluff Middle School
Lake Charles, LA

Tony Spoores
Switzerland County Middle School
Vevay, IN

Nancy A. Stearns
Switzerland County Middle School
Vevay, IN

Kari Vogel
Princeton Middle School
Princeton, MN

Alison Welch
Wm. D. Slider Middle School
El Paso, TX

Linda Workman
Parkway Northeast Middle School
Creve Coeur, MO

Teacher Advisory Board

The Teacher Advisory Board gave the authors, editorial staff, and design team feedback on the content and design of the Student Edition. They provided valuable input in the development of *Glencoe ⓘScience.*

Frances J. Baldridge
Department Chair
Ferguson Middle School
Beavercreek, OH

Jane E. M. Buckingham
Teacher
Crispus Attucks Medical
Magnet High School
Indianapolis, IN

Elizabeth Falls
Teacher
Blalack Middle School
Carrollton, TX

Nelson Farrier
Teacher
Hamlin Middle School
Springfield, OR

Michelle R. Foster
Department Chair
Wayland Union
Middle School
Wayland, MI

Rebecca Goodell
Teacher
Reedy Creek Middle School
Cary, NC

Mary Gromko
Science Supervisor K–12
Colorado Springs District 11
Colorado Springs, CO

Randy Mousley
Department Chair
Dean Ray Stucky
Middle School
Wichita, KS

David Rodriguez
Teacher
Swift Creek Middle School
Tallahassee, FL

Derek Shook
Teacher
Floyd Middle Magnet School
Montgomery, AL

Karen Stratton
Science Coordinator
Lexington School District One
Lexington, SC

Stephanie Wood
Science Curriculum Specialist,
K–12
Granite School District
Salt Lake City, UT

Online Guide

connectED.mcgraw-hill.com

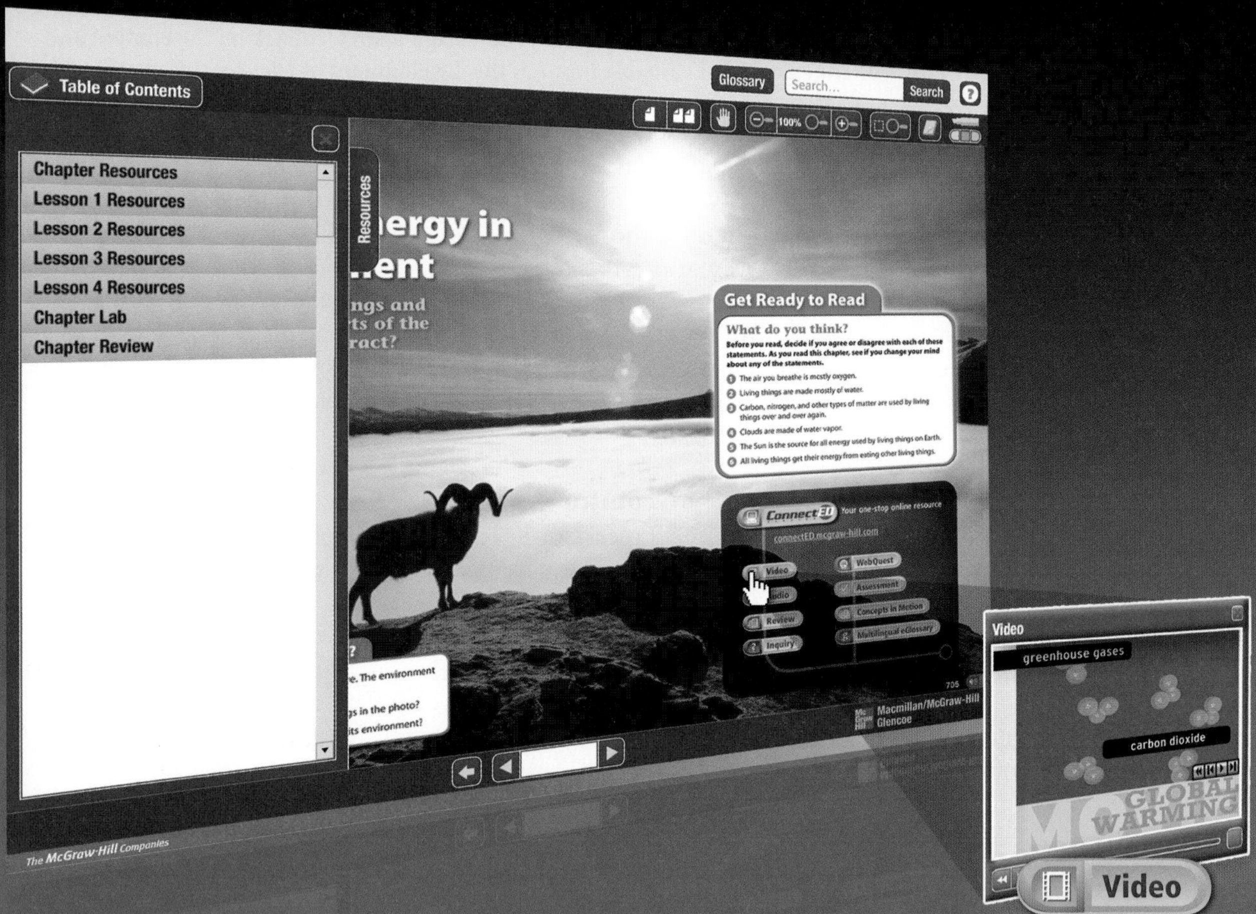

Your Digital Science Portal

See the science in real life through these exciting

Click the link and you can listen to the text while you

Try these interactive tools to help you review

Explore concepts through hands–on and virtual labs.

These web-based challenges relate the concepts you're learning

The icons in your online student edition link you to interactive learning opportunities. Browse your online student book to find more.

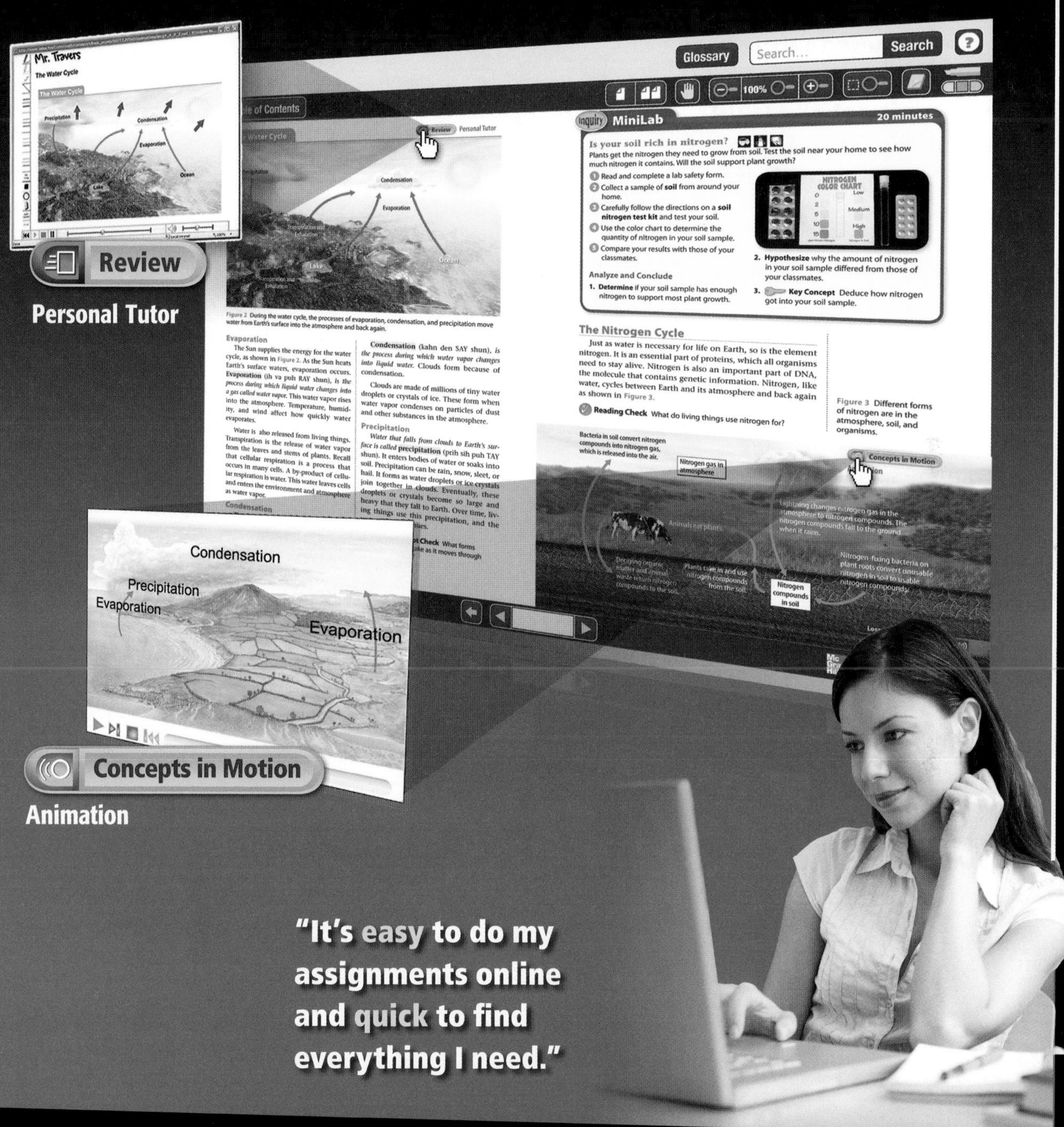

"It's easy to do my assignments online and quick to find everything I need."

Assessment

Check how well you understand the concepts with online quizzes and practice questions.

Concepts in Motion

The textbook comes alive with animated explanations of important concepts.

Multilingual eGlossary

Read key vocabulary in 13 languages.

Treasure Hunt

Your science book has many features that will aid you in your learning. Some of these features are listed below. You can use the activity at the right to help you find these and other special features in the book.

- THE BIG IDEA can be found at the start of each chapter.
- The Reading Guide at the start of each lesson lists **Key Concepts,** vocabulary terms, and online supplements to the content.
- ConnectED icons direct you to online resources such as animations, personal tutors, math practices, and quizzes.
- Inquiry Labs and Skill Practices are in each chapter.
- Your FOLDABLES® help organize your notes.

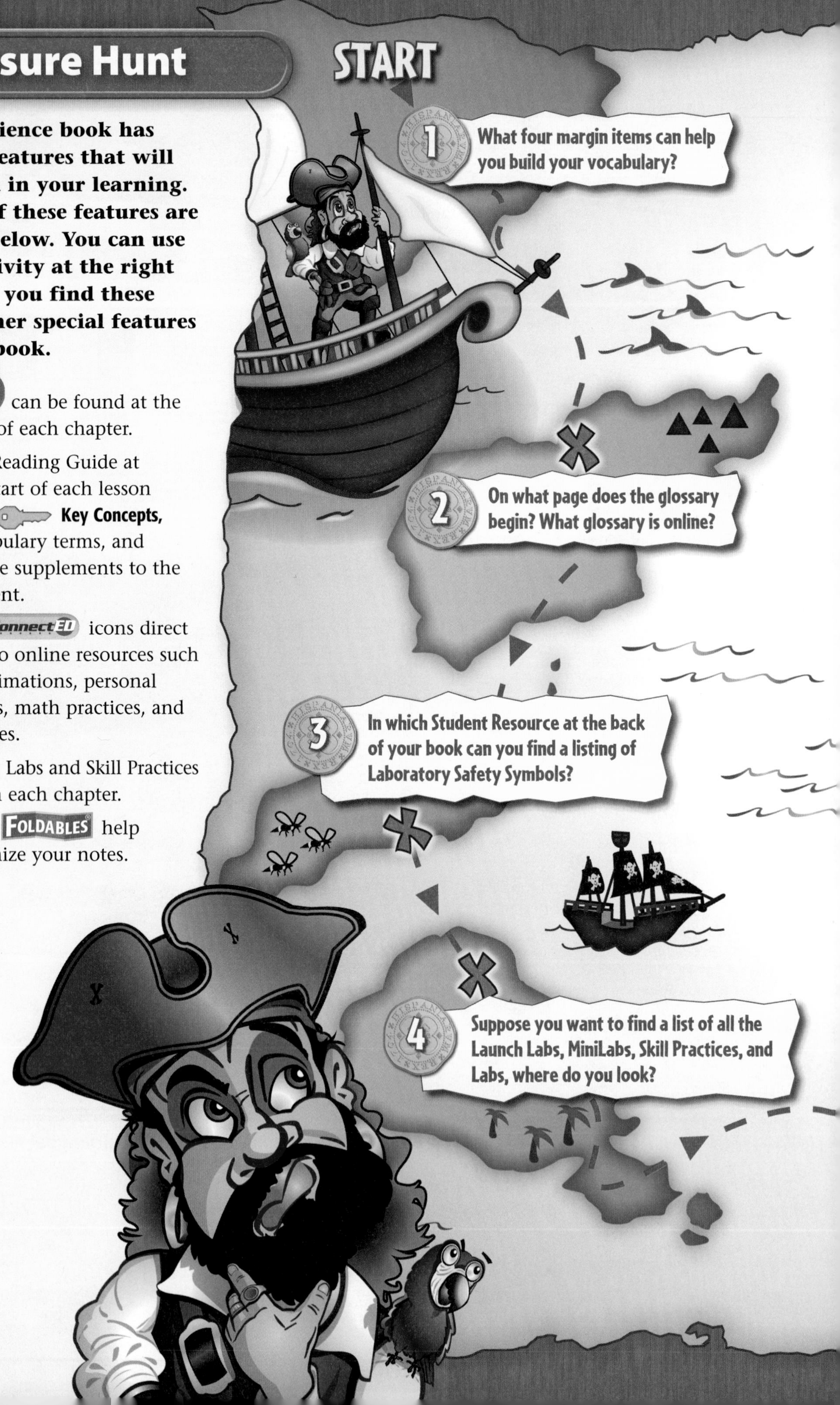

START

1. What four margin items can help you build your vocabulary?

2. On what page does the glossary begin? What glossary is online?

3. In which Student Resource at the back of your book can you find a listing of Laboratory Safety Symbols?

4. Suppose you want to find a list of all the Launch Labs, MiniLabs, Skill Practices, and Labs, where do you look?

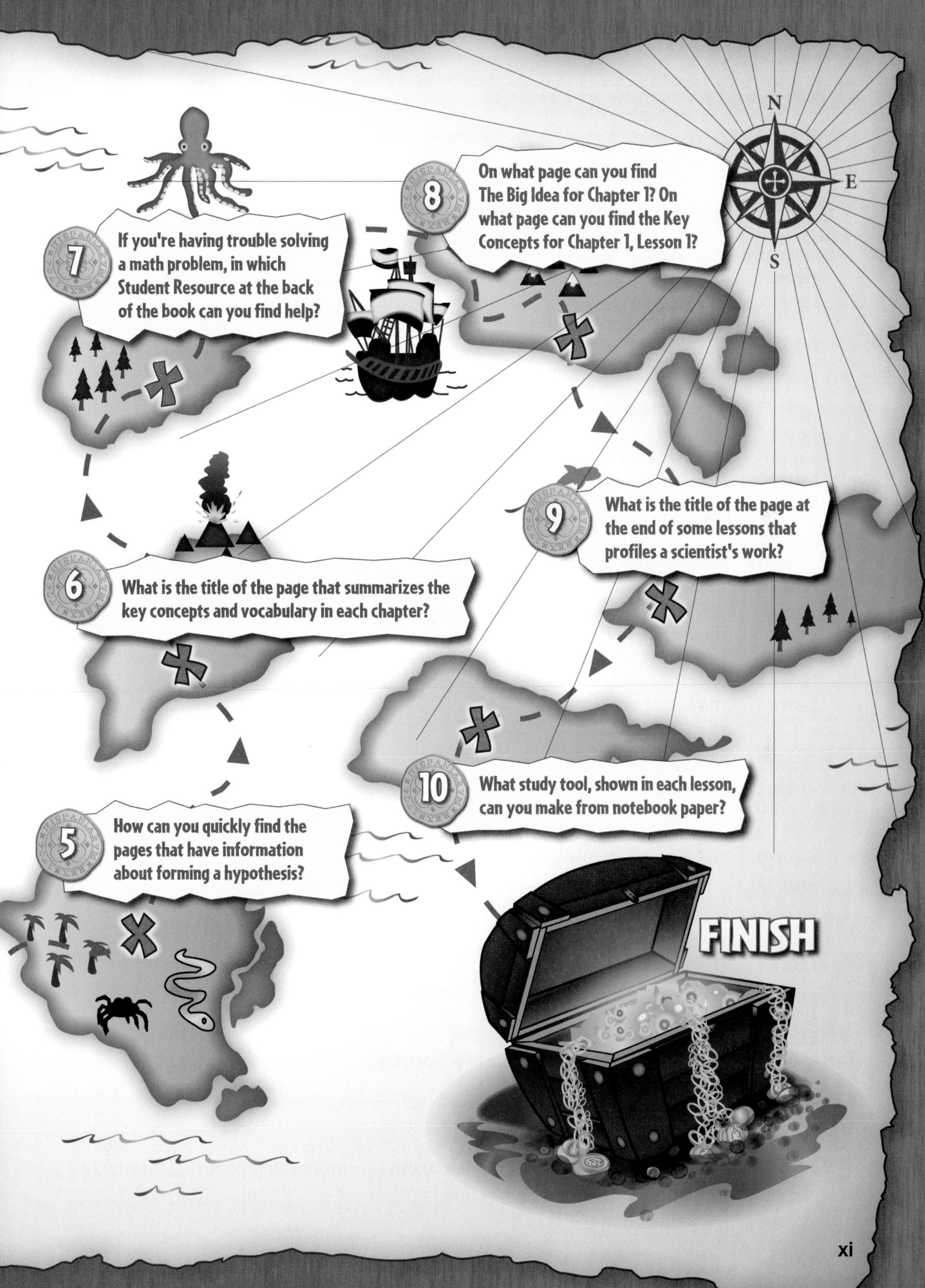
N
E
S
8
On what page can you find The Big Idea for Chapter 1? On what page can you find the Key Concepts for Chapter 1, Lesson 1?
7
If you're having trouble solving a math problem, in which Student Resource at the back of the book can you find help?
9
What is the title of the page at the end of some lessons that profiles a scientist's work?
6
What is the title of the page that summarizes the key concepts and vocabulary in each chapter?
10
What study tool, shown in each lesson, can you make from notebook paper?
5
How can you quickly find the pages that have information about forming a hypothesis?
FINISH

Table of Contents

Student Resources

Inquiry

TABLE OF CONTENTS

Inquiry Launch Labs

Inquiry MiniLabs

Inquiry Skill Practice

Inquiry Labs

Features

How It Works

Science & Society

Unit 4

Human Body Systems

1800 · 1850 · 1900 · 1950

1818
The first well-documented case of a person-to-person blood transfusion is performed. British obstetrician James Blundell transfuses 4 oz. of blood from a man to his wife who had just given birth.

1823
German surgeon Christian Bünger performs the first autograft, replacing skin on a man's nose with some from his thigh.

1905
Dr. Eduard Zirm performs the first successful cornea transplant on patient Alois Glogar.

1954
The first successful organ transplant between living relatives (a kidney transplant between twins) is completed by Dr. Joseph Murray and Dr. David Hume in Boston.

1967
The first successful liver and heart transplants are completed. Dr. Thomas Starzl performs the liver transplant in Denver, Colorado. Dr. Christiaan Barnard performs the heart transplant in Cape Town, South Africa.

Turn the dial a bit, and we see the respiratory and digestive systems in great detail!

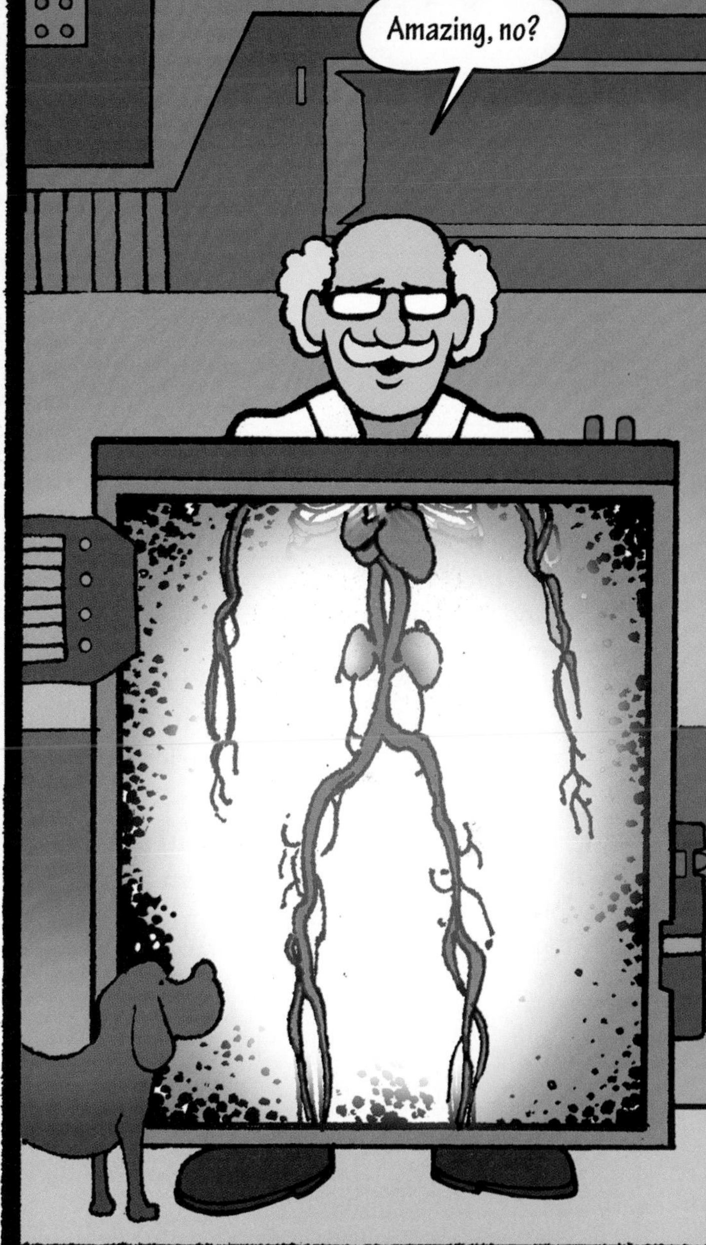

1960 1980 2000

1968
The Uniform Anatomical Gift Act establishes the Uniform Donor Card as a legal document that allows anyone 18 years of age or older to legally donate their organs upon death.

1984
The National Organ Transplant Act (NOTA) establishes a nationwide computer registry, authorizes financial support for organ procurement organizations, and prohibits the buying or selling of organs in the United States.

2001
Due to widespread use of advanced surgical techniques and higher success rates for surgeries, the number of living donors passes the number of deceased donors.

 **Inquiry**

Visit ConnectED for this unit's **STEM** activity.

Systems

A **system** is a collection of parts that influence or interact with one another. For example, the human body is a large system made up of many smaller subsystems, such as the ones shown in **Figure 1.**

Like the human body system, complex systems often contain smaller, or less complex, subsystems. The parts of each subsystem interact among themselves, as well as with other subsystems. Each subsystem has a different purpose, but they interact to keep the larger system working properly.

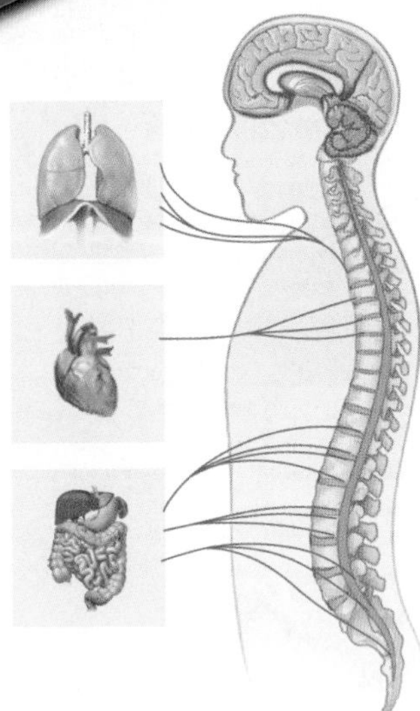

▲ **Figure 1** The nervous system, the respiratory system, the circulatory system, and the digestive system are subsystems of a larger system—the human body.

Parts of a System

Systems and subsystems often attempt to achieve a goal. For example, the nervous system, a subsystem of the human body system, regulates your body temperature, as shown in **Figure 2**. Systems and subsystems often are described in terms of their input, processing, and output.

Input is the matter, energy, or information that enters a system. When you exercise, one input to your nervous system is thermal energy. The input, or thermal energy, is detected by special brain and skin cells called receptors.

Processing is the changing of the input to achieve a goal. The hypothalamus processes the input from receptors. It sends electrical signals, carried by the nerves, to other parts of the body. The signals tell the body it is warmer than it should be.

Output is the material, energy, or information that leaves a system. Outputs from the nervous system include sweat, goose bumps, and shivers, all of which can change body temperature.

Figure 2 The nervous system is responsible for regulating body temperature. ▼

Input: Thermal energy released by contracting muscles is detected by receptors.

Processing: Signals from receptors are sent to the brain. The brain then signals glands in the skin to produce sweat.

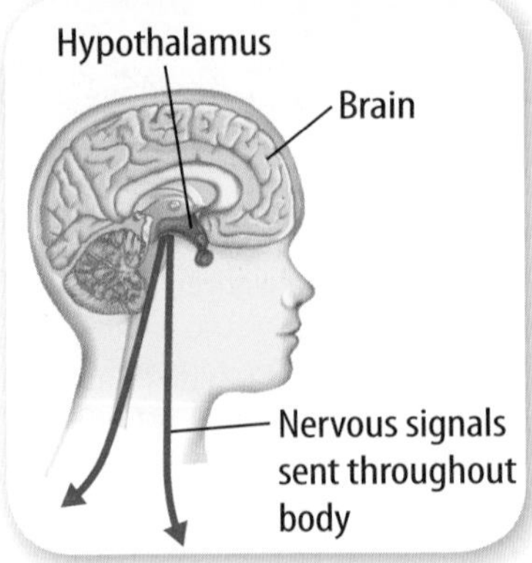

Output: Sweat forms on the skin. Then, it cools the body as it evaporates.

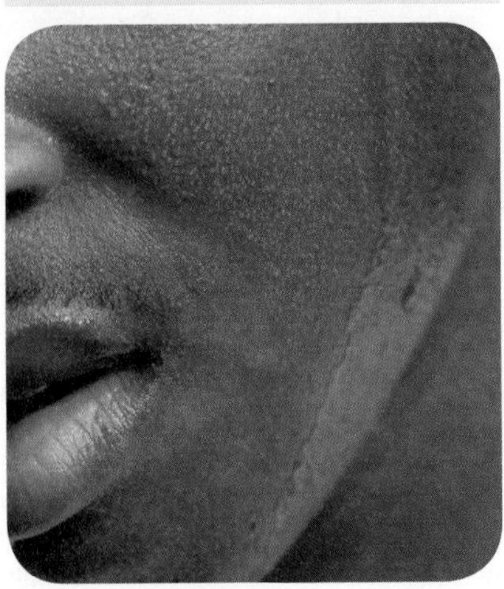

Figure 3 After a hard race, decreasing levels of carbon dioxide in a rower's blood act as feedback to the nervous system. This feedback signals the rower's brain to restore her breathing to a normal rate.

Feedback in Systems

Many systems use feedback, or information, to monitor and regulate input, process, and output. For example, when you exercise, as shown in **Figure 3,** your muscles produce carbon dioxide as a waste product. Receptors detect this input—high levels of carbon dioxide in your blood. The brain processes this information and signals your nervous system to increase breathing. When you breathe harder and faster, you take in more oxygen and the levels of carbon dioxide in your blood decrease. Once this change is detected by receptors, your brain signals your nervous system to return to normal breathing.

Cooperation, Order, and Change

Body subsystems work together in specific ways to regulate temperature, remove waste from your blood, and respond to other changes in your body. A failure in one subsystem affects other subsystems. For example, if a bad cold causes your lungs to become congested, your respiratory system cannot efficiently exchange oxygen for carbon dioxide. Therefore, muscle cells no longer receive enough oxygen to function normally. As a result, you easily become tired and have trouble catching your breath when you exercise.

Systems Thinking

Thinking in terms of systems might change the way you make choices. For example, some people think that if they reduce the amount they eat, they will lose weight. However, protein is necessary for your muscular system to function properly. Without an input of protein, muscle tissues begin to break down. Someone who does not eat enough protein might not lose weight because they become weak and tired and stop exercising. Thinking about the interactions of the systems and subsystems in your body can lead to decisions that help you achieve long-term goals.

Can you analyze a system?

The human body system has many subsystems that respond to environmental changes. How do these subsystems interact?

1. Identify an environmental change that you experience every day.
2. Research one or more subsystems of your body that regulate this change.
3. Design a chart to show the inputs, the processing, the outputs, and the feedback of one of your subsystems.
4. Identify and include another subsystem with which your subsystem interacts.

Analyze and Conclude

1. **Infer** Can a subsystem in the body operate without interacting with one or more other subsystems? Why or why not?
2. **Draw Conclusions** Why do injuries to the brain so often threaten a person's life?
3. **Demonstrate** Use a Venn diagram to show how the subsystem you chose interacts with two other systems in the body.

Chapter 14

Structure and Movement

How do bones, muscles, and skin help maintain the body's homeostasis?

Inquiry Why would this never work?

You could not play soccer if your body were made of only bones, only muscles, or only skin. Bones, muscles, and skin work together and help you move and carry out processes you need in order to live.

- What would happen to your body if you had no skin? No muscles?
- What functions do you think bones, muscles, and skin have?
- How do bones, muscles, and skin work together to help your body function?

Get Ready to Read

What do you think?

Before you read, decide if you agree or disagree with each of these statements. As you read this chapter, see if you change your mind about any of the statements.

1. Bones protect internal organs.
2. Bones do not change during a person's lifetime.
3. The same type of muscle that moves bones also pumps blood through the heart.
4. Muscles cannot push bones.
5. Skin helps regulate body temperature.
6. Skin is made of two layers of tissue.

Lesson 1

The Skeletal System

Reading Guide

Key Concepts
ESSENTIAL QUESTIONS

- What does the skeletal system do?
- How do the parts of the skeletal system work together?
- How does the skeletal system interact with other body systems?

Vocabulary

skeletal system p. 487
cartilage p. 490
periosteum p. 490
joint p. 491
ligament p. 491
arthritis p. 492
osteoporosis p. 492

Multilingual eGlossary

Video BrainPOP®

Inquiry How do bones move so easily?

Bones work like mechanical joints, enabling you to turn, pivot, twist, and bend. However, they don't do this alone. Other parts of your skeletal system work with the bones and help you move.

Launch Lab

10 minutes

How are bones used for support?

If you have ever watched the construction of a building, you might have seen a wood or steel frame being used to provide support. In a similar way, bones support your body and the organs inside it.

1. Read and complete a lab safety form.
2. Using pieces of **clay,** try to build a tower taller than your partner's. One person should use a **wooden dowel** to support his or her tower.
3. Measure and record the height of both towers. Find the class average for the height of towers with and without dowels.

Think About This

1. Were towers with or without dowels generally higher?
2. **Key Concept** What do you think your body would be like if you had no bones?

Functions of the Skeletal System

Squeeze your hands and arms. The hard parts you feel are parts of your skeleton. When you think of your skeleton, you might think of bones, like those shown in **Figure 1.** These bones support your body and help you move. Your skeleton is part of your skeletal system and contains more than 200 bones. *The* **skeletal system** *contains bones as well as other structures that connect and protect the bones and that support other functions in the body.* In this lesson, you will learn how all parts of the skeletal system work together.

Support

Can you imagine trying to stack jiggling blocks of gelatin 1 m high? You would probably have a hard time because gelatin does not have any support structures inside it. Without bones, your body would be similar to gelatin. Bones provide support. They help you sit up, stand, and raise your arm over your head to ask a question.

Reading Check How do bones act as a support system?

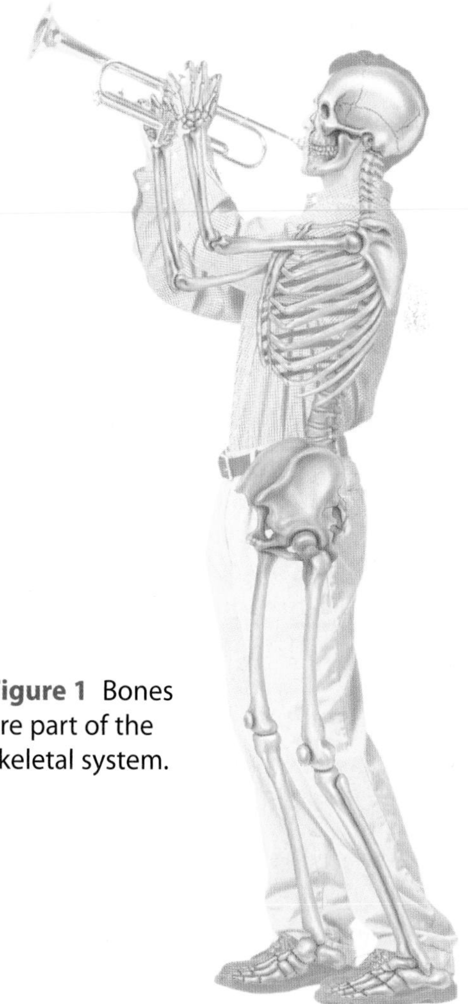

Figure 1 Bones are part of the skeletal system.

Science Use v. Common Use

tissue

Science Use a group of the same type of cells that perform a specific function within an organism

Common Use a soft, thin paper

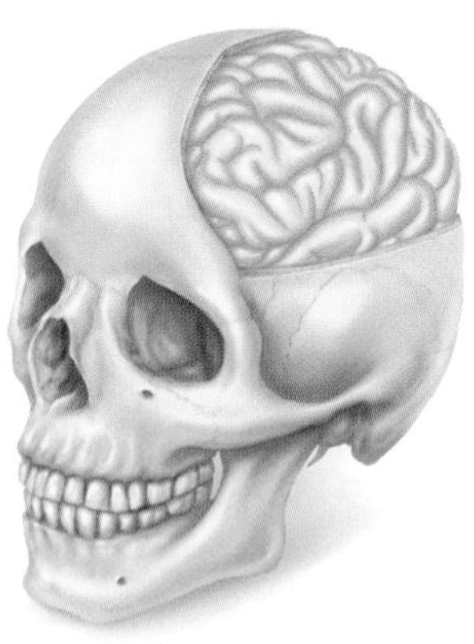

Figure 2 The skull protects the soft tissue of the brain.

Movement

The skeletal system enables different parts of the body to move in different ways, like when a person kicks a soccer ball, as shown in the beginning of this chapter. Bones can move because they are attached to muscles. The skeletal system and the muscular system work together and move your body.

Protection

Feel your head and then feel your stomach. Your stomach is softer than your head. The hard, rigid structure you feel in your head is your skull, shown in **Figure 2.** It protects the soft, fragile **tissue** of your brain from damage. Other bones protect the spinal cord, heart, lungs, and other internal organs.

Production and Storage

Another function of bones is to produce and store materials needed by your body. Red blood cells are produced inside your bones. Bones store fat and calcium. Calcium is needed for strong bones and for many cellular processes. When the body needs calcium, it is released from bones into the blood.

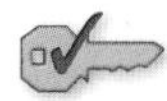 **Key Concept Check** What are the major functions of the skeletal system?

Inquiry MiniLab

15 minutes

How does the skeleton protect organs?

The hard bones of your skeleton protect some of the soft tissues and organs of your body.

1. Read and complete a lab safety form.
2. Place one piece of **bubble wrap** in a **plastic bag** and another piece of bubble wrap in a **plastic jar.**
3. Firmly squeeze the plastic bag for five seconds, popping as many bubbles as you can. Remove the bubble wrap and count how many bubbles are popped. Record the data in your Science Journal.
4. Repeat step 3 with the bubble wrap in the plastic jar.

Analyze and Conclude

1. **Analyze** your data to determine which container provided more protection for the bubble wrap. Explain your answer.
2. **Infer** which part of your skeletal system the plastic jar models in this experiment.
3. 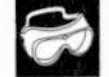 **Key Concept** How does the skeleton provide protection for internal organs?

Structure of Bones

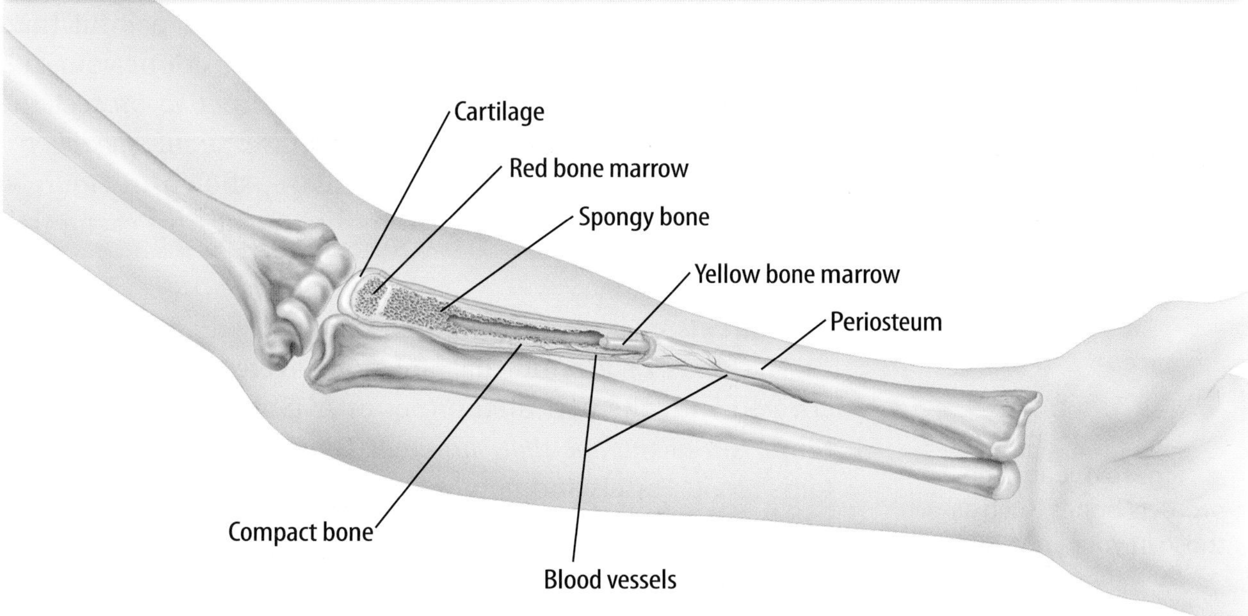

Figure 3 A bone contains many types of tissue.

Structure of Bones

A bone is an organ composed of living tissue. There are two main types of bone tissue: compact and spongy. A bone also contains other types of tissue, as you can see in the arm bone shown in **Figure 3.**

Compact Bone Tissue

The hard, outer portions of bones are made of compact bone tissue. Compact bone tissue is a dense web of fibers. A long bone, like the arm bone shown in **Figure 3,** is made mostly of compact bone tissue. However, in its ends, a long bone contains a different kind of tissue called spongy bone tissue.

Spongy Bone Tissue

The small holes in spongy bone tissue make it look like a sponge. Because of these holes, spongy bone is less dense than compact bone. A short bone, like one in your wrist, is mostly spongy bone tissue.

Bone Marrow

The insides of most bones contain a soft tissue called bone marrow (MER oh). There are two types of bone marrow. Red bone marrow is the tissue where red blood cells are made. It is found in the spongy ends of long bones and in some flat bones, such as the ribs. Yellow bone marrow stores fat and is found inside the longest part of long bones.

Reading Check What is the difference between red bone marrow and yellow bone marrow?

FOLDABLES®

Fold a sheet of paper in half to make a horizontal half book. Use your book to illustrate and describe the structure of bones.

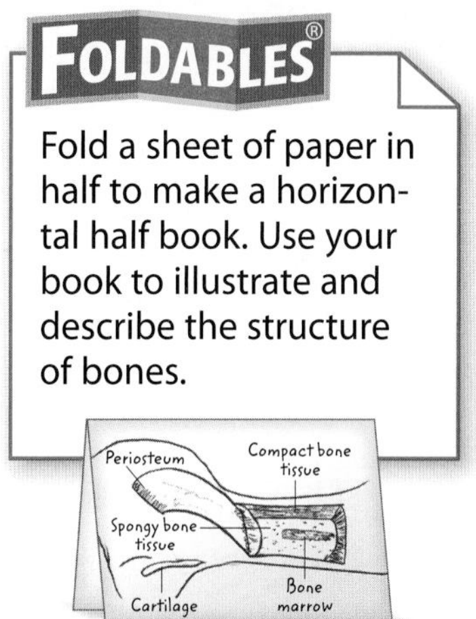

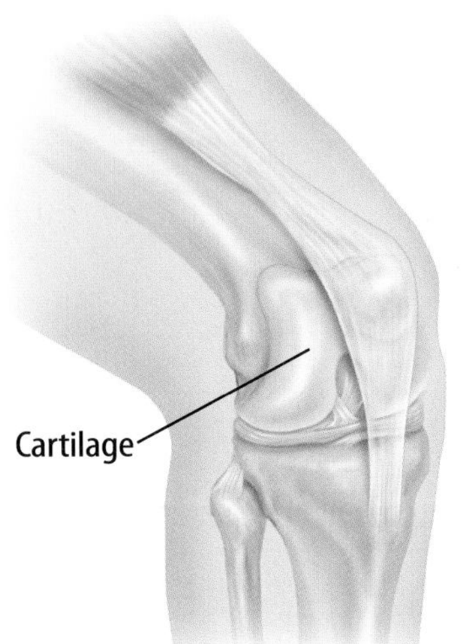

▲ **Figure 4** Cartilage protects bones, such as those in your knee joint.

Cartilage

Have you ever fallen on a hard floor made of tiles or concrete? Falling on a hard surface is usually more painful than falling on a soft surface like a carpet, which cushions your fall. A special tissue in your body acts like a carpet to protect the skeletal system. **Cartilage** (KAR tuh lihj) *is a strong, flexible tissue that covers the ends of bones.* Cartilage, shown in **Figure 4,** prevents the surfaces of bones from rubbing against each other and reduces friction.

Periosteum

The parts of a bone that are not covered in cartilage are covered with the periosteum (per ee AHS tee um). *The* **periosteum** *is a membrane that surrounds bone.* This thin tissue contains blood vessels and nerves as well as cells that produce new bone tissue. Periosteum nourishes bones and helps them function and grow properly, as well as heal after injury.

WORD ORIGIN

ligament
from Latin *ligare,* means "to bind, tie"

Formation of Bones

Before you were born, your skeleton was made mostly of cartilage. During your infancy and childhood, the cartilage was gradually replaced by bone, as shown in **Figure 5.** The long bones in children and young teens have regions of bone growth that produce new bone cells. These regions are called growth plates. A growth plate produces cartilage that is then replaced by bone tissue. A growth plate, shown in **Figure 5,** is the weakest part of an adolescent bone. Growth continues until adulthood, when most of the cartilage has turned to bone.

Figure 5 Bone gradually replaces cartilage as children grow. ▼

Bone Development

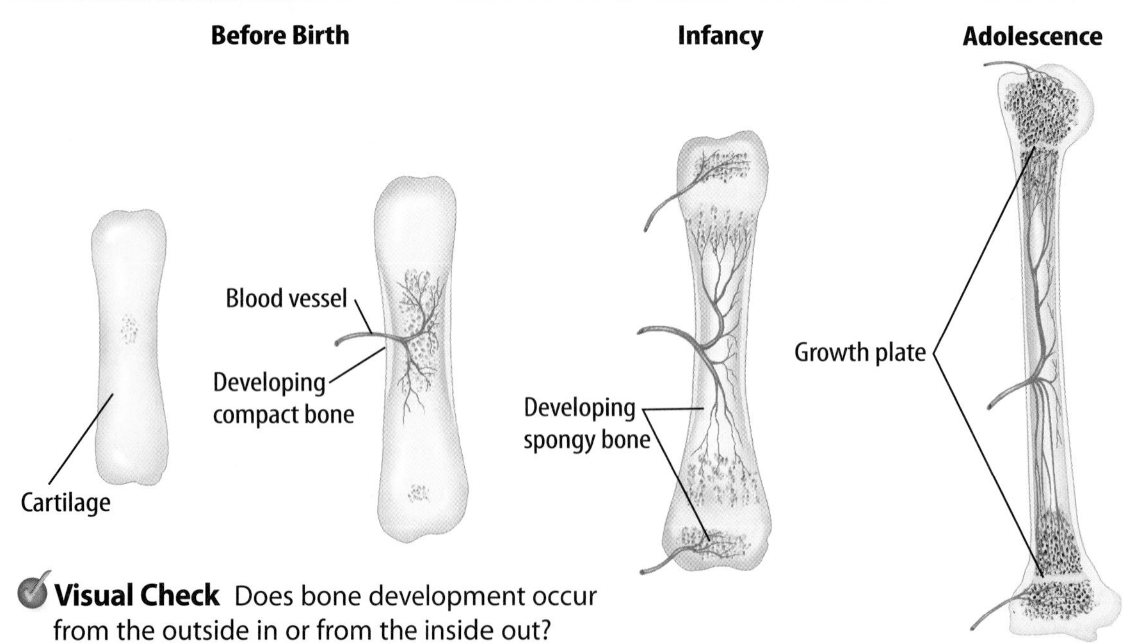

Visual Check Does bone development occur from the outside in or from the inside out?

Joints

Your bones do not function independently. They work together at places called joints. *A* **joint** *is where two or more bones meet.* Joints provide flexibility and enable the skeleton to move. *Bones are connected to other bones by tissues called* **ligaments** (LIH guh munts). When the bones in joints move, ligaments stretch and keep the bones from shifting away from each other.

Ligaments connect bones at joints, but they do not protect bones. As you read earlier, cartilage protects the ends of bones, reducing friction between them so the bones move easily. Your skeletal system contains two types of joints—immovable joints and movable joints.

Immovable Joints

Some parts of your skeleton are made of bones that connect but do not move. These are immovable joints. Your skull contains several immovable joints.

Movable Joints

You are able to move your hand or bend your body because of movable joints. The body's movable joints allow for a wide range of motion. Three main types of movable joints and the ligaments that hold them together are shown in **Table 1.**

Key Concept Check How do ligaments and cartilage help the skeletal system function?

Table 1 Types of Movable Joints Concepts in Motion Interactive Table

Joint	Description	Example
Ball and socket (Ligaments)	allows bones to move and rotate in nearly all directions	hips and shoulders
Hinge	allows bones to move back and forth in a single direction	fingers, elbows, knees
Pivot	allows bones to rotate	neck, lower arm below the elbow

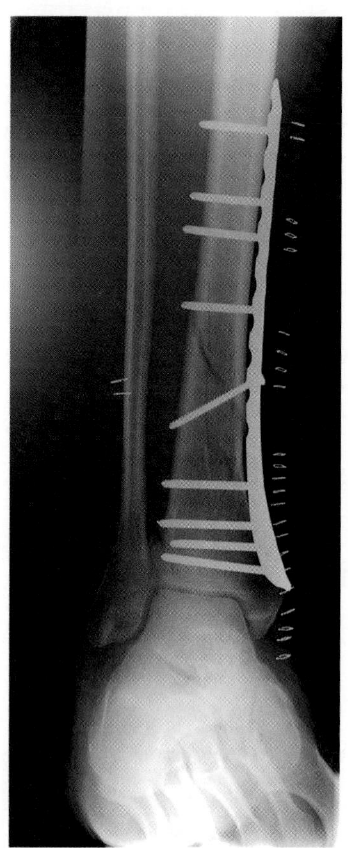

▲ **Figure 6** This X-ray was taken after screws and plates were placed in the bone to hold it together.

Bone Injuries and Diseases

Because bones are made of living tissues, they are at risk of injury and disease. A wooden board is hard and strong, but if enough pressure is applied to it, the board will break. The same is true for bones.

Broken Bones

A break in a bone is called a fracture (FRAK chur). Broken bones are able to repair themselves, but it is a slow process. A broken bone must be held together while it heals, just as you hold two glued objects together while the glue dries. Often, a person wears a cast to keep broken bones in place. Sometimes metal plates and screws like those in **Figure 6** hold bones together while they heal.

Arthritis

You are able to move because your skeleton bends and rotates at joints. If the joints become irritated, it can be painful to move. **Arthritis** (ar THRI tus) *is a disease in which joints become irritated or inflamed, such as when cartilage in joints is damaged or wears away.* Arthritis is most common in adults, but it can also affect children.

Osteoporosis

Another common bone disease is **osteoporosis** (ahs tee oh puh ROH sus), *which causes bones to weaken and become brittle.* Anyone can develop osteoporosis, but it is most common in women over the age of 50. Osteoporosis can change a person's skeleton and cause fractures, as shown in **Figure 7.**

Osteoporosis

Figure 7 Osteoporosis can weaken the skeletal system over time.

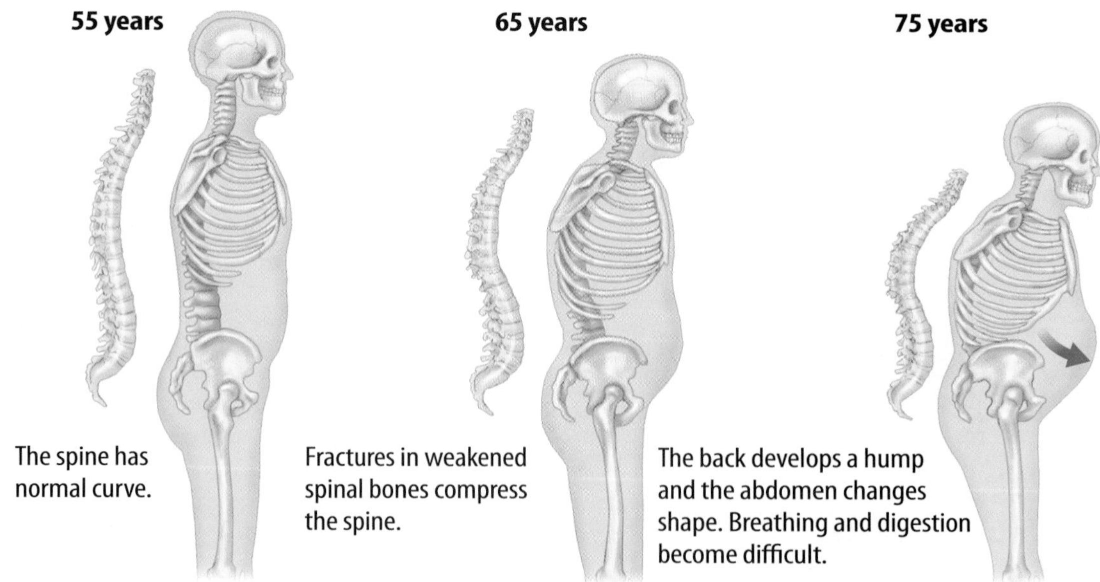

Figure 8 Exercises that place weight on bones or work bones against gravity help keep the skeletal system healthy.

Healthy Bones

One of the best ways to keep bones healthy is to exercise. Certain types of exercise, such as those shown in **Figure 8,** place weight on bones. This strengthens bones and builds new bone tissue. Without exercise, bones weaken and lose mass.

A balanced diet also keeps bones healthy. Bones especially need calcium and vitamin D. Calcium makes bones strong. It is also necessary for cellular processes in blood, nerve, and muscle cells. If you do not have enough calcium in your diet, your body will use the calcium stored in your bones. If your body uses too much of the stored calcium, your bones can become weak. Vitamin D is also important because it helps the body use calcium.

The Skeletal System and Homeostasis

You might recall that **homeostasis** is an organism's ability to maintain a stable internal environment. Homeostasis requires that all body systems function properly together. Because bones supply calcium to your nerves, muscles, and heart, a healthy skeletal system is important in maintaining your body's homeostasis.

REVIEW VOCABULARY
homeostasis
the ability of an organism to maintain a stable internal environment

Bones also help you respond to unpleasant stimuli, such as a mosquito bite. Working together with muscles, bones enable you to move away from unpleasant stimuli or danger, or even swat a mosquito.

 **Key Concept Check** How does the skeletal system help the body maintain homeostasis?

Lesson 1 Review

Assessment Online Quiz

Inquiry Virtual Lab

Visual Summary

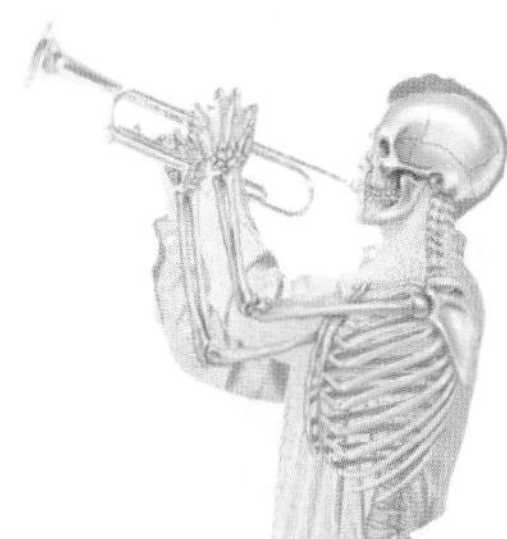

The skeletal system consists of all the bones, ligaments, and cartilage in the body.

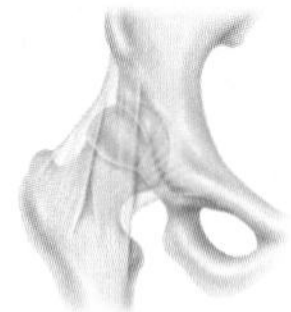

Joints provide flexibility and enable the skeleton to move.

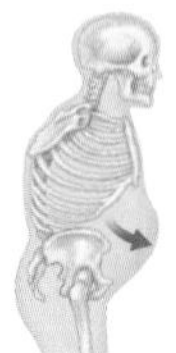

Osteoporosis is a disease that causes bones to weaken and become brittle.

FOLDABLES®

Use your lesson Foldable to review the lesson. Save your Foldable for the project at the end of the chapter.

Use Vocabulary

1. Strong, flexible tissue that covers the ends of bones is called ________.
2. Bones connect to other bones at ________.
3. A person with ________ has irritated joints.

Understand Key Concepts 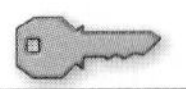

4. Which is NOT a part of the periosteum?
   - **A.** blood vessels
   - **B.** bone cells
   - **C.** bone marrow
   - **D.** nerves
5. **Distinguish** between cartilage and ligaments.
6. **Give** an example of how the skeletal system interacts with the nervous system.

Interpret Graphics

7. **Determine** the type of joint shown in the figure below.

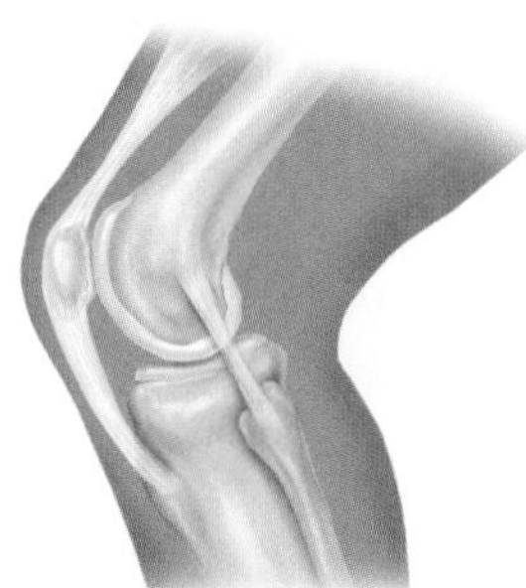

8. **Organize** Copy the graphic organizer below and, in the ovals, list the four major functions of the skeletal system.

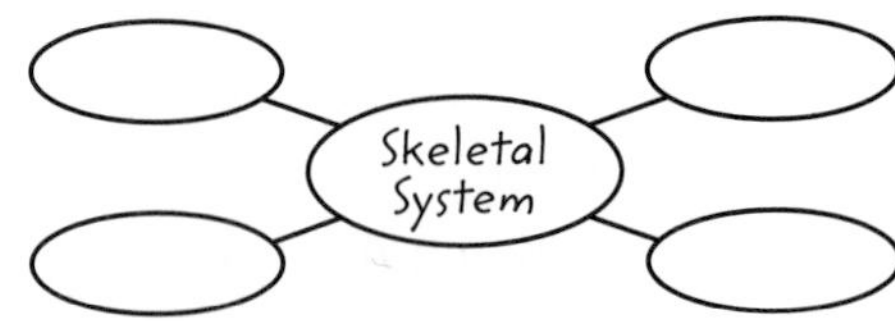

Critical Thinking

9. **Summarize** how bones help the body maintain homeostasis.
10. **Evaluate** why it is better for an adult's skull to have immovable joints and not movable joints.

What do you think NOW?

You first read the statements below at the beginning of the chapter.

1. Bones protect internal organs.
2. Bones do not change during a person's lifetime.

Did you change your mind about whether you agree or disagree with the statements? Rewrite any false statements to make them true.

A Bionic Arm

HOW IT WORKS

How brains control mechanical arms

Imagine what your arms would be like without muscles. They would simply swing from your shoulders like pendulums. You would not be able to control them. For many years prosthetic, or artificial, arms looked real, but they didn't work like real arms. Recently scientists have developed a bionic, or mechanical, arm. Signals from the patient's brain control it.

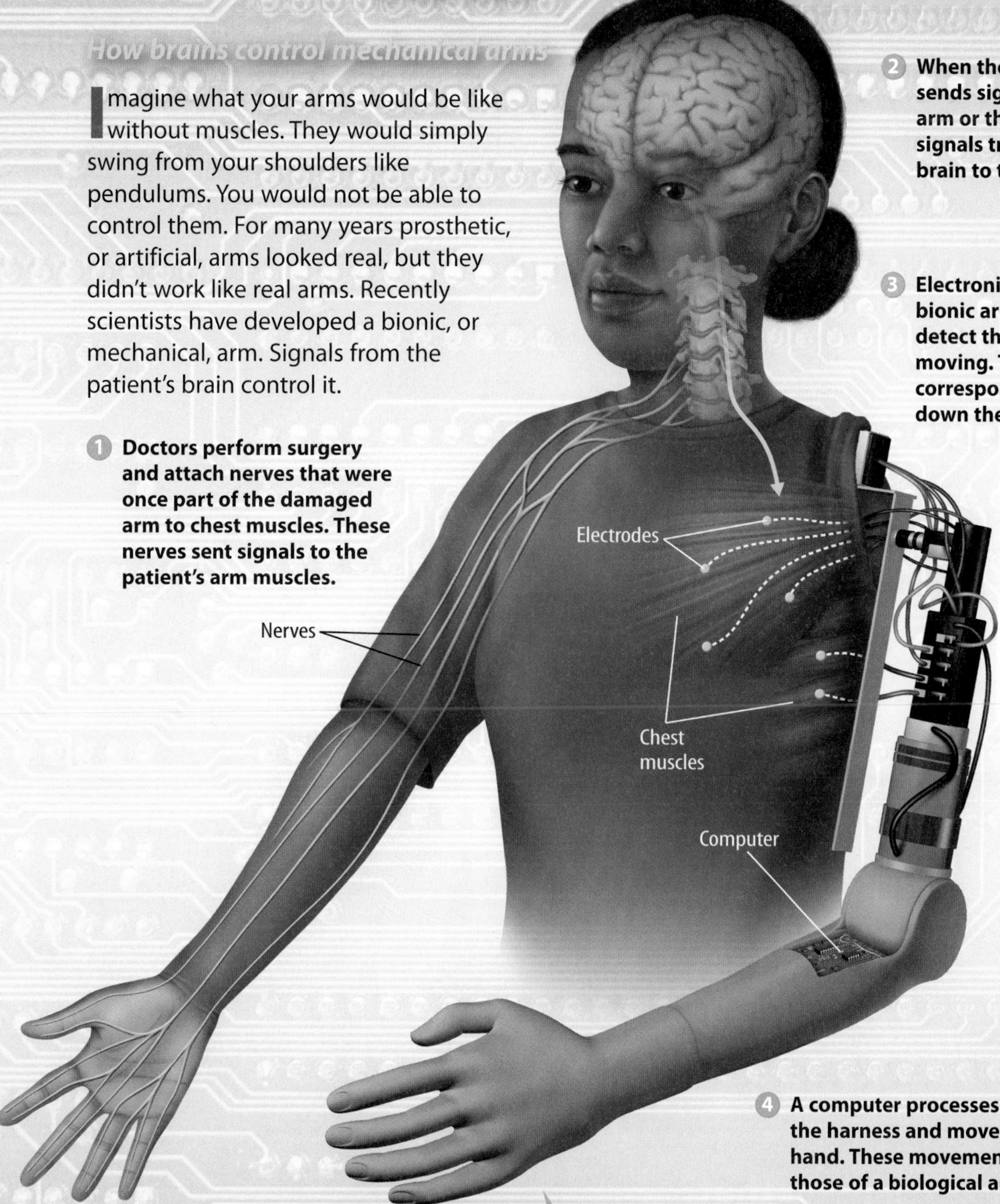

1 Doctors perform surgery and attach nerves that were once part of the damaged arm to chest muscles. These nerves sent signals to the patient's arm muscles.

2 When the patient's brain sends signals to move the arm or the hand, the signals travel from the brain to the chest muscles.

3 Electronic sensors in the bionic arm's harness detect the chest muscle moving. The sensors send corresponding signals down the bionic arm.

4 A computer processes the signals from the harness and moves the arm and hand. These movements are similar to those of a biological arm and hand.

It's Your Turn

RESEARCH AND REPORT In science-fiction films, some characters have bionic body parts such as ears and eyes. Are any other bionic body parts in development? Summarize your findings in a paragraph.

Lesson 2

Reading Guide

Key Concepts
ESSENTIAL QUESTIONS

- What does the muscular system do?
- How do types of muscle differ?
- How does the muscular system interact with other body systems?

Vocabulary

muscle p. 497
skeletal muscle p. 499
voluntary muscle p. 499
cardiac muscle p. 500
involuntary muscle p. 500
smooth muscle p. 500

Multilingual eGlossary

Video

- Science Video
- What's Science Got to do With It?

The Muscular System

Inquiry How can she do that?

This person is a contortionist (kun TOR shuh nist)—someone who can bend and flex his or her body in unusual ways. The contortionist pictured here is able to stay in this position because her muscular system can hold her body in place.

Launch Lab

15 minutes

Can you control all your muscles?

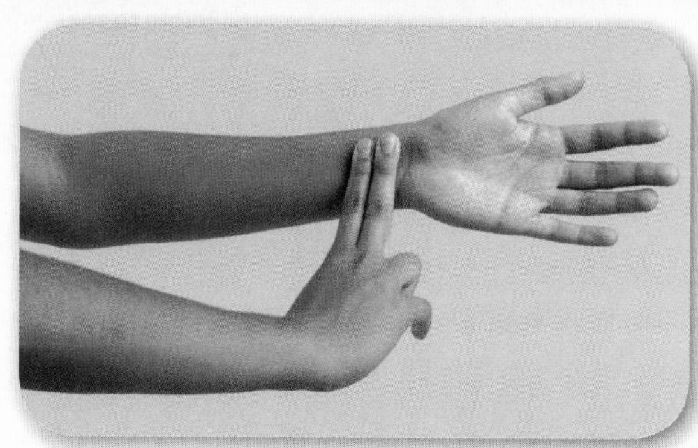

Can you feel your heart beating? Your body contains many muscles that you can control by thinking about them. However, not all types of muscle are controllable.

1. Shake hands with another student. Did you have to think about this action?
2. Rest your index and middle fingers on the thumb side of your wrist until you can feel your pulse. Can you change the speed of your pulse by thinking about it?

Think About This

1. Make a list of the muscles in your body that you can consciously control. What are their functions?
2. **Key Concept** Think of other muscles in your body, besides your heart, that work without you thinking about them. How do the functions of these muscles differ from the ones you consciously control?

Functions of the Muscular System

What comes to mind when you think of muscles? Maybe you think about the muscles in your arms or legs that enable you to carry books or run fast. Movement is an important function of the muscular system. But muscles are also important for protection, stability, and maintaining body temperature.

Although muscles have different functions, all muscle tissues are made of cells that contract. *A* **muscle** *is made of strong tissue that can contract in an orderly way.* When a muscle contracts, the cells of the muscle become shorter, as shown in **Figure 9.** When the muscle relaxes, the cells return to their original length.

Movement

Many of your muscles are attached to bone and enable your skeleton to move. Bones move when these muscles contract. This movement can be fast, such as when you run, or slow, like when you stretch.

You also have many muscles in your body that are not attached to bones. The contractions in these muscles cause blood and food to move throughout your body. They also cause your heart to beat and the hair on your arms to stand on end when you get goose bumps.

Key Concept Check What is one major function of the muscular system?

Figure 9 A muscle cell works by contracting and relaxing.

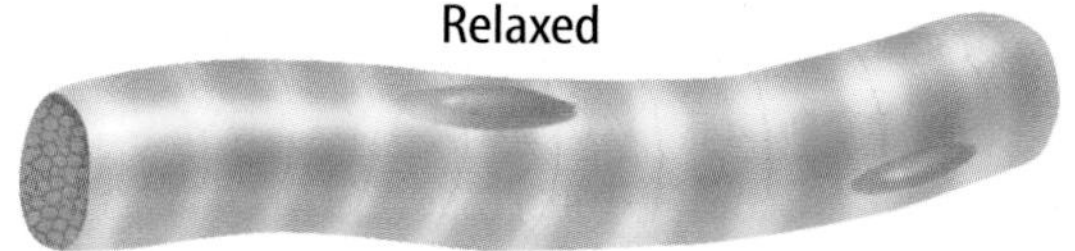

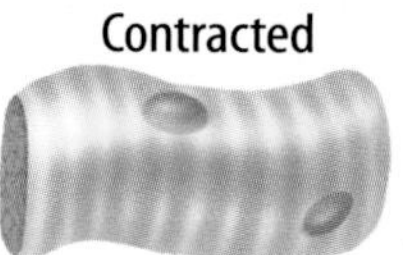

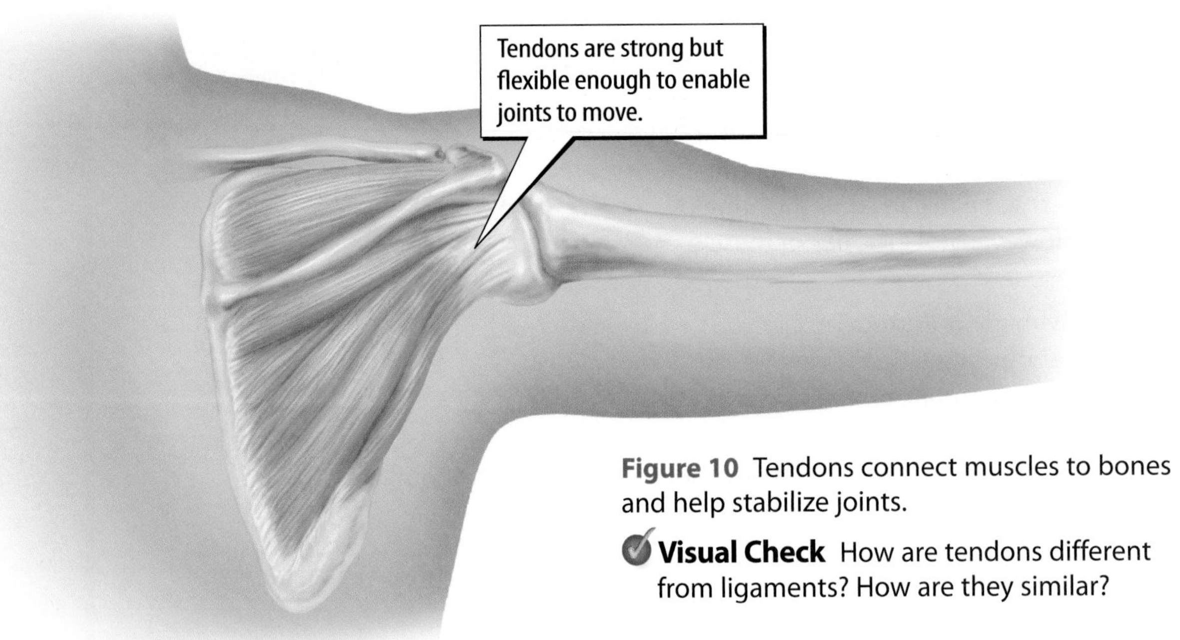

Figure 10 Tendons connect muscles to bones and help stabilize joints.

Visual Check How are tendons different from ligaments? How are they similar?

Stability

Muscles that are attached to bones support your body and help you balance. If your body becomes unstable, such as when you trip and lose your balance, your muscles pull in different directions and you regain your balance.

Tendons attach muscles to bones, as shown in **Figure 10.** Look at the back of your hand and move your fingers. Do you see cordlike structures moving under your skin? These are tendons. You can feel the Achilles tendon above either of your heels. Tendons work with muscles and keep joints in place when your body moves. Tendons also help hold your body in a proper posture, or shape.

Figure 11 Muscles cover the skeleton. ▼

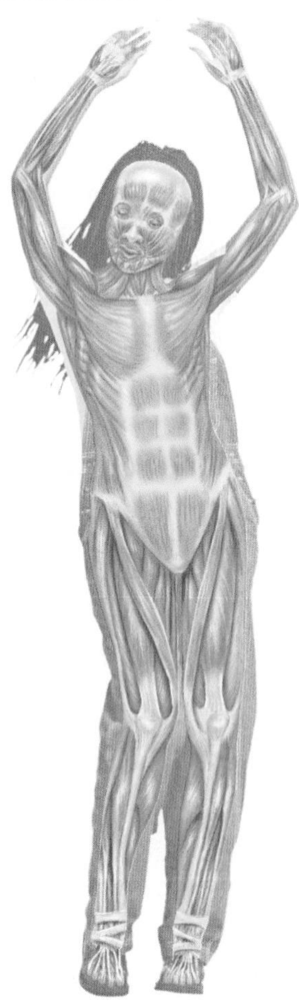

Protection

Muscles protect your body. As shown in **Figure 11,** muscles cover most of your skeleton. Muscles also cover most of the organs inside your body. Muscles are like a layer of padding. They surround your abdomen, chest, and back and protect your internal organs.

Temperature Regulation

Have you ever been in a cold environment and started shivering? Shivering is when muscles contract rapidly and change chemical energy to thermal energy. The thermal energy helps raise your body's temperature. This is important because a human's body temperature must stay around 37°C in order for the body to function properly. Muscles also change chemical energy to thermal energy during exercise. This is why you feel warm after physical activity.

Types of Muscles

Your body has three different types of muscles: skeletal, cardiac, and smooth. Each of these muscle tissues is specialized for a different function.

Skeletal Muscle

The type of muscle that attaches to bones is **skeletal muscle.** Skeletal muscles, as shown in **Figure 12,** are also called **voluntary muscles,** *which are muscles that you can consciously control.* Have you ever played with a puppet on a string? You controlled how the puppet moved. In a similar way, you control how skeletal muscles move. The contractions of skeletal muscles can be quick and powerful, such as when you run fast. However, contracting these muscles for long periods of time can exhaust or cramp them.

How Skeletal Muscles Work Skeletal muscles work by pulling on bones. Because muscles cannot push bones, they must work in pairs. **Figure 12** illustrates how an arm's biceps (BI seps) and triceps (TRI seps) muscles work as a pair.

Reading Check Why must skeletal muscles work in pairs?

Changes in Skeletal Muscles Your skeletal muscles can change throughout your lifetime. If you exercise, your muscle cells increase in size and the entire muscle becomes larger and stronger.

FOLDABLES®

Fold a sheet of paper into thirds to make a tri-fold book. Use it to organize your notes on the three types of muscle tissue.

Skeletal muscle
Cardiac muscle
Smooth muscle

Concepts in Motion Animation

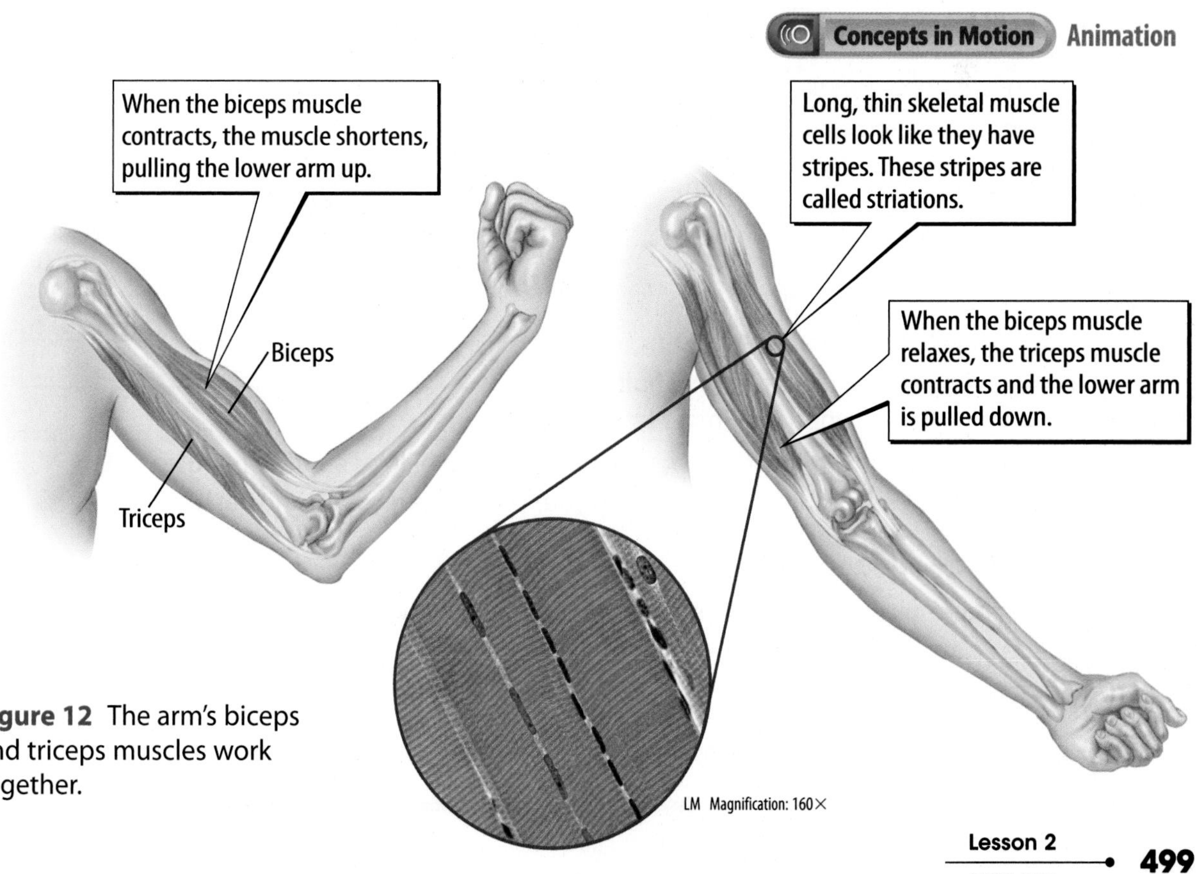

Figure 12 The arm's biceps and triceps muscles work together.

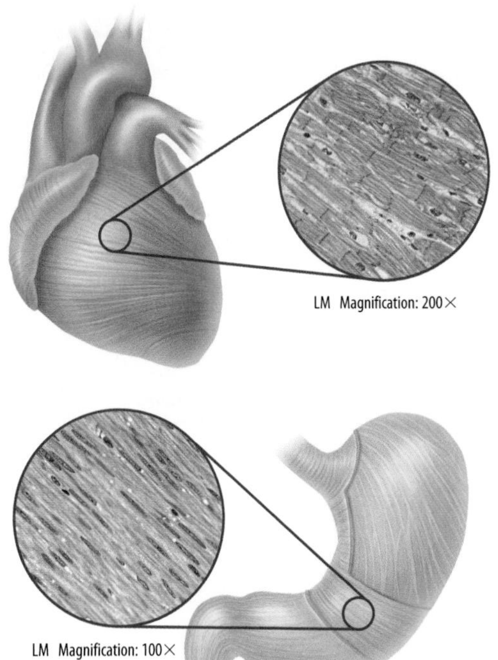

Figure 13 The heart is made of cardiac muscle. The stomach is lined with three layers of smooth muscle.

Cardiac Muscle

Your heart is made of **cardiac** (KAR dee ak) **muscles,** *which are found only in the heart.* A cardiac muscle is a type of **involuntary muscle,** *which is muscle you cannot consciously control.* When cardiac muscles contract and relax, they pump blood through your heart and through vessels throughout your body. Cardiac muscle cells, shown in **Figure 13,** have branches with discs at the ends. The discs send signals to other cardiac cells and they all contract at nearly the same time.

Smooth Muscle

Blood vessels and many organs, such as the stomach shown in **Figure 13,** are lined with smooth muscles. **Smooth muscles** *are involuntary muscles named for their smooth appearance.* Contraction of smooth muscles helps move material through the body, such as food in the stomach, and controls the movement of blood through vessels.

Key Concept Check What are the three types of muscles?

Inquiry MiniLab — 20 minutes

How strong are your hand muscles?

Have you ever done hand exercises? Do your hand muscles tire?

1. Read and complete a lab safety form.
2. Copy the data table to the right in your Science Journal.
3. Choose a partner. While your partner keeps time with a **stopwatch,** count how many times you can squeeze a **tennis ball** in one minute. Record your data in the table.
4. Immediately count how many times you can squeeze the ball for another minute. Again record your data in the table. Switch roles so your partner can collect the data.

	Number of Squeezes	
Student Name	First Minute	Second Minute

Analyze and Conclude

1. **Compare** the results from the first minute to the second minute for both partners.
2. **Design** an experiment to determine whether daily practice would increase the number of squeezes per minute.
3. **Key Concept** Evaluate the importance of doing exercises to keep muscles strong.

Healthy Muscles

Recall that a good diet keeps your bones healthy. Your muscles benefit from a healthy diet, too. All muscles require energy to contract. This energy comes from the food you eat. Eating a diet full of nutrients such as protein, fiber, and potassium can help keep muscles strong.

Exercise also helps keep muscles healthy. Muscle cells decrease in size and strength without exercise, as shown in **Figure 14.** Decreased muscle strength can increase the risk of heart disease and bone injuries. It can also make joints less stable.

Word Origin

cardiac
from Greek *kardia,* means "heart"

Figure 14 Muscles lose size, strength, and mobility if they are not exercised.

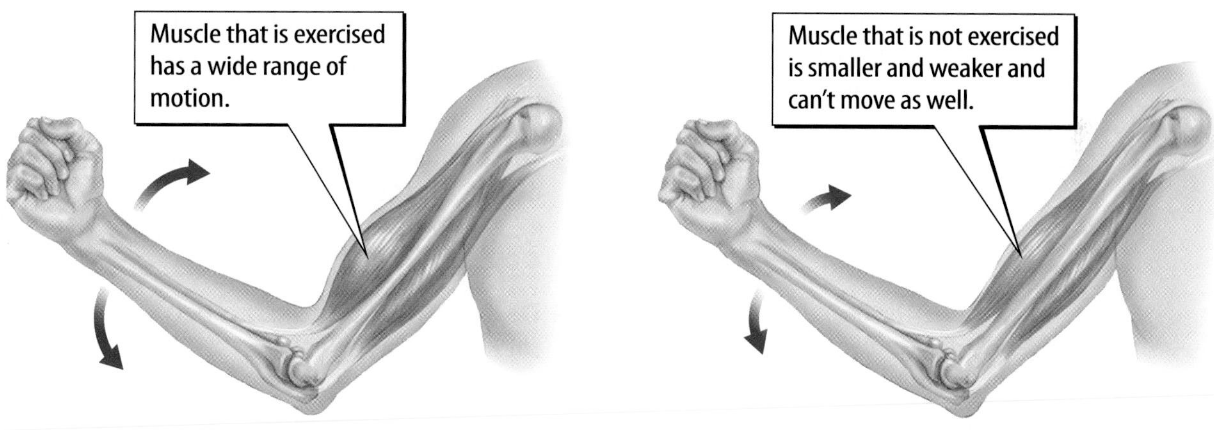

The Muscular System and Homeostasis

There are many ways the muscular system helps your body maintain homeostasis. You are probably sitting in a room where the temperature is somewhere between 21°C and 27°C. However, your body temperature is around 37°C. Your body must stay at this temperature to function well. As you have read, muscle contractions **convert** chemical energy to thermal energy and keep your body warm.

When you exercise, your cells need more oxygen and release more waste, such as carbon dioxide. The cardiac muscles of your heart help maintain homeostasis by contracting more often. When it contracts faster, the heart pumps more blood, and more oxygen is carried to the cells.

Academic Vocabulary

convert
(verb) to change something into a different form

 **Key Concept Check** How do muscles help maintain homeostasis in the body?

Lesson 2 Review

Assessment Online Quiz

Visual Summary

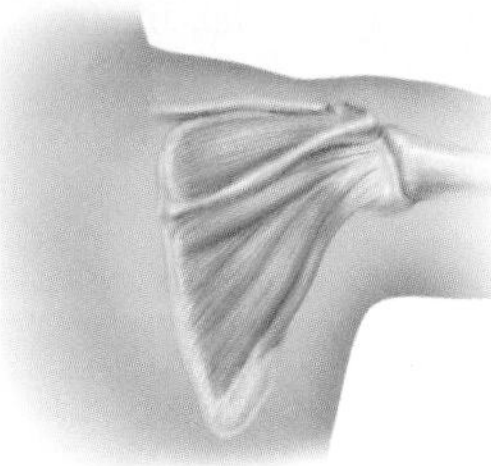

Your muscular system is made of different types of muscles. Skeletal muscles attach to bone and are muscles you can control.

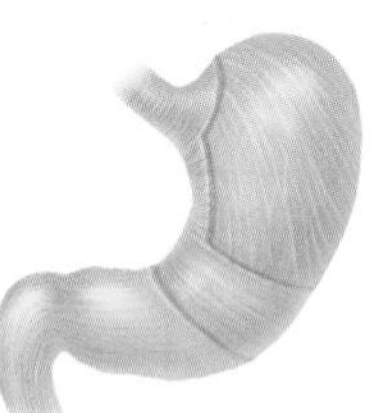

Smooth muscles line blood vessels and many internal organs.

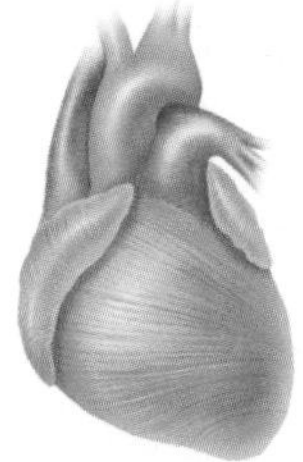

The heart is made of cardiac muscle.

FOLDABLES®

Use your lesson Foldable to review the lesson. Save your Foldable for the project at the end of the chapter.

What do you think NOW?

You first read the statements below at the beginning of the chapter.

3. The same type of muscle that moves bones also pumps blood through the heart.

4. Muscles cannot push bones.

Did you change your mind about whether you agree or disagree with the statements? Rewrite any false statements to make them true.

Use Vocabulary

1. Muscle that pumps blood through the body is called ________.
2. A person is unable to control the contractions of ________.
3. Strong tissue that contracts is called ________.

Understand Key Concepts 

4. Which is a voluntary muscle?
   A. biceps **C.** blood vessel
   B. stomach **D.** small intestine
5. **Explain** how the muscular system regulates body temperature.
6. **Distinguish** between a cardiac muscle and a smooth muscle.

Interpret Graphics

7. **Describe** the relationship between the two muscles shown.

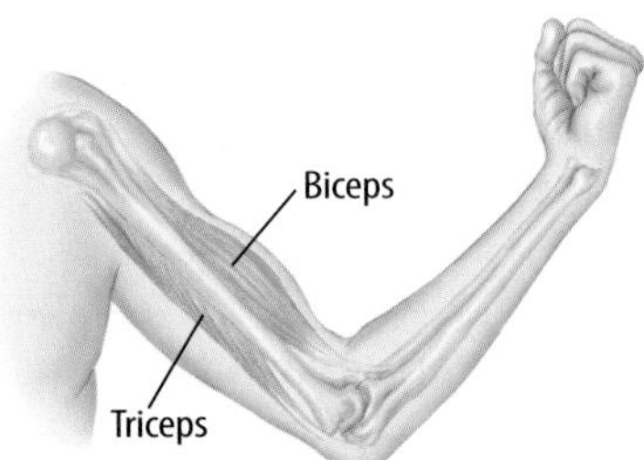

8. **Identify** Copy and fill in the graphic organizer below to identify three functions of the muscular system.

Critical Thinking

9. **Predict** what would happen if the smooth muscles in a person's body could not contract.
10. **Assess** the importance of exercise for muscle health.

Skill Practice Observe

25 minutes

How do the three types of muscle cells compare?

As you have read in this chapter, there are three types of muscle cells. Skeletal muscle cells are voluntary muscle cells and can be controlled. Smooth and cardiac muscle cells are involuntary muscle cells. An organism does not need to think about their movement. These three types of muscle cells have different characteristics and functions in the body. In this activity, you will examine cells of each type and observe their similarities and differences.

Materials

compound microscope

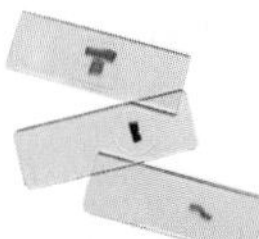

muscle cell slides

Safety

Learn It

Scientists can learn a great deal about organisms or objects through **observation.** A scientist might note the characteristics that several things have in common. He or she might also observe characteristics that make something unique. Observations help scientists classify things and understand how they function.

Try It

1. Read and complete a lab safety form.
2. Copy the table below into your Science Journal.
3. Place a prepared slide of skeletal muscle cells on the stage of a compound light microscope.
4. In your table, describe the cells' shape and color. Note any differences among them.
5. Locate the nucleus in one of the cells. Where is it in the cell? Do any cells have more than one nucleus?
6. Describe the arrangement of the cells in the tissue. Do they have a pattern or are they randomly arranged?
7. Repeat steps 3 through 6 for the prepared cardiac muscle and smooth muscle slides.

Apply It

8. **Compare and Contrast** How do the characteristics of each cell type differ?
9. **Key Concept** Based on your observations and on what you have read, what characteristics of each cell type might relate to its function?

Observations of Three Types of Muscle Cells			
Cell Characteristics	Skeletal	Smooth	Cardiac
Shape			
Nucleus location			
Arrangement patterns			

Lesson 3

The Skin

Reading Guide

Key Concepts
ESSENTIAL QUESTIONS

- What does the skin do?
- How do the three layers of skin differ?
- How does the skin interact with other body systems?

Vocabulary

integumentary system p. 505
epidermis p. 507
melanin p. 507
dermis p. 507
bruise p. 508

Multilingual eGlossary

Color-enhanced SEM Magnification: unknown

Inquiry What is this?

You might think this is a picture of a landscape on another planet. However, this image shows what your skin looks like under a microscope. Your skin sheds its top layer of cells constantly as new cells are made to take their place.

Launch Lab

5 minutes

How does your skin protect your body?

Your skin is your body's first line of defense. When you touch something with your fingers you can instantly tell if it is hot or cold. Are all parts of your body equally sensitive?

1. Read and complete a lab safety form.
2. Touch the back of your hand with an **ice cube** in a **plastic bag.**
3. Now do the same to the back of your knee.

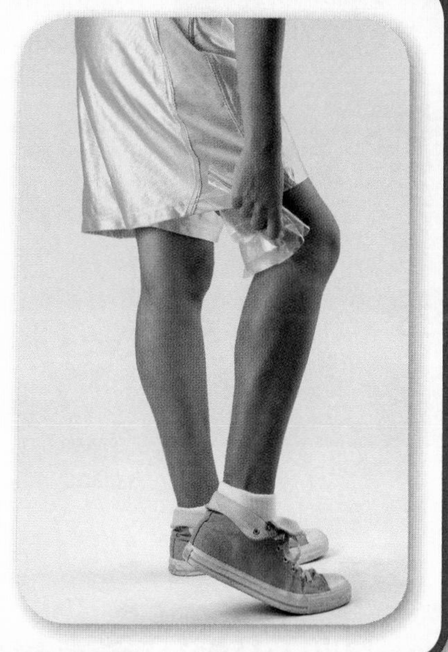

Analyze and Conclude

1. Which area was more sensitive to cold?
2. How do you think the skin senses temperature?
3. **Key Concept** How does sensitivity to temperature protect the body?

Functions of the Skin

Touch your fingertips, your arm, and your face. The soft tissue you feel is the outermost layer of your skin. Skin is the largest organ of your body. It is part of the **integumentary** (ihn teh gyuh MEN tuh ree) **system,** *which includes all the external coverings of the body, including the skin, nails, and hair.* Most parts of the integumentary system are shown in **Figure 15.** Like bones and muscles, skin serves many different functions in your body.

Word Origin

integumentary
from Latin *integere,* means "to cover"

Protection

When you look at yourself in a mirror, you cannot see the bones, muscles, or other parts of your skeletal and muscular systems. Instead, you see your skin. Skin covers your bones and muscles and protects them from the external environment. It keeps your body from drying out in sunlight and wind. Skin also protects the cells and tissues under the skin from damage. Skin is the first line of defense against dirt, bacteria, viruses, and other substances that might enter your body.

Reading Check What would happen to your body if you had no skin?

Figure 15 The integumentary system includes skin, nails, and hair.

Figure 16 During exercise, sweat evaporates and blood vessels enlarge. This releases thermal energy.

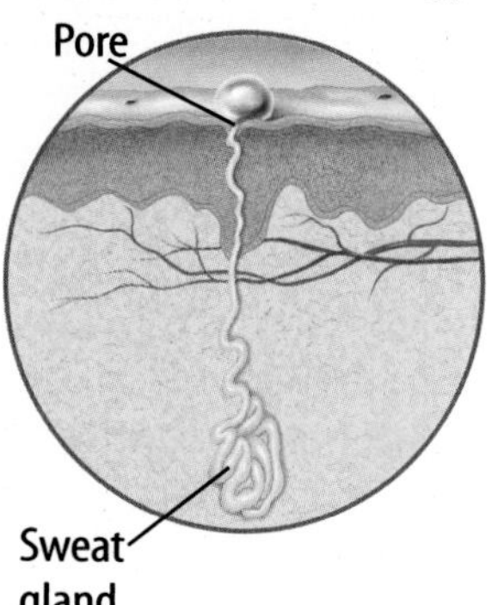

Sensory Response

Close your eyes, and feel the surface of your desk and the objects on top. Even with your eyes closed, you can tell the difference between the desk, a book, paper, and pencils. This is because your skin has special cells called sensory receptors that detect texture. Sensory receptors also detect temperature and sense pain. The more sensory receptors there are in an area of skin, the more sensitive it is.

Temperature Regulation

Skin helps control body temperature. When you exercise, sweat comes from tiny holes, or pores, on the skin's surface, as shown in **Figure 16.** Sweating is one way skin lowers your body temperature. As sweat evaporates, excess thermal energy leaves the body and the skin cools.

Another way that skin lowers body temperature is by releasing thermal energy from blood vessels. Has your face ever turned red while exercising? The girl in **Figure 16** has a red face because the blood vessels near the skin's surface dilated, or enlarged. This increases the surface area of the blood vessels and releases more thermal energy.

Key Concept Check How does skin regulate body temperature?

Production of Vitamin D

If your skin is exposed to sunlight, it can make vitamin D. Your body needs vitamin D to help it absorb calcium and phosphorous and to promote the growth of bones. Your skin is not the only source of vitamin D. Vitamin D is usually added to milk and is found naturally in certain types of fish.

Elimination

Normal cellular processes produce waste products. The skin helps to eliminate these wastes. Water, salts, and other waste products are removed through the pores. This occurs all the time, but you might only notice it when you sweat during exercise.

Inquiry MiniLab — 10 minutes

Why are you sweating?

You might think that you sweat only when you exercise, but your body also sweats when you are at rest.

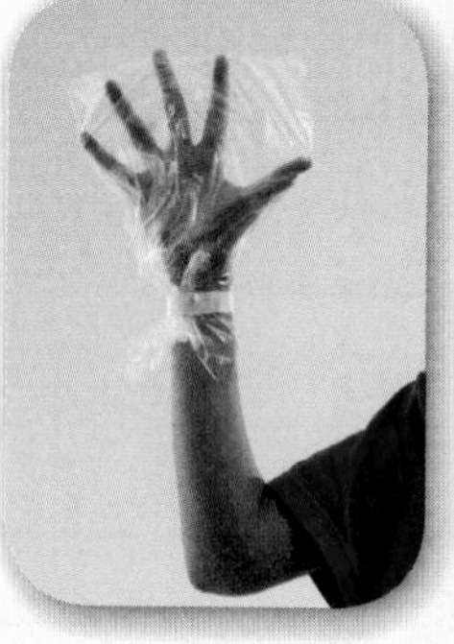

1. Read and complete a lab safety form.
2. Wipe your hands dry on a **paper towel.**
3. Place your non-writing hand in a **one-quart plastic bag.** With a partner's help, tape it around your wrist.
4. Observe the inside of the bag after 5 minutes. Record your observations in your Science Journal.
5. Observe the inside of the bag again after 10 minutes. Record your observations.

Analyze and Conclude

1. **Describe** what happened to the inside of the bag after 10 minutes.
2. **Explain** what might have caused this change.
3. **Key Concept** How does skin help regulate body temperature?

Skin Structures

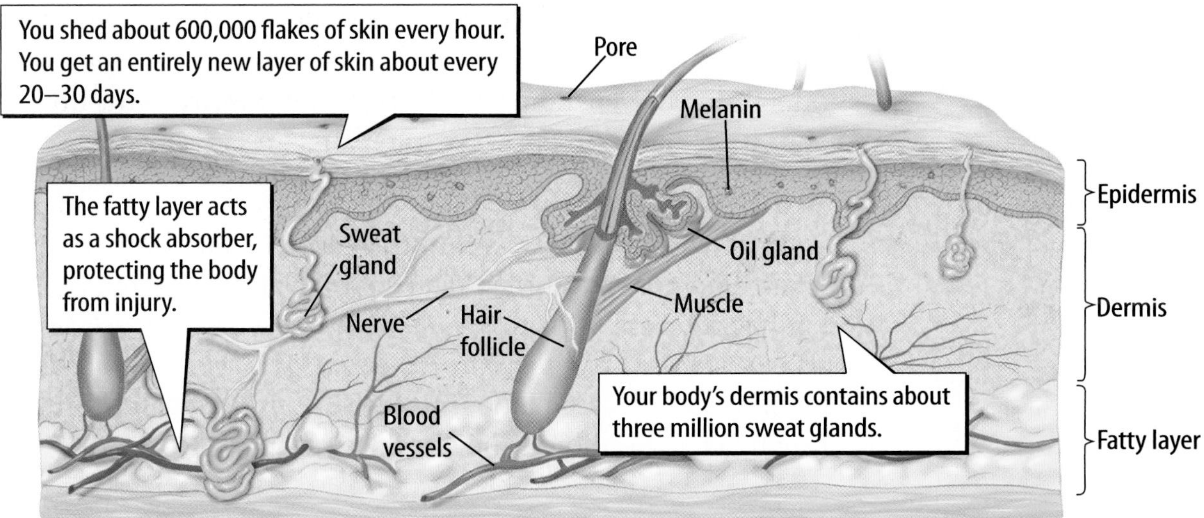

Figure 17 Skin has three layers, each with different functions.

Structures of the Skin

The skin that you see and feel on your body is the outermost layer of skin. Below it are two other layers of skin. These layers, shown in **Figure 17,** differ in structure and function.

Epidermis

The **epidermis** (eh puh DUR mus) *is the outermost layer of skin and the only layer in direct contact with the outside environment.* The epidermis is tough but thin. The epidermis on your eyelids is thinner than a sheet of paper. Cells of the epidermis are constantly shed and replaced by new cells.

One important function of the epidermis is the production of melanin (MEH luh nun). **Melanin** *is a pigment that protects the body by absorbing some of the Sun's damaging ultraviolet rays.*

Dermis

Below the epidermis is the dermis. *The* **dermis** *is a thick layer of skin that gives skin strength, nourishment, and flexibility.* The dermis contains sweat glands, blood vessels, nerves, hair follicles, and muscles, as shown in **Figure 17.** When the muscles in the dermis contract, you get goose bumps.

Fatty Layer

The innermost layer of skin insulates the body, acts as a protective padding, and stores energy. This layer is sometimes called the fatty layer. It can be very thin or very thick, depending on its location on the body.

 **Key Concept Check** How do the skin's three layers differ?

Math Skills 

Using Proportions

The ratios $\frac{5}{1}$ and $\frac{25}{5}$ are equivalent, so they can be written as the proportion $\frac{5}{1} = \frac{25}{5}$. When ratios form a proportion, the cross products are equal. In the above proportion, $5 \times 5 = 25 \times 1$. You can use cross products to find a missing term. For example, if each 1 cm^2 of skin contains 300 pores, how many pores are there in 5 cm^2 of skin?

$$\frac{1\ \text{cm}^2}{300\ \text{pores}} = \frac{5\ \text{cm}^2}{n\ \text{pores}}$$

$$1 \times n = 300 \times 5;$$

$$n = 1500\ \text{pores}$$

Practice

The palm of the hand has about 500 sweat glands per 1 cm^2. How many sweat glands would there be on a palm measuring 7 cm by 8 cm?

- Math Practice
- Personal Tutor

Skin Injuries and Repair

You have probably fallen down and injured your knees or other parts of your body. You might also have damaged your skin. Because skin is exposed to the outside environment, it is often injured. Depending on the type and severity of the injury, your body has different ways to repair skin.

Bruises

Have you ever bumped into the edge of a table and noticed later that your skin turned red or purple and blue? You probably had a bruise. *A* **bruise** *is an injury where blood vessels in the skin are broken, but the skin is not cut or opened.* The broken blood vessels release blood into the surrounding tissue. Bruises usually change color as they heal, as shown in **Figure 18.**

Figure 18 Bruises change color as they heal due to chemical changes in the blood under the skin's surface.

A new bruise is red because blood pools under the skin.

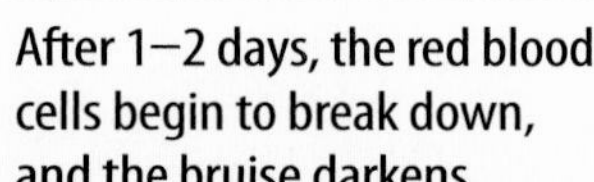

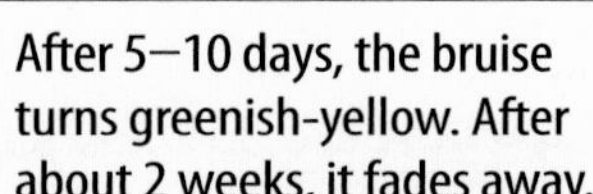

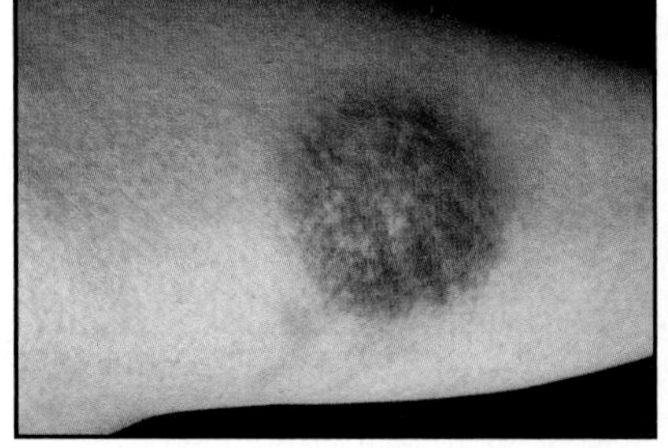

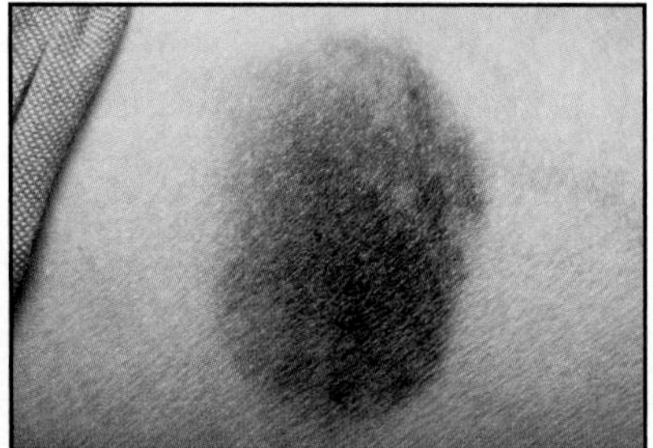

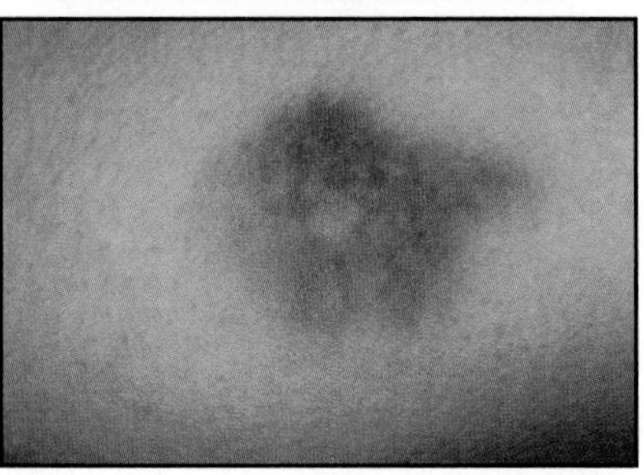

Cuts

When you break one or more layers of skin, it is called a cut. Cuts often cut blood vessels, too. The released blood will usually thicken and form a scab over the cut. The scab prevents dirt and other outside substances from entering the body. Skin heals by producing new skin cells that eventually repair the cut. Some cuts are too large to heal naturally. If that happens, stitches might be needed to close the cut while it heals.

FOLDABLES®

Make a horizontal half book from a sheet of notebook paper. Use it to record information about the different types of skin injuries and how the body repairs them.

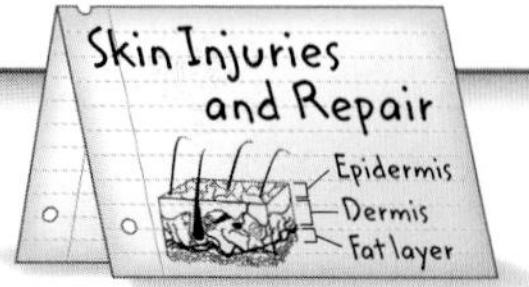

Burns

Have you ever injured your skin by contact with hot water or hot food? If so, then you had a burn. You might think that burns only occur from touching hot objects. However, burns can also be caused by touching extremely cold objects, chemicals, radiation (such as sunlight), electricity, or friction (rubbing). **Table 2** describes the three degrees, or levels, of burns.

Reading Check What can cause your skin to burn?

Table 2 Degrees of Burns

Burn	Symptoms	Damage
First-degree burn	Damages top layer of skin Symptoms: pain, redness, swelling Usually heals in 5–7 days without scarring	
Second-degree burn	Damages top two layers of skin Symptoms: pain, redness, swelling, blistering Usually heals in 2–6 weeks with some scarring	
Third-degree burn	Damages all three layers of skin and sometimes the tissue below skin Symptoms: black or white charred skin, may be temporarily numb due to damaged nerves Heals over months with scarring; might require surgery	

Table 2 Burns are classified by the depth of damage to the skin. There are three different degrees, or levels, of burns.

Visual Check If swelling and blisters appear on the surface of a burned area of skin, what degree burn would it be?

▲ **Figure 19** Protecting your skin from sunlight is important for maintaining skin health.

Healthy Skin

Just as diet and exercise are important to healthy skeletal and muscular systems, good life choices can keep your skin functioning well.

One important thing you can do for your skin is protect it from sunlight. The ultraviolet rays in sunlight can cause permanent damage to the skin, including wrinkles, dry skin, and skin cancer. Like the person in **Figure 19,** you can protect your skin by using sunscreen. You can also wear protective clothing and avoid outdoor activities during the middle of the day.

Another way to keep your skin healthy is to eat a balanced diet. You can also apply lotion to your skin to help keep it moist and use gentle soaps to clean it.

The Skin and Homeostasis

Skin helps your body maintain homeostasis in many ways. You read that the skin can make vitamin D and that it protects the body from outside substances. These functions of the skin help regulate the body's internal environment.

The skin also works with other body systems to maintain homeostasis. You read earlier how the skin and the circulatory system help cool the body when it becomes overheated. If the body becomes cold, blood vessels constrict, or narrow, reducing thermal energy loss. As you can see in **Figure 20,** the skin also works with the nervous and muscular systems to help the body react to stimuli.

Key Concept Check How does the skin interact with other body systems to help maintain homeostasis?

Figure 20 The skin, the nervous system, and the muscular system work together to maintain homeostasis. ▼

Reaction to Stimuli

Review Personal Tutor

Lesson 3 Review

Assessment Online Quiz

Visual Summary

The integumentary system consists of all the external coverings of the body.

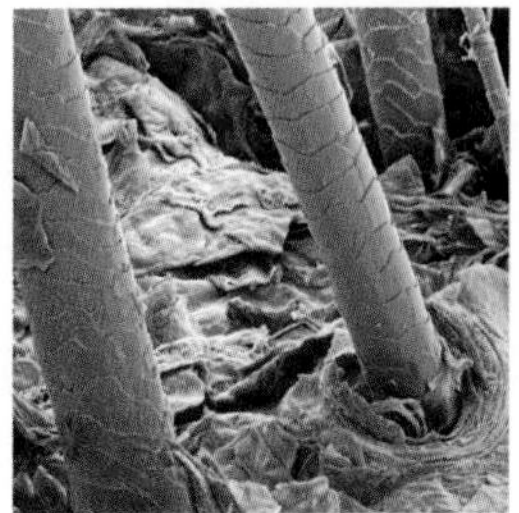

The epidermis is the outermost layer of skin.

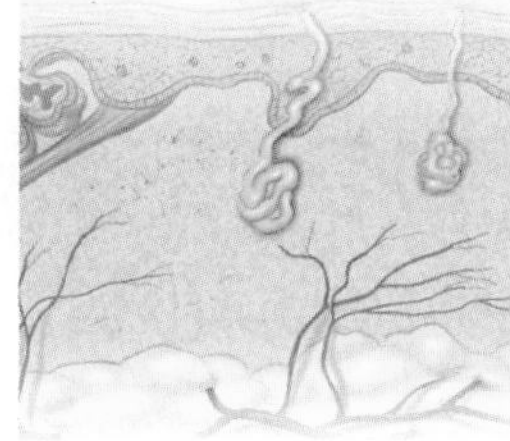

The dermis is the layer of skin that contains sweat glands, blood vessels, nerves, hair follicles, and muscles.

FOLDABLES®

Use your lesson Foldable to review the lesson. Save your Foldable for the project at the end of the chapter.

What do you think NOW?

You first read the statements below at the beginning of the chapter.

5. Skin helps regulate body temperature.

6. Skin is made of two layers of tissue.

Did you change your mind about whether you agree or disagree with the statements? Rewrite any false statements to make them true.

Use Vocabulary

1. A pigment that absorbs ultraviolet rays is called ________.
2. The ________ is the layer that nourishes the skin.

Understand Key Concepts 

3. Which is the innermost layer of skin?
   - **A.** dermis
   - **B.** epidermis
   - **C.** fatty layer
   - **D.** melanin
4. **Give** an example of how the skin regulates body temperature.
5. **Compare** the dermis and epidermis. Explain the function of each.

Interpret Graphics

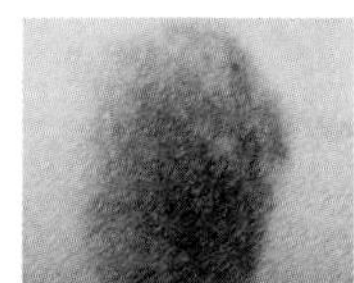

6. **Estimate** the age of the bruise shown on the right.
7. **Organize Information** Copy and fill in the table below with details about the three degrees of burns.

First-degree burn	
Second-degree burn	
Third-degree burn	

Critical Thinking

8. **Evaluate** the importance of your skin to your homeostasis.

Math Skills

Review Math Practice

9. The palms of the hands and soles of the feet have the highest concentration of sweat glands—about $500/cm^2$. How many sweat glands would there be on the sole of a foot measuring 10 cm × 27 cm?

45 minutes

Dissect a Chicken Wing

Materials

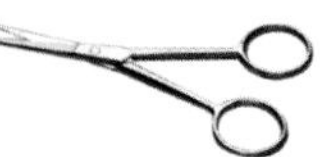

dissecting scissors

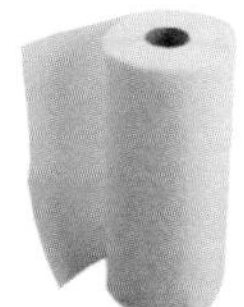

paper towel

forceps

dissection tray

chicken wing

Safety

A chicken wing has many structures similar to those of a human arm. As in humans, these structures perform different functions. How do these structures work together? How are they like your arm?

Question

How do the skin, muscles, and bones of a chicken help support the chicken, enable it to move, and regulate its internal environment?

Procedure

1. Read and complete a lab safety form.
2. Obtain a boiled chicken wing and pat it dry with a paper towel. Flex the wing at its joints and notice what happens.
3. Insert the scissors under the skin at the large end and cut the skin down the length of the wing. Look at the underside of the skin. Does it look any different from the outside?

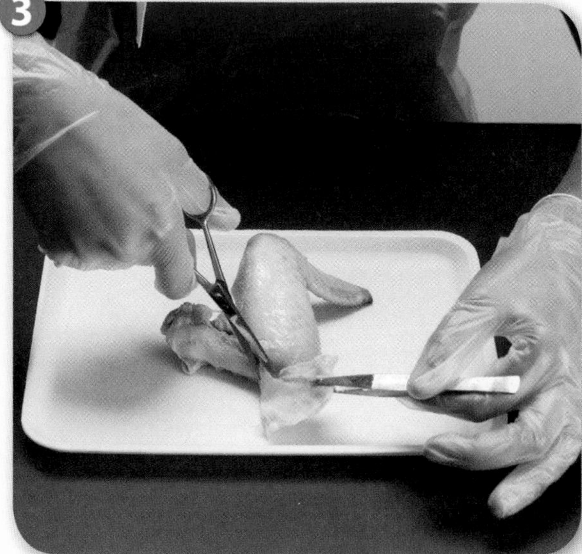

4. Carefully peel the skin off the wing. Identify the muscles—the pink or white tissues between the skin and the bone.
5. Examine the outer surface of the muscles. Identify the tendons, which are shiny white tissues at either end of the muscles. Carefully cut one of the tendons and pull the muscle back.
6. Remove enough muscle from the wing to expose a bone. Look for veins and arteries, which are thin, dark, rubbery tissues.
7. Examine the surface of the bone, including the ends and the middle. Try to separate it from another bone. Look for cartilage and ligaments holding the bones together. Try to identify the type of joint you see.

Analyze and Conclude

8 **Describe** how the wing changed when you bent and then released it. Why do you think this happened?

9 **Think Critically** Compare the joints of the chicken wing to joints in your own arm and shoulder.

10 **The Big Idea** Explain how the different parts of the chicken wing work together and how they help the chicken maintain its homeostasis.

Communicate Your Results

Prepare a poster illustrating the observations that you made as you dissected the chicken wing. Include a diagram of the chicken wing and label the structures that you identified.

Extension

How do the structure and function of chicken wings differ from those of human arms? Design an experiment that would help you determine what structures human arms and chicken wings do not share, and how the functions of each of these structures differ. Include your hypothesis, any materials you will use, and any safety guidelines that should be observed.

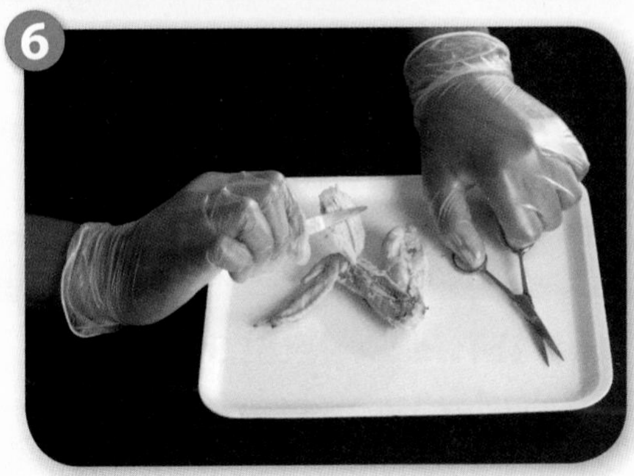

Lab Tips

- ☑ As you dissect, look for evidence of injuries or abnormalities the chicken might have. Share your observations with your classmates.
- ☑ Cut skin and other tissues only with the chicken wing resting in the dissection pan.

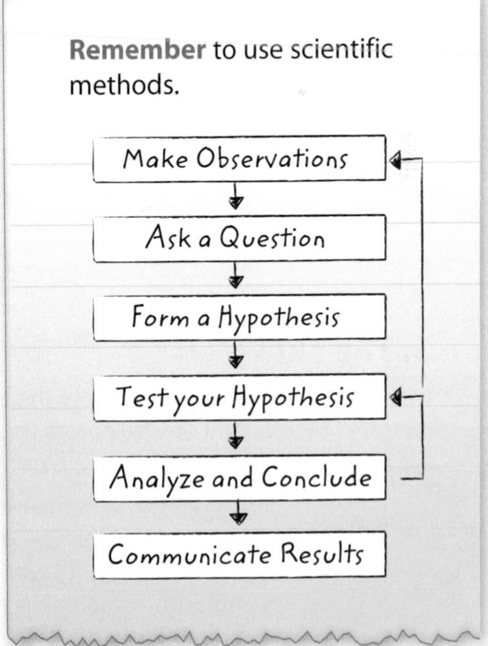

Chapter 14 Study Guide

The skeletal system, the muscular system, and the skin work together to move, support, and protect the body and maintain its homeostasis.

Key Concepts Summary

Lesson 1: The Skeletal System

- The **skeletal system** supports the body, helps it move, and protects internal organs. Bones store fat and calcium and make red blood cells.
- Bones are protected by **cartilage** and are connected to other bones by **ligaments** at **joints.**
- The skeletal system works with the muscular system and the skin to protect and support the body and enable it to move.

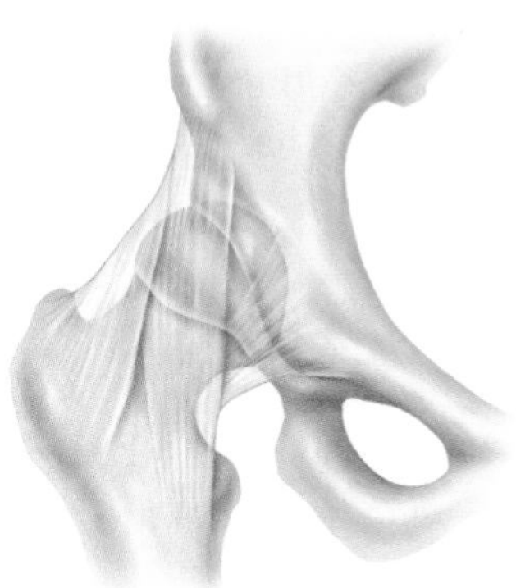

Vocabulary

skeletal system p. 487
cartilage p. 490
periosteum p. 490
joint p. 491
ligament p. 491
arthritis p. 492
osteoporosis p. 492

Lesson 2: The Muscular System

- **Muscles** support and stabilize the skeleton, enable bones and organs to move, protect the body, and regulate temperature.
- You can consciously control **skeletal,** or **voluntary, muscles.** You cannot consciously control **involuntary muscles,** which include the **cardiac muscle** that pumps blood and the **smooth muscles** that move food and blood.
- The muscular system interacts with other body systems to protect, support, and move the body.

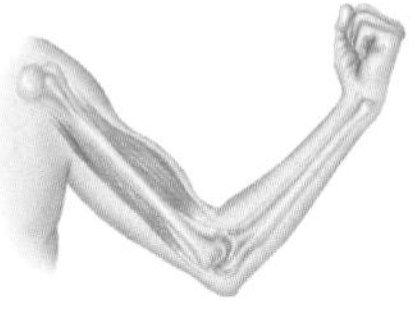

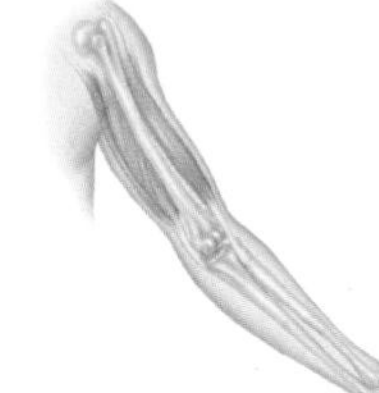

Vocabulary

muscle p. 497
skeletal muscle p. 499
voluntary muscle p. 499
cardiac muscle p. 500
involuntary muscle p. 500
smooth muscle p. 500

Lesson 3: The Skin

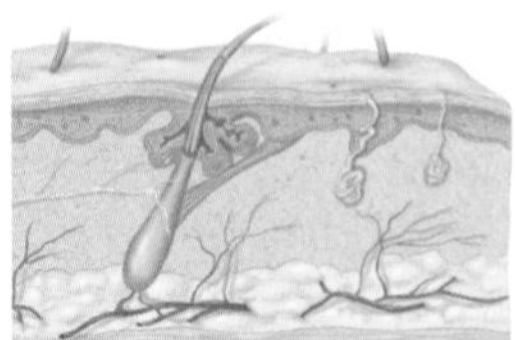

- The skin protects the body, regulates temperature, contains receptors that respond to stimuli, makes vitamin D, and helps eliminate waste.
- The skin has three layers. The **epidermis** is the outermost layer. The **dermis** is below, and the fat layer is the inner layer. The epidermis produces **melanin,** which help protect the body from ultraviolet rays.
- The skin works together with the skeletal and muscular systems to protect and support the body.

Vocabulary

integumentary system p. 505
epidermis p. 507
melanin p. 507
dermis p. 507
bruise p. 508

Review
- Personal Tutor
- Vocabulary eGames
- Vocabulary eFlashcards

FOLDABLES® Chapter Project

Assemble your lesson Foldables as shown to make a Chapter Project. Use the project to review what you have learned in this chapter.

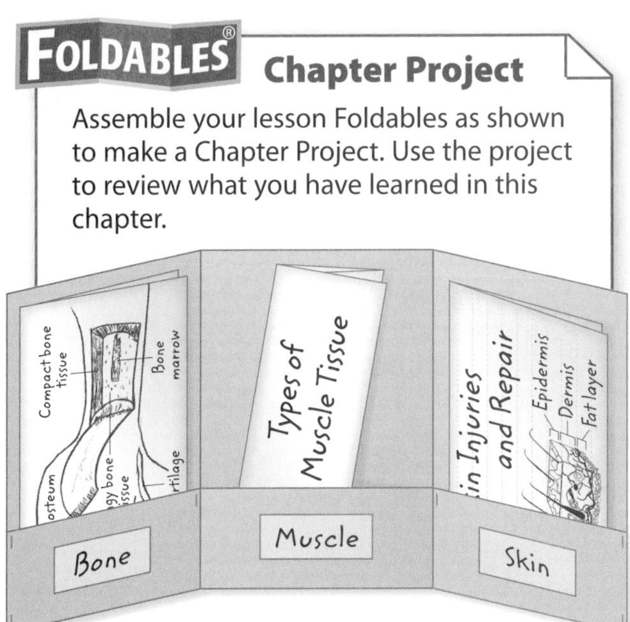

Use Vocabulary

Explain the differences and similarities between the vocabulary words in each of the following sets.

1. arthritis, osteoporosis
2. ligament, joint
3. cartilage, periosteum
4. skeletal muscle, cardiac muscle
5. voluntary muscle, involuntary muscle
6. skeletal system, muscular system
7. epidermis, dermis
8. integumentary system, muscular system
9. cut, bruise

Link Vocabulary and Key Concepts

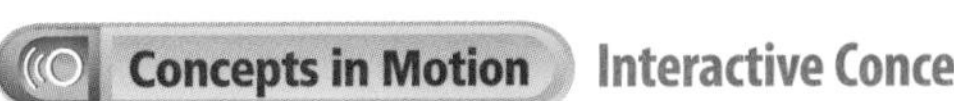

Copy this concept map, and then use vocabulary terms from the previous page and other terms from the chapter to complete the concept map.

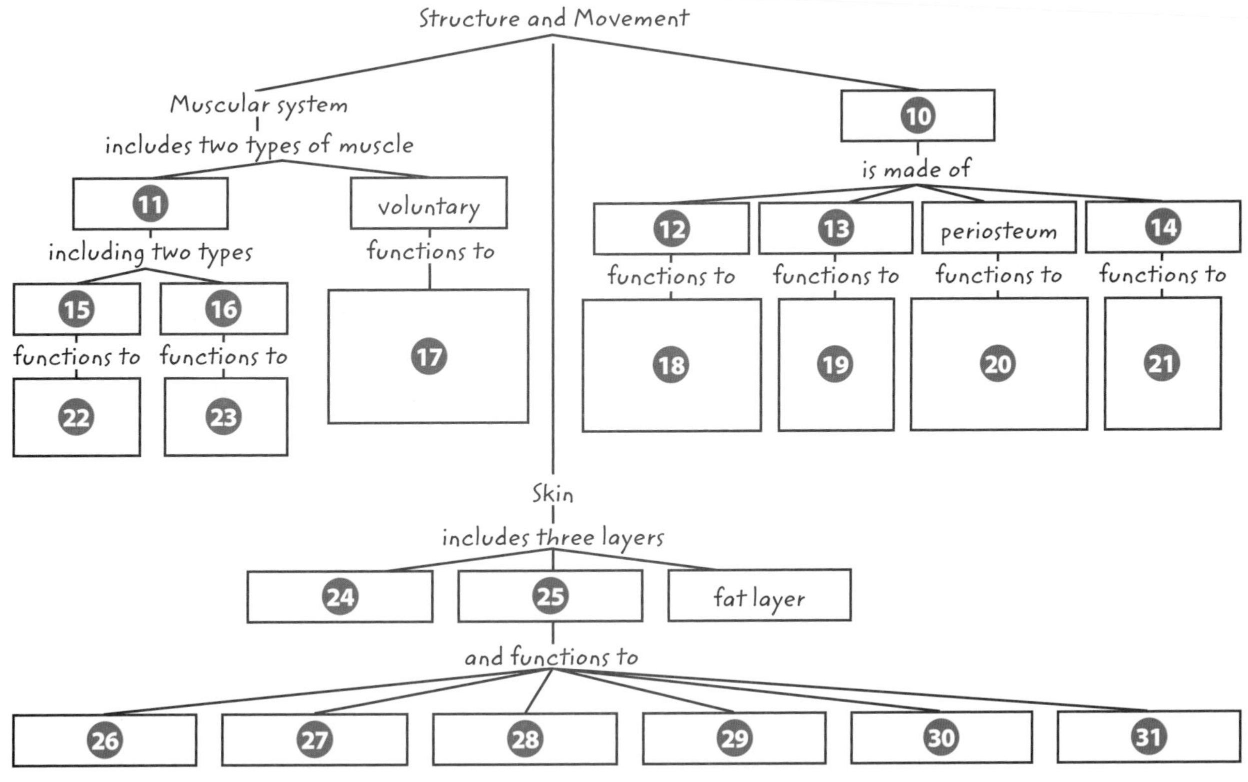

Chapter 14 Review

Understand Key Concepts

1. What is the arrow pointing to in the figure below?

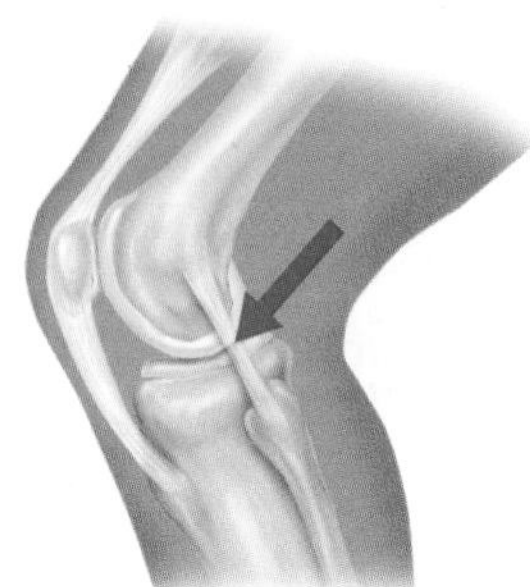

   A. cartilage
   B. ligament
   C. periosteum
   D. tendon

2. Which is NOT produced or stored inside bones?
   A. calcium
   B. fat
   C. melanin
   D. red blood cells

3. Which organ is NOT protected by the skeletal system?
   A. brain
   B. heart
   C. lungs
   D. skin

4. Which prevents bones from rubbing against each other?
   A. cartilage
   B. compact bone
   C. ligament
   D. spongy bone

5. What happens to a muscle when it contracts?
   A. The muscle lengthens.
   B. The muscle pushes on a bone.
   C. The muscle pushes on another muscle.
   D. The muscle shortens.

6. Which type of muscle is responsible for movement of the arms and legs?
   A. cardiac muscle
   B. involuntary muscle
   C. skeletal muscle
   D. smooth muscle

7. A contraction of what type of muscle is not consciously controlled?
   A. arm muscle
   B. involuntary muscle
   C. skeletal muscle
   D. voluntary muscle

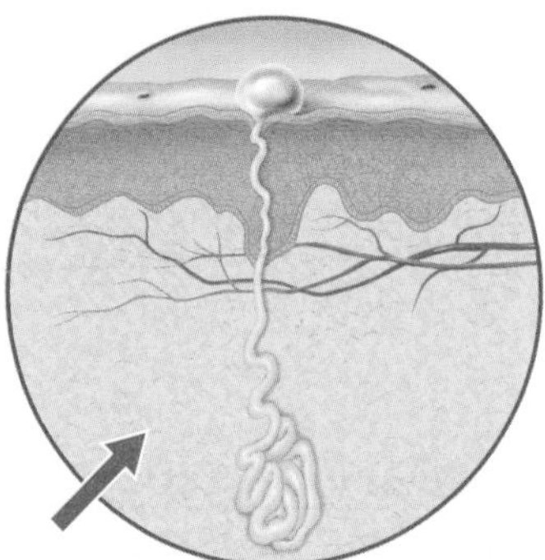

8. To which layer of skin is the arrow in the figure above pointing?
   A. dermis
   B. epidermis
   C. fat layer
   D. melanin

9. What is one function of sweating?
   A. allows the body to move
   B. enables calcium to be absorbed
   C. increases body temperature
   D. lowers body temperature

10. Which vitamin is made in the skin?
   A. vitamin A
   B. vitamin B
   C. vitamin C
   D. vitamin D

Assessment

Online Test Practice

Critical Thinking

11 **Explain** how the muscular system and the skin regulate body temperature.

12 **Compare** spongy bone and compact bone.

13 **Assess** the importance of calcium in your diet, including the consequences of not getting enough.

14 **Give an example** of how the skeletal system maintains homeostasis.

15 **Predict** how your daily life would change if your skin contained no melanin.

16 **Discuss** why muscles must work in pairs.

17 **Evaluate** how the skin's ability to continually produce new skin cells helps protect your body.

18 **Identify** the kind of joint shown below. Give an example of where it is found in the body and what type of movement it enables the body to do. Record what structures you can identify and what system they belong to.

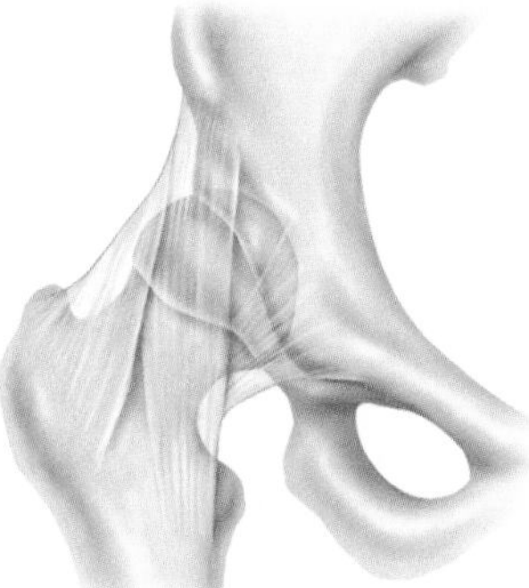

Writing in Science

19 **Write** a paragraph about how bones grow, starting before birth. Hypothesize why children and teens are in greater danger of breaking a bone than adults. Include a main idea, supporting details and examples, and a concluding sentence.

REVIEW THE BIG IDEA

20 How do the skeletal system, the muscular system, and the skin help the body maintain homeostasis? Explain how each system works independently to maintain homeostasis, and give an example of how two of them work together.

21 The photo below shows two soccer players—one using just muscles, the other using just bones. Explain why this would not be possible in real life.

Math Skills

Math Practice

Use Proportions

22 A particular person produces about 95 mg of sweat per minute. How much sweat will the person produce in 1 hour?

23 There is approximately 1.8 m^2 of skin on the average human body. (1 m^2 =10,000 cm^2)

A. What is the surface area of the human body in cm^2?

B. If the body averages 277 sweat glands per square centimeter, about how many sweat glands would the human body contain?

Standardized Test Practice

Record your answers on the answer sheet provided by your teacher or on a sheet of paper.

Multiple Choice

1 Which is part of the integumentary system?

A fingernail

B lung

C muscle

D nerve

Use the diagram below to answer question 2.

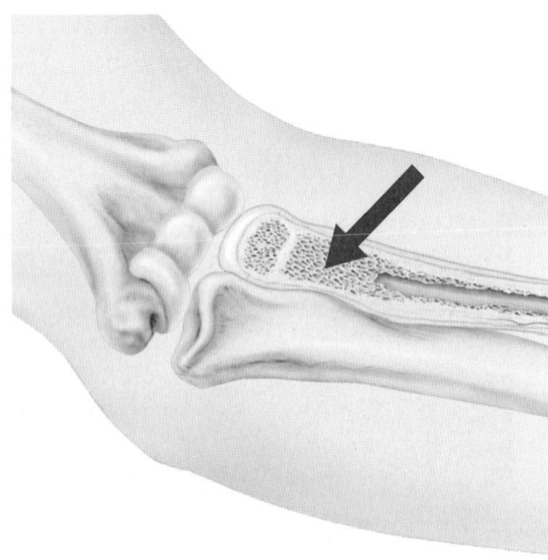

2 What does the arrow in the figure point to?

A cartilage

B compact bone

C spongy bone

D yellow marrow

3 Which does NOT help to control a person's body temperature?

A contracting muscles rapidly

B enlarging blood vessels

C producing melanin

D sweating through pores

4 Which is a characteristic of skeletal muscles?

A They are involuntary.

B They are smooth muscle.

C They contain discs.

D They have striations.

5 Where is cardiac muscle?

A heart

B legs

C lungs

D stomach

Use the diagram below to answer question 6.

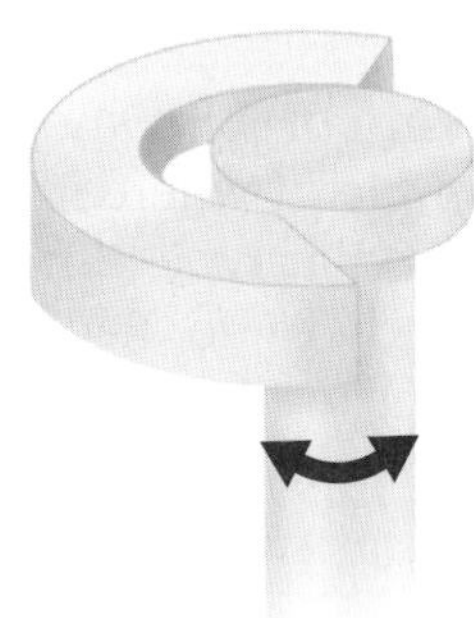

6 Where is this type of joint in the human body?

A finger

B knee

C neck

D shoulder

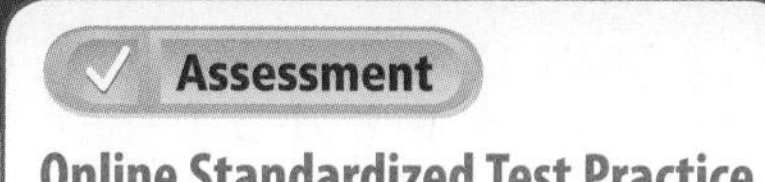

Use the figure below to answer questions 7 and 8.

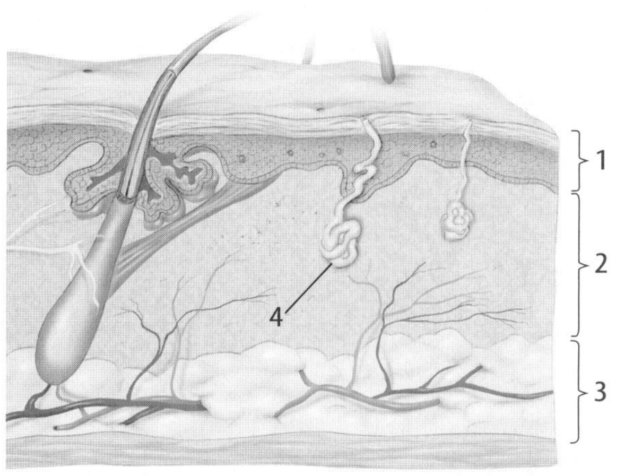

7 Where is melanin produced?

A 1

B 2

C 3

D 4

8 Which area gives the skin strength, flexibility, and nourishment?

A 1

B 2

C 3

D 4

9 Which bone material stores fat?

A cartilage

B compact bone

C spongy bone

D yellow bone marrow

Constructed Response

Use the image below to answer questions 10–13.

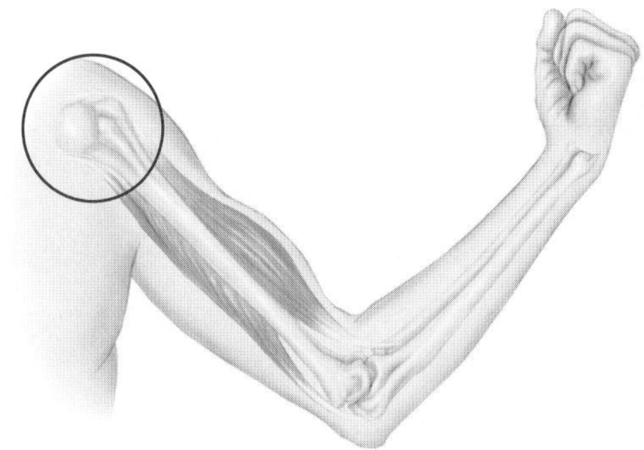

10 What joint type is circled in the diagram? What other area of the body contains the same kind of joint?

11 How do ligaments and tendons function in this type of joint?

12 What kind of muscle is attached to the joint shown? Explain your answer. How does this type of muscle differ from the muscle in a person's stomach?

13 How are the functions of muscle and skin similar? Explain how each performs these functions.

NEED EXTRA HELP?													
If You Missed Question...	1	2	3	4	5	6	7	8	9	10	11	12	13
Go to Lesson...	3	1	2, 3	2	2	1	3	3	1	1	1, 2	2	2, 3

Chapter 15

Digestion and Excretion

THE BIG IDEA **How do the digestive and excretory systems help maintain the body's homeostasis?**

Inquiry Why So Long?

This image shows parts of the digestive system. The small intestine is the structure that looks like a tangled-up rope. The small intestine can be up to 6 m long.

- Why do you think the small intestine is so long?
- What do you think the function of the small intestine is?
- How might the digestive system help your body maintain homeostasis?

Get Ready to Read

What do you think?

Before you read, decide if you agree or disagree with each of these statements. As you read this chapter, see if you change your mind about any of the statements.

1. An activity such as sleeping does not require energy.
2. All fats in food should be avoided.
3. Digestion begins in the mouth.
4. Energy from food stays in the digestive system.
5. Several human body systems work together to eliminate wastes.
6. Blood contains waste products that must be removed from the body.

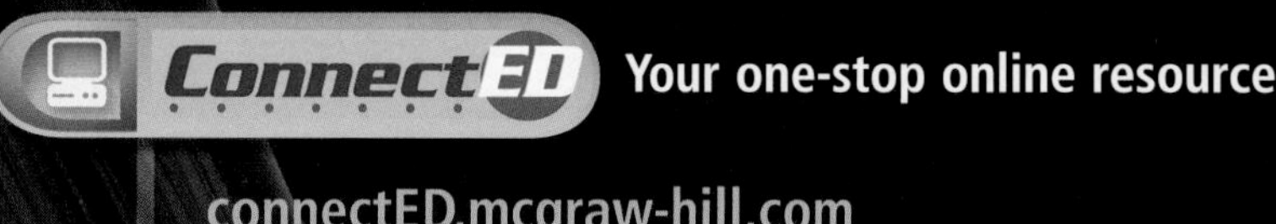

Video

Audio

Review

Inquiry

WebQuest

Assessment

Concepts in Motion

Multilingual eGlossary

Lesson 1

Nutrition

Reading Guide

Key Concepts
ESSENTIAL QUESTIONS

- Why do you eat?
- Why does your body need each of the six groups of nutrients?
- Why is eating a balanced diet important?

Vocabulary

Calorie p. 523
protein p. 524
carbohydrate p. 524
fat p. 525
vitamin p. 525
mineral p. 525

 Multilingual eGlossary

 Video Science Video

Inquiry Time for Lunch?

This photo shows fried moth larvae on a banana leaf. It might not look appetizing, but it contains nutrients your body needs for energy and growth. Nutrients are in many different foods, from a cheeseburger to a fried insect.

Launch Lab

20 minutes

How much energy is in an almond?

⚠ Food allergy

Food contains energy. Is there enough energy in an almond to boil water?

1. Read and complete a lab safety form.
2. Place a small amount of **clay** in a **shallow baking dish.** Straighten a **metal paper clip.** Insert one end into an unshelled **almond.** Anchor the other end in the clay.
3. Place a **25-mL test tube** in a **test-tube clamp.** Add 10 mL of **water** to the test tube.
4. Have your partner light the almond with a **long wooden match** until the almond starts burning on its own.
5. Gently swirl the test tube at an angle over the flame until the almond completely burns. Record your observations in your Science Journal.

⚠ *Point the test tube away from fellow students.*

Think About This

1. What happened to the water? Why did this happen?
2. **Key Concept** What do you think happens to your body when you eat an almond?

Why do you eat?

How do you decide what to eat or when to eat? Although you can survive for weeks without food, you might become hungry within hours of your last meal. Hunger is your body's way of telling you that it needs food. Why does your body need food? Food provides your body with the energy and nutrients it needs to survive.

Energy

Every activity you do, such as riding a bike or even sleeping, requires energy. Your digestive system processes food and releases energy that is used for cellular processes and all activities that you do.

The amount of energy in food is measured in Calories. *A* **Calorie** *(Cal) is the amount of energy it takes to raise the temperature of 1 kg of water by 1°C.* How much energy do foods contain? Each food is different. One grape contains 2 Cal, but a slice of cheese pizza has 220 Cal. All foods give your body energy to use.

The amount of energy a person needs depends on several factors, such as weight, age, activity level, and gender. For example, a person with a mass of 68 kg usually burns more Calories than a person with a mass of 45 kg. Playing soccer requires more energy than playing a video game. How does the food you eat supply you with energy? The energy comes from nutrients.

Nutrients

Food is made of nutrients—substances that provide energy and materials for cell development, growth, and repair. The types and amounts of nutrients a person needs depend on age, gender, and activity level. Toddlers need more fat in their diets than older children do. Women need more calcium and iron than men do. Active people need more protein. Next, you'll read about the six groups of nutrients and their roles in maintaining your health.

Key Concept Check Why do you eat?

Groups of Nutrients

The six groups of nutrients are proteins, carbohydrates, fats, vitamins, minerals, and water. Each nutrient has a different function in the body. To be healthy, you need foods from each group every day.

Proteins

Most of the tissues in your body are made of proteins. *A* **protein** *is a large molecule that is made of amino acids and contains carbon, hydrogen, oxygen, nitrogen, and sometimes sulfur.* Proteins have many functions, such as relaying signals between cells, protecting against disease, providing support to cells, and speeding up chemical reactions. All of these functions are needed to maintain homeostasis, or the regulation of an organism's internal condition regardless of changes in its environment.

Word Origin
protein
from Greek *proteios,* means "the first quality"

Combinations of 20 different amino acids make up the proteins in your body. Your cells can make more than half of these amino acids. The remaining amino acids must come from the foods that you eat. Some foods that are good sources of protein are shown in **Figure 1.**

Reading Check How does your body obtain amino acids that cannot be made in cells?

Figure 1 Good sources of protein include red meat, eggs, beans, and peanuts. Good sources of carbohydrates include red beans, fruits, and vegetables.

Proteins

Carbohydrates

Visual Check Describe a lunch that is high in proteins and carbohydrates.

Carbohydrates

What do pasta, bread, and potatoes have in common? They are all foods that have high levels of carbohydrates (kar boh HI drayts). **Carbohydrates** *are molecules made of carbon, hydrogen, and oxygen atoms and are usually the body's major source of energy.* They are commonly in one of three forms—starches, sugars, or fibers. All of them are made of sugar molecules that are linked together like a chain. It is best to eat foods that contain carbohydrates from whole grains because they are easier to digest. Also shown in **Figure 1** are some foods that are high in carbohydrates.

Fats

You might think that fats in food are bad for you. But, you need a certain amount of fat in your diet and on your body to stay healthy. **Fats,** *also called lipids, provide energy and help your body absorb vitamins.* They are a major part of cell membranes. Body fat helps to insulate against cold temperatures. Most people get plenty of fat in their diet, so deficiencies in fats are rare. But too much fat in your diet can lead to health problems. Only about 25–35 percent of the Calories you consume should be fats.

Fats are often classified as either saturated or unsaturated. A diet high in saturated fats can increase levels of cholesterol, which can increase the risk of heart disease. Most of the fat in your diet should come from unsaturated fats, such as those shown in **Figure 2.**

Figure 2 Fish, nuts, and liquid vegetable oils contain unsaturated fats.

Fats

Vitamins

Has anyone ever told you to eat certain foods because you need vitamins? **Vitamins** *are nutrients that are needed in small amounts for growth, regulation of body functions, and prevention of some diseases.* You can obtain most of the vitamins you need by eating a well-balanced diet. If you do not consume enough of one or more vitamins, then you might develop symptoms of vitamin deficiency. The symptoms depend on which vitamin you are lacking. **Table 1** lists some vitamins people need in their diet.

Reading Check Why do you need vitamins in your diet?

Minerals

In addition to vitamins, you also need other nutrients called minerals. **Minerals** *are inorganic nutrients—nutrients that do not contain carbon—that help the body regulate many chemical reactions.* Similar to vitamins, if you do not consume enough of certain minerals, you might develop a mineral deficiency. **Table 1** also lists some minerals that you need in your diet.

Concepts in Motion Interactive Table

Table 1 Vitamins and Minerals

Vitamin	Good Sources	Health Benefit
Vitamin B_2 (riboflavin)	milk, meats, vegetables	helps release energy from nutrients
Vitamin C	oranges, broccoli, tomatoes, cabbage	growth and repair of body tissues
Vitamin A	carrots, milk, sweet potatoes, broccoli	enhances night vision, helps maintain skin and bones
Mineral	**Good Sources**	**Health Benefit**
Calcium	milk, spinach, green beans	builds strong bones and teeth
Iron	meat, eggs, green beans	helps carry oxygen throughout the body
Zinc	meat, fish, wheat/grains	aids protein formation

Table 1 Vitamins and minerals are essential for maintaining a healthy body.

Visual Check What foods are good sources of vitamin A?

Water

You might recall that your body is mostly water. You need water for chemical reactions to occur in your body. Your body takes in water when you eat or drink. However, you lose water when you sweat, urinate, and breathe. To stay healthy, it is important to replace the water that your body loses. If you exercise, live in a warm area, or become sick, your body loses more water. When lost water is not replaced, you might become dehydrated. Symptoms of dehydration include thirst, headache, weakness, dizziness, and little or no urination.

 **Key Concept Check** Why does your body need nutrients?

FOLDABLES®

Fold a sheet of paper into a chart with three columns and two rows. Use your chart to organize information about the major food groups and to list examples of each.

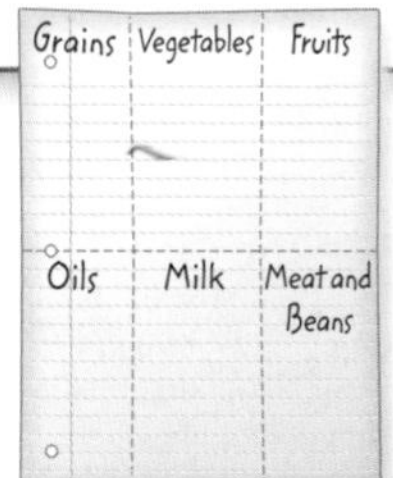

Healthful Eating

Imagine walking through a grocery store. Each aisle in the store contains hundreds of different foods. With so many choices, it's difficult to choose foods that are part of a healthful diet. Healthful eaters need to be smart shoppers. They make grocery lists beforehand and buy products that are high in nutrients. Nutritious foods come from the major food groups, which include grains, vegetables, fruits, oils, milk products, and meats and beans.

Inquiry MiniLab

25 minutes

What nutrients are in foods?

Food provides your body with nutrients and Calories. Each nutrient is important and has its own function in the body.

1. Using the materials provided by your teacher, search for foods that contain a high amount of your assigned nutrient.
2. Find the number of items for your nutrient that your teacher has assigned.
3. Once you have found the appropriate number of items, form a group with other students who were assigned the same nutrient.
4. As a group, make a chart listing your food items. Show the amount of your assigned nutrient present in each item. Share your chart with the class.

Analyze and Conclude

1. **Classify** the foods studied by all groups according to their nutrient value. Which foods were high in proteins? Fats? Carbohydrates?
2. **Explain** the function each nutrient has in the body.
3. **Key Concept** Describe what might happen if your body did not get enough of a particular nutrient.

Table 2 Daily Recommended Amounts of Each Food Group for 9–13-Year-Olds

Food Group	Daily Amount males, 9–13 years old	Daily Amount females, 9–13 years old	Examples of Foods
Grains	6-ounce equivalents	5-ounce equivalents	whole-wheat flour, rye bread, brown rice
Vegetables	2 1/2 cups	2 cups	broccoli, spinach, carrots
Fruits	1 1/2 cups	1 1/2 cups	apples, strawberries, oranges
Fats	5 teaspoons or less	5 teaspoons or less	canola oil, olive oil, avocados
Milk	3 cups	3 cups	milk, cheese, yogurt
Meat and beans	5 ounces or less	5 ounces or less	fish, beans, lean beef, lean chicken

A Balanced Diet

A healthful diet includes carbohydrates, proteins, fats, vitamins, minerals, and water. But how do you know how much of each food group you should eat? **Table 2** lists the daily recommended amounts of each food group for 9–13-year-olds.

The nutrient-rich foods that you choose might be different from the nutrient-rich foods eaten by people in China, Kenya, or Mexico. People usually eat foods that are grown and produced regionally. Regardless of where you live, eating a balanced diet ensures that your body has the nutrients it needs to function.

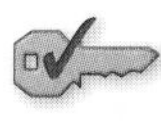

Key Concept Check Why is eating a balanced diet important?

Food Labels

What foods would you buy to follow the recommended guidelines in **Table 2?** Most grocery stores sell many varieties of bread, milk, meat, and other types of food. How would you know what nutrients these foods contain? You can look at food labels, such as the one in **Figure 3.** Food labels help you determine the amount of protein, carbohydrates, fats, and other substances in food.

Figure 3 A food label lists a food's nutrients per serving, not per container.

Visual Check List the nutrients in this food product.

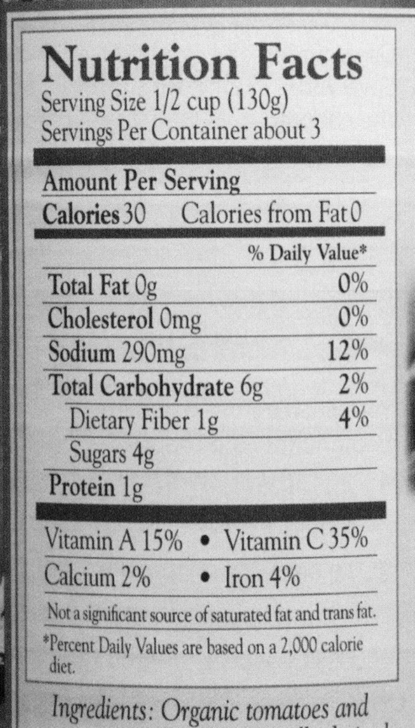

Lesson 1 Review

Assessment Online Quiz
Inquiry Virtual Lab

Visual Summary

People eat food to obtain the energy their bodies need to function.

Proteins are one of the six groups of nutrients.

Evaluating food labels can help you eat a balanced diet.

FOLDABLES®

Use your lesson Foldable to review the lesson. Save your Foldable for the project at the end of the chapter.

What do you think NOW?

You first read the statements below at the beginning of the chapter.

1. An activity such as sleeping does not require energy.
2. All fats in food should be avoided.

Did you change your mind about whether you agree or disagree with the statements? Rewrite any false statements to make them true.

Use Vocabulary

1. Nutrients made of long chains of amino acids are ________.
2. The major source of energy in your diet comes from ________.
3. The amount of energy in food is measured in ________.

Understand Key Concepts 

4. **Explain** why it is important to consume vitamins.
5. Which nutrient helps your body absorb vitamins?
   A. carbohydrate
   B. fat
   C. mineral
   D. protein
6. **Give an example** of when you might need to drink more water than usual.

Interpret Graphics

7. **Calculate** How many grams of carbohydrates are in three servings of this food?

Sodium 290mg	12%
Total Carbohydrate 6g	2%
Dietary Fiber 1g	4%
Sugars 4g	
Protein 1g	

8. **Summarize** Copy and fill in the graphic organizer below to identify the six groups of nutrients.

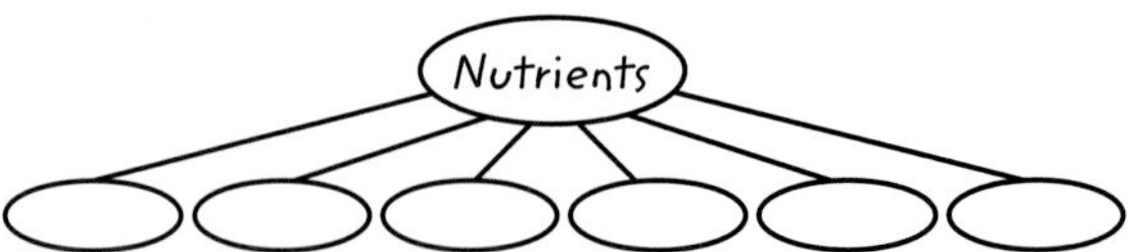

Critical Thinking

9. **Plan** a meal that contains a food from each of the six food groups.
10. **Analyze** One serving of a certain food contains 370 Cal, 170 Cal from fat, and 12 g of saturated fat (60% of the daily value). Is this food a good choice for a healthful lifestyle? Why or why not?

Inquiry **Skill Practice** **Compare and Contrast** 45 minutes

How do foods compare?

Materials

brown paper grocery bag

permanent marker

food samples

small plastic cup

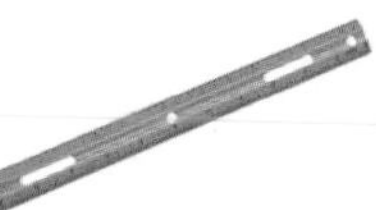

metric ruler

Safety

As you have learned, not all foods are alike. Knowing about different types of nutrients will help you make good food choices. Foods with a lot of fat often taste good but might not be healthful for you in large amounts. It is important to be able to identify foods with different fat contents in order to have a balanced diet. How do these different foods compare?

Learn It

Observations can be analyzed by noting the similarities and differences between two or more objects or events that you observe. You **compare** objects or events by seeing how they are similar. You **contrast** objects or events by looking for differences.

4

Try It

1. Read and complete a lab safety form.
2. Create a data table like the one below in your Science Journal.
3. Use a permanent marker and a plastic cup to draw seven circles on a large piece of a brown paper grocery bag.
4. Obtain one each of the seven food items your teacher has provided. Label each circle with the name of the food to be tested.
5. Place one piece of the labeled food in each circle.
6. Allow the foods to sit for 30 minutes.
7. Remove the foods and properly dispose of them. Record in the table whether the food left a greasy mark, a wet mark, or no mark. Also record the diameter of the mark.
8. Dispose of the used grocery bag as directed by your teacher.

Apply It

9. **Compare and contrast** the marks produced by the foods. Describe both their appearances and their sizes.
10. **Infer** Which items left a greasy mark on the paper bag? How are these foods alike?
11. **Key Concept** Why is it important to eat a variety of foods every day?

Product	Type of Mark	Diameter (cm)

Lesson 2

The Digestive System

Reading Guide

Key Concepts
ESSENTIAL QUESTIONS

- What does the digestive system do?
- How do the parts of the digestive system work together?
- How does the digestive system interact with other systems?

Vocabulary

digestion p. 531
mechanical digestion p. 532
chemical digestion p. 532
enzyme p. 532
esophagus p. 534
peristalsis p. 534
chyme p. 535
villi p. 536

Multilingual eGlossary

Video BrainPOP®

Inquiry Under the Sea?

These colorful projections look like something you might see on the ocean floor, but they are found in your body. They line the walls of the small intestine, which is part of your digestive system. What do you think these projections do?

Launch Lab

15 minutes

Which dissolves faster?

Has anyone ever told you to take small bites and chew your food thoroughly? The size of chewed food particles can affect how quickly food is digested. Similarly, the size of a sugar particle can affect how fast it dissolves in water.

1. Read and complete a lab safety form.
2. Add the contents of one serving package of **granulated sugar** to a **500-mL beaker** containing 300 mL of **warm water.**
3. Gently stir the contents of the beaker with a **plastic spoon.** Have your partner use a **stopwatch** to time how long it takes the sugar to dissolve. Record the time in your Science Journal.
4. Add a **sugar cube** to another **500-mL beaker** containing 300 mL of warm water.
5. Repeat step 3.

Think About This

1. Which dissolved faster—the granulated sugar or the sugar cube?
2. Why do you think particle size affects the rate at which sugar dissolves?
3. **Key Concept** How might food particle size affect how quickly food is digested?

Functions of the Digestive System

Suppose you ate a cheeseburger and a pear for lunch. What happens to the food after it is eaten?

As soon as the food enters your mouth, it begins its journey through your digestive system. No matter what you eat, your food goes through four steps—ingestion, digestion, absorption, and elimination. All four steps happen in the organs and tissues of the digestive system in the following order:

- Food is ingested. Ingestion is the act of eating, or putting food in your mouth.
- Food is digested. **Digestion** *is the mechanical and chemical breakdown of food into small particles and molecules that your body can absorb and use.*
- Nutrients and water in the food are absorbed, or taken in, by cells. Absorption occurs when the cells of the digestive system take in small molecules of digested food.
- Undigested food is eliminated. Elimination is the removal of undigested food and other wastes from your body.

Key Concept Check What does the digestive system do?

WORD ORIGIN
digestion
from Latin *digestus,* means "to separate, divide"

Types of Digestion

Before your body can absorb nutrients from food, the food must be broken down into small molecules by digestion. There are two types of digestion—mechanical and chemical. *In* **mechanical digestion**, *food is physically broken into smaller pieces.* Mechanical digestion happens when you chew, mash, and grind food with your teeth and tongue. Smaller pieces of food are easier to swallow and have more surface area, which helps with chemical digestion. *In* **chemical digestion**, *chemical reactions break down pieces of food into small molecules.*

Enzymes

Chemical digestion cannot occur without substances called enzymes (EN zimez). **Enzymes** *are proteins that help break down larger molecules into smaller molecules. Enzymes also speed up, or catalyze, the rate of* **chemical reactions.** Without enzymes, some chemical reactions would be too slow or would not occur at all.

REVIEW VOCABULARY

chemical reaction
process in which a compound is formed or broken down

There are many kinds of enzymes. Each one is specialized to help break down a specific molecule at a specific location.

 **Reading Check** What are enzymes?

20 minutes

How can you model digestion?

Your saliva contains enzymes to help with digestion. You can use radishes to model the effect saliva has on food. Radishes and saliva contain the same enzyme.

1. Read and complete a lab safety form.
2. Place a small amount of **cooked rice** into two **100-mL beakers.**
3. Add a small amount of **grated radish** to one beaker and stir well with a **plastic spoon.**
4. Let the rice sit for 5 minutes.
5. Use a **dropper** to add three drops of **iodine tincture solution** to the rice in each beaker. Record the color of the rice in your Science Journal.

Analyze and Conclude

1. **Compare** the colors of the rice in the two beakers after the iodine was added.
2. **Infer** Iodine reacts with starches. Starches are made up of sugar molecules. Infer what happened to the starches in the rice when an enzyme was added.
3. **Key Concept** Summarize the role enzymes play in digestion.

The Role of Enzymes in Digestion

Nutrients in food are made of different molecules, such as carbohydrates, proteins, and fats. Many of these molecules are too large for your body to use. But, because these molecules are made of long chains of smaller molecules joined together, they can be broken down into smaller pieces.

The digestive system produces enzymes that are specialized to help break down each type of food molecule. For example, the enzyme amylase helps break down carbohydrates. The enzymes pepsin and papain help break down proteins. Fats are broken down with the help of the enzyme lipase. **Figure 4** illustrates how an enzyme helps break down food molecules into smaller pieces.

Notice in **Figure 4** that the food molecule breaks apart, but the enzyme itself does not change. Therefore, the enzyme can immediately be used to break down another food molecule.

Reading Check What happens to an enzyme after it helps break down a food molecule?

Organs of the Digestive System

In order for your body to use the nutrients in the foods you eat, the nutrients must pass through your digestive system. Your digestive system has two parts: the digestive tract and the other organs that help the body break down and absorb food. These organs include the tongue, salivary glands, liver, gallbladder, and pancreas.

The digestive tract extends from the mouth to the anus. It has different organs connected by tubelike structures. Each of these organs is specialized for a certain function.

Recall the cheeseburger and pear mentioned at the beginning of this lesson. Where do you think digestion of this food begins?

Figure 4 An enzyme helps break down food molecules into smaller pieces.

Step 1
An enzyme attaches to a food particle.

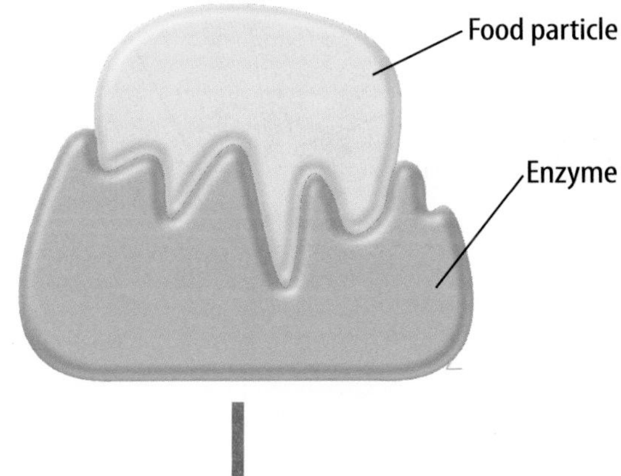

Step 2
The enzyme speeds up a chemical reaction that breaks down the food particle.

Step 3
The enzyme releases the broken-down food particle.

The Digestive System

Concepts in Motion Animation

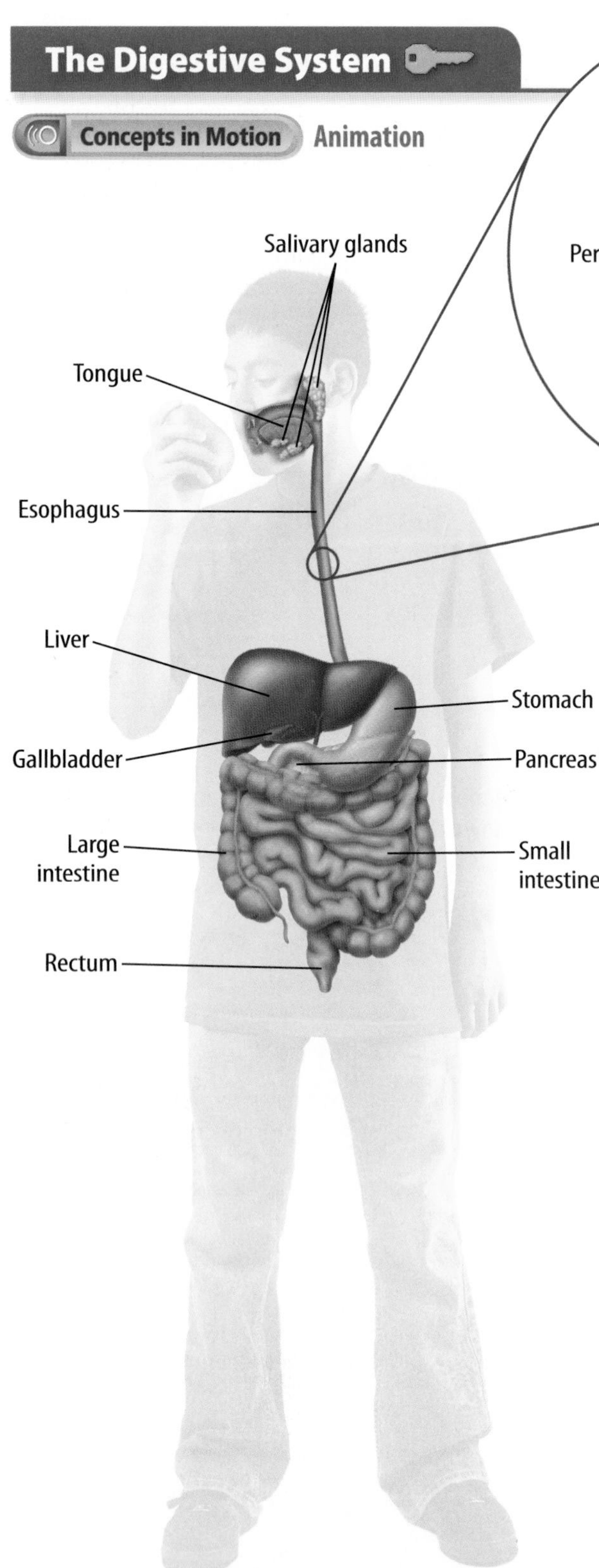

Figure 5 The digestive system includes the organs of the digestive tract, as well as other organs such as the tongue, salivary glands, liver, gallbladder, and pancreas.

Visual Check Which organ connects the mouth to the stomach?

The Mouth

You can follow the path food takes through your digestive tract in **Figure 5.** Mechanical digestion of food, such as a pear or a cheeseburger, begins in your mouth. Your teeth and tongue mechanically digest food as you chew. But even before chewing begins, your mouth prepares for digestion.

Your salivary (SA luh ver ee) glands produce saliva (suh LI vuh) at the very thought of food. They produce more than 1 L of saliva every day. Saliva contains an enzyme that helps break down carbohydrates, such as those found in a hamburger bun. Saliva also contains substances that neutralize acidic foods. It also contains a slippery substance that makes food easier to swallow.

The Esophagus

After you swallow a bite of your food, it enters your esophagus (ih SAH fuh gus). *The* **esophagus** *is a muscular tube that connects the mouth to the stomach. Food moves through the esophagus and the rest of the digestive tract by waves of muscle contractions, called* **peristalsis** (per uh STAHL sus).

Peristalsis is similar to squeezing a tube of toothpaste. When you squeeze the bottom of the tube, toothpaste is forced toward the top of the tube. As muscles in the esophagus contract and relax, partially digested food is pushed down the esophagus and into the stomach.

The Stomach

Once your partially digested food leaves the esophagus, it enters the stomach. The stomach is a large, hollow organ. One function of the stomach is to temporarily store food. This allows you to go many hours between meals. The stomach is like a balloon that can stretch when filled. An adult stomach can hold about 2 L of food and liquids.

Reading Check Why is the stomach's ability to store food beneficial?

Another function of the stomach is to aid in chemical digestion. As shown in **Figure 6,** the walls of the stomach are folded. These folds enable the stomach to expand and hold large amounts of food. In addition, the cells in these folds produce chemicals that help break down proteins. For example, the stomach contains an acidic fluid called gastric juice. Gastric juice makes the stomach acidic. Acid helps break down some of the structures that hold plant and animal cells together, like the cells in hamburger meat, lettuce, tomatoes, and pears. Gastric juice also contains pepsin, an enzyme that helps break down proteins in foods into amino acids. Food and gastric juices mix as muscles in the stomach contract through peristalsis. As food mixes with gastric juice in the stomach, it forms *a thin, watery liquid called* **chyme** (KIME).

FOLDABLES®

Make a shutterfold book to illustrate the organs of the digestive system. Use it to record information about their functions.

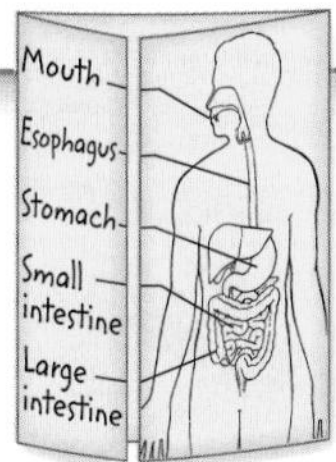

Figure 6 The stomach temporarily stores food and aids in chemical digestion.

Visual Check Where does food go after it leaves the stomach?

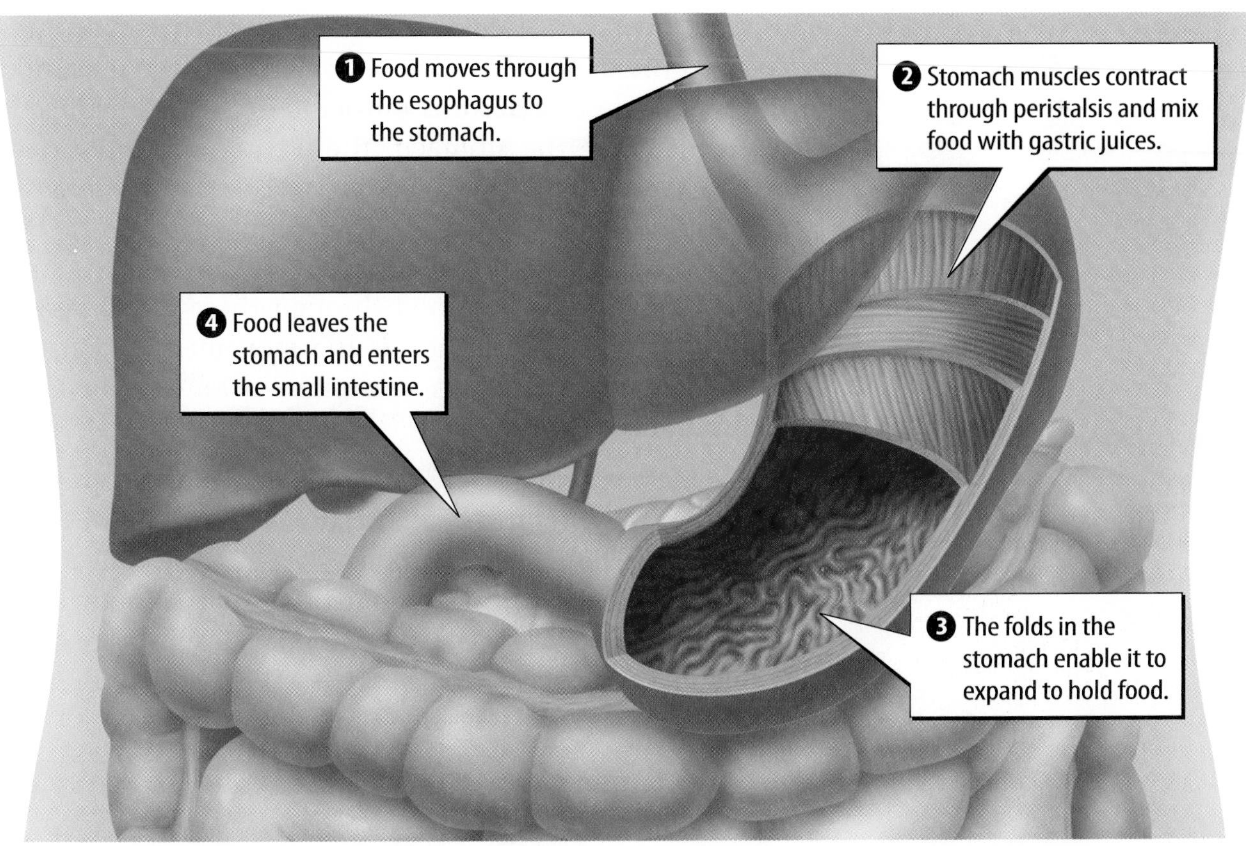

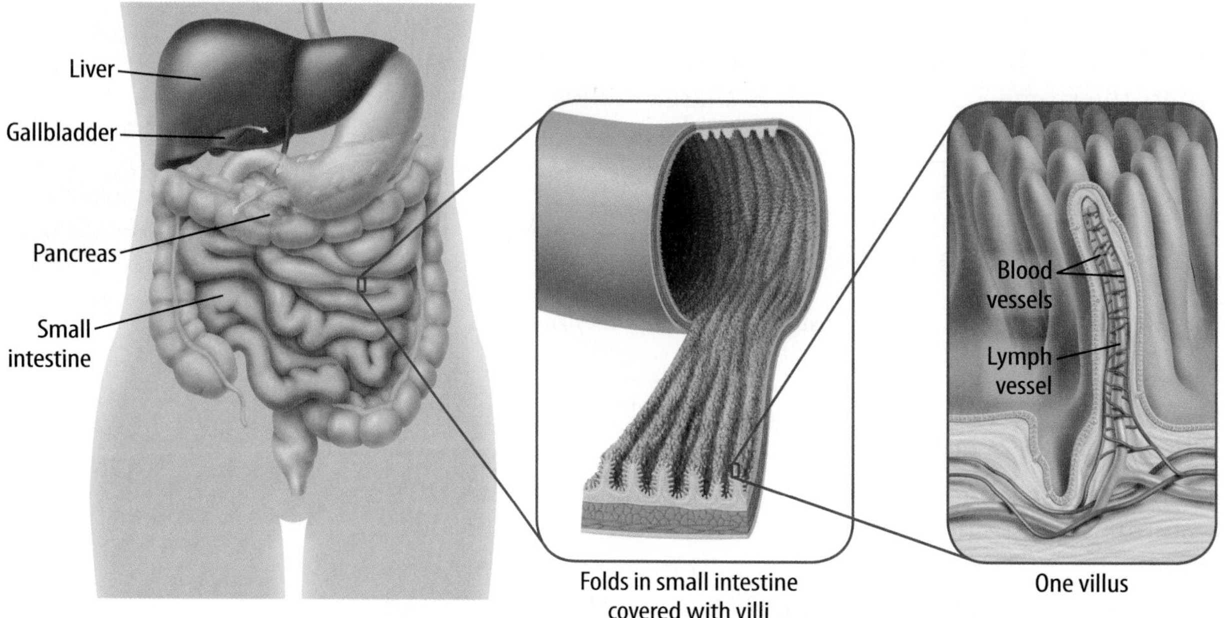

Figure 7 The walls of the small intestine are covered with villi that help move nutrients into the blood.

The Small Intestine

Chemical digestion of your cheeseburger and pear begins in the mouth and stomach. But most chemical digestion occurs in the small intestine. The small intestine is a long tube connected to the stomach. It is where chemical digestion and nutrient absorption occur. The small intestine is named for its small diameter—about 2.5 cm. It is about 7 m long.

Chemical digestion of proteins, carbohydrates, nucleic acids, and fats takes place in the first part of the small intestine, called the duodenum (doo uh DEE num). The remainder of the small intestine absorbs nutrients from food. Notice in **Figure 8** that, like the stomach, the wall of the intestine is folded. *The folds of the small intestine are covered with fingerlike projections called* **villi** (VIH li) (singular, villus). Notice also that each villus contains small blood vessels. Nutrients in the small intestine diffuse into the blood through these blood vessels. You might recall that diffusion is the movement of particles from an area of higher concentration to an area of lower concentration.

Science Use v. Common Use

substance

Science Use matter that has a particular chemical makeup

Common Use a fundamental quality

The pancreas and the liver, shown in **Figure 7,** produce **substances** that enter the small intestine and help with chemical digestion. The pancreas produces an enzyme called amylase that helps break down carbohydrates and a substance that neutralizes stomach acid. The liver produces a substance called bile. Bile makes it easier to digest fats. The gallbladder stores bile until it is needed in the small intestine.

Key Concept Check What organs work together to help with chemical digestion?

Figure 8 The bacteria shown here live in the large intestine. Without them, your food would not be digested well.

Visual Check Cocci bacteria are spherical, bacilli bacteria are rod-shaped, and spirilla bacteria are spiral-shaped. Which type of bacteria is shown in the photo?

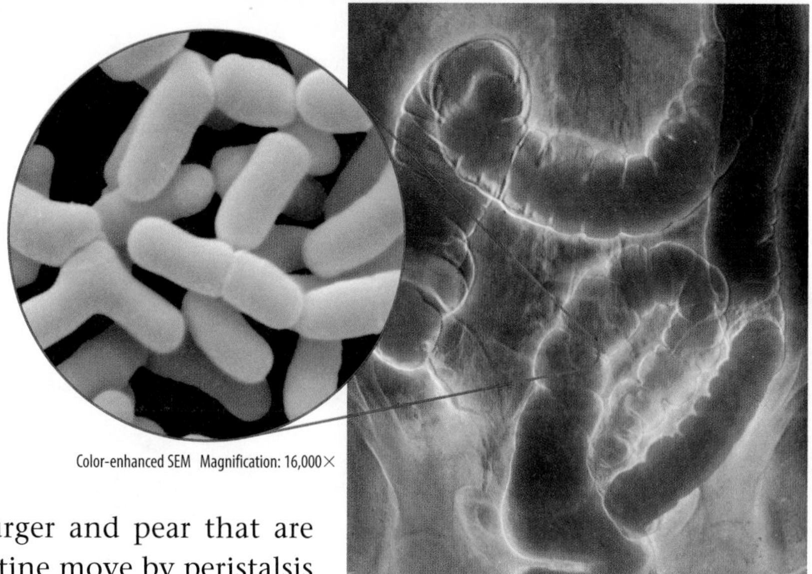

Color-enhanced SEM Magnification: 16,000×

The Large Intestine

The parts of your cheeseburger and pear that are not absorbed in the small intestine move by peristalsis into the large intestine, also called the colon. The large intestine, shown in **Figure 8,** has a larger diameter (about 5 cm) than the small intestine. However, at about 1.5 m long, it is much shorter than the small intestine.

Most of the water in ingested foods and liquids is absorbed in the small intestine. As food travels through the large intestine, even more water is absorbed. Materials that pass through the large intestine are the waste products of digestion. The waste products become more solid as excess water is absorbed. Peristalsis continues to force the remaining semisolid waste material into the last section of the large intestine, called the rectum. Muscles in the rectum and anus control the release of this semisolid waste, called feces (FEE seez).

Bacteria and Digestion

You might think that all bacteria are harmful. However, some bacteria have an important role in the digestive system. Bacteria, such as the ones shown in **Figure 8,** digest food and produce important vitamins and amino acids. Bacteria in the intestines are essential for proper digestion.

The Digestive System and Homeostasis

Recall that nutrients from food are absorbed in the small intestine. The digestive system must be functioning properly for this absorption to occur. These nutrients are necessary for other body systems to maintain homeostasis. For example, the blood in the circulatory system absorbs the products of digestion. The blood carries the nutrients to all other body systems, providing them with materials that contain energy.

Key Concept Check What might happen to other body systems if the digestive system did not function properly?

Math Skills 

Use Percentages

A percentage is a ratio that compares a number to 100. For example, the total length of the intestines is about 8.5 m. That value represents 100%. If the rectum is 0.12 m long, what percentage of the intestines is made up of the rectum?

The ratio is $\frac{0.12 \text{ m}}{8.5 \text{ m}}$.

Find the equivalent decimal for the ratio.

$$\frac{0.12 \text{ m}}{8.5 \text{ m}} = 0.014$$

Multiply by 100.

$$0.014 \times 100 = 1.4\%$$

Practice

The total length of the intestines is about 8.5 m. If the small intestine is 7.0 m long, what percentage of the intestines is made up of the small intestine?

- Math Practice
- Personal Tutor

Lesson 2 Review

Visual Summary

Enzymes in the digestive system break down food so nutrients can be absorbed by your body.

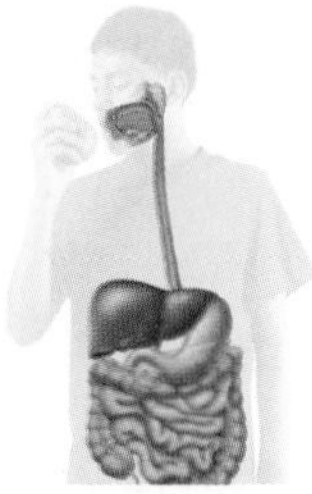

Food moves through the digestive tract by waves of peristalsis.

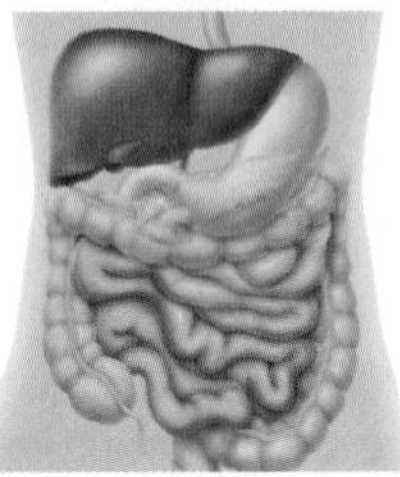

The liver and the pancreas produce substances that help with chemical digestion.

FOLDABLES®

Use your lesson Foldable to review the lesson. Save your Foldable for the project at the end of the chapter.

What do you think NOW?

You first read the statements below at the beginning of the chapter.

3. Digestion begins in the mouth.

4. Energy from food stays in the digestive system.

Did you change your mind about whether you agree or disagree with the statements? Rewrite any false statements to make them true.

Use Vocabulary

1. **Define** *enzyme* in your own words.
2. **Distinguish** between absorption and digestion.

Understand Key Concepts 

3. Where is the first place digestion occurs?
   A. mouth
   B. stomach
   C. large intestine
   D. small intestine
4. **Compare** the functions of the stomach and the small intestine.
5. **Give an example** of how the digestive system affects other body systems.

Interpret Graphics

6. **Explain** How do structures like the one to the right affect digestion?

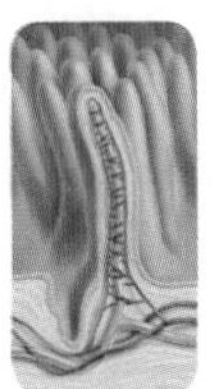

7. **Organize Information** Copy and fill in the graphic organizer below to show how food moves through the digestive tract.

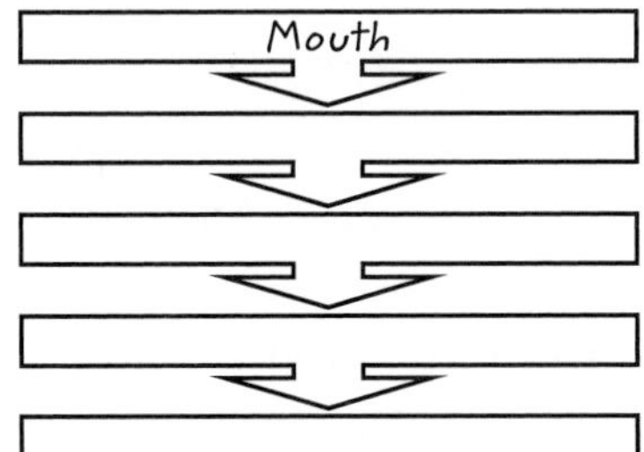

Critical Thinking

8. **Infer** what would happen if food passed more quickly than normal through the digestive system.

Math Skills

Math Practice

9. If the total length of the intestines is 8.5 m and the large intestine is 1.5 m long, what percentage of the intestines is made up of the small intestine?

Are digestive bacteria related to obesity?

SCIENCE & SOCIETY

Bacteria percentages might affect your health.

The worldwide rate of obesity greatly concerns medical and health professionals. New research reveals a possible link between bacteria in the human digestive tract and the risk of being overweight.

Your digestive system is home to between 10 and 100 trillion bacteria. That's ten times the number of cells in your body! Certain bacteria are necessary, however, for the digestion of food. Without "friendly" bacteria, you could eat all you wanted, but the food would pass through your intestines mostly undigested.

Recent studies suggest there might be a link between the bacteria in the human digestive tract and obesity. Some people have a type of bacteria that causes them to absorb more calories than normal from their food. They gain more weight than people with a different type of bacteria. In general, obese humans have a lower percentage of a group of bacteria called Bacteroidetes (BAK-tear-oid-dee-teez) and more of a group of bacteria called Firmicutes (fir-MIC-cu-teez). It is not clear whether Firmicutes bacteria make people obese, or whether obese people have more of this type of bacteria. But evidence supports the idea that changing the bacteria in someone's intestines and stomach—by means of diet or medications—might be an important weapon in the fight against obesity.

Additional research is needed to understand any link between digestive bacteria and obesity. But it is an exciting possibility that managing the bacteria in the digestive tract could be a new way to improve human health.

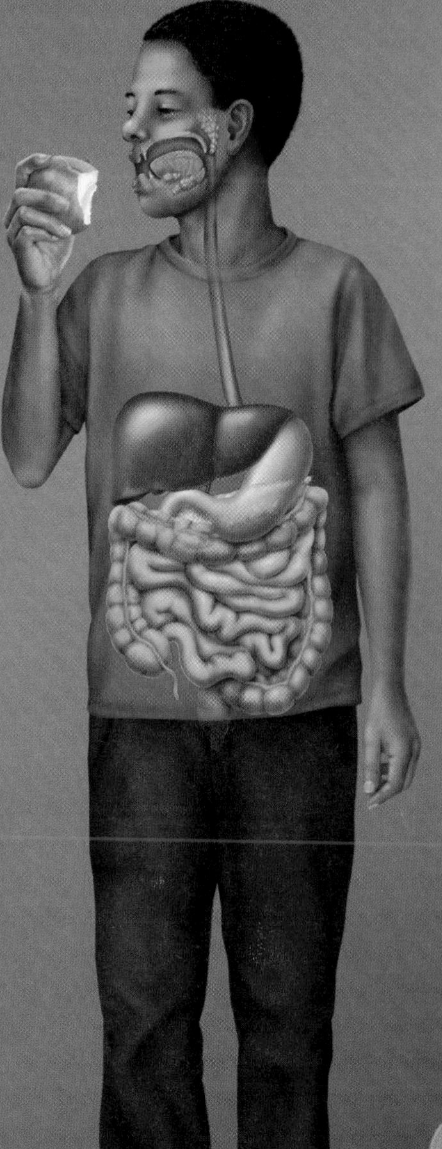

It's Your Turn

RESEARCH Find out more about the role of bacteria in human health. Research how the bacteria in your digestive tract help to regulate your immune system.

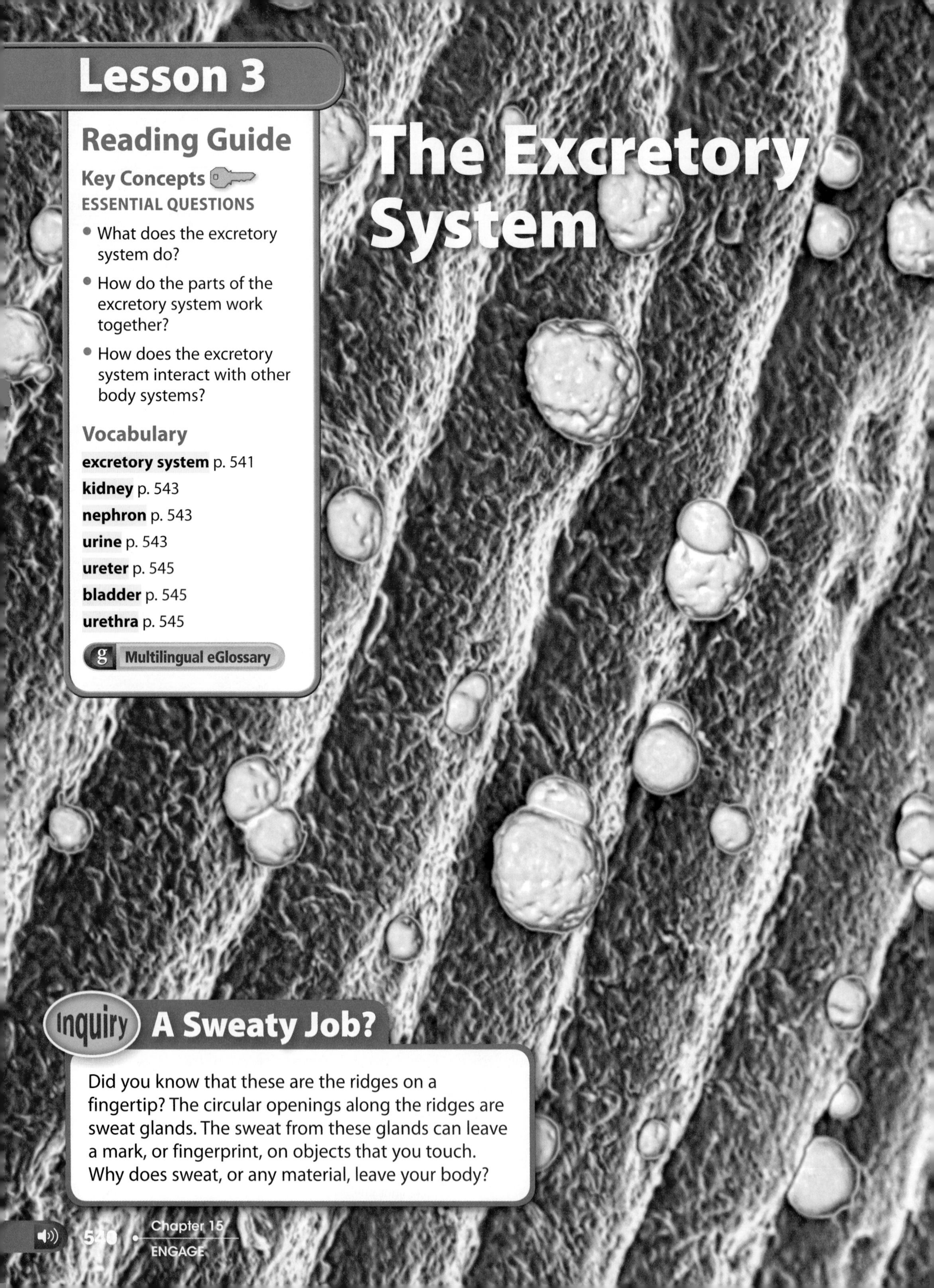

Lesson 3

The Excretory System

Reading Guide

Key Concepts
ESSENTIAL QUESTIONS

- What does the excretory system do?
- How do the parts of the excretory system work together?
- How does the excretory system interact with other body systems?

Vocabulary

g Multilingual eGlossary

Inquiry A Sweaty Job?

Did you know that these are the ridges on a fingertip? The circular openings along the ridges are sweat glands. The sweat from these glands can leave a mark, or fingerprint, on objects that you touch. Why does sweat, or any material, leave your body?

Launch Lab

10 minutes

What happens when you breathe out? 

Look again at the photo of the fingertip on the previous page. The sweat glands in your skin are one way substances leave your body. Do substances also leave your body when you breathe out?

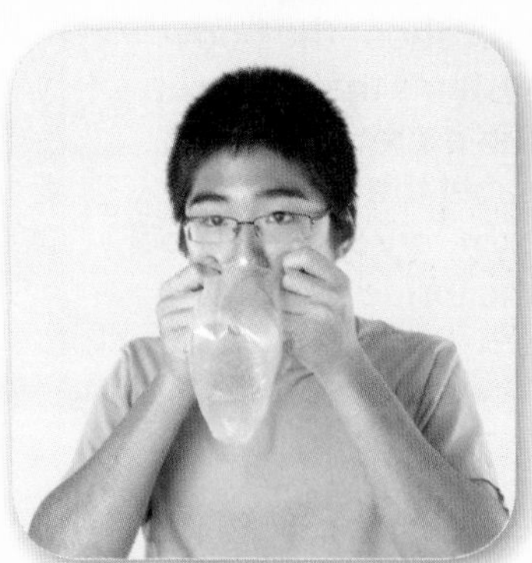

1. Read and complete a lab safety form.
2. Take a deep breath and hold it.
3. Breathe out through your mouth into a **plastic bag.** Leave a small opening to allow some of the air to leave the bag as you blow into it.
4. Remove the bag from around your mouth. Let the air escape from the bag, but do not push the sides of the bag together.
5. Using the same plastic bag, repeat steps 2–4 three more times.
6. Observe the contents of the bag. Record your observations in your Science Journal.

Think About This

1. Did the plastic bag look different after you breathed into it? Explain.
2. What do you think was in the plastic bag at the end of the activity?
3. **Key Concept** Based on your observations, do you think the respiratory system is part of the excretory system? Explain.

Functions of the Excretory System

You have read about the nutrients in food that are necessary to maintain health. You have also read how the digestive system processes that food. However, your body doesn't use all the food that you ingest. The unused food parts are waste products. What happens to the wastes? They are processed by the excretory system. *The* **excretory system** *collects and eliminates wastes from the body and regulates the level of fluid in the body.*

Collection and Elimination

Your home probably has several places where waste is collected. You might have a trash can in the kitchen and another one in the bathroom. The furnace has an air filter that removes and collects dust from the air. Similarly, your body also collects wastes. The digestive system collects waste products in the intestines. The circulatory system collects waste products in the blood.

When the trash cans in your home fill up, you must take the trash outside. The same is true of the waste in your body. If waste is not removed, or eliminated, from your body, it could become toxic and damage your organs. You'll read about the different body systems that eliminate waste later in this lesson.

Regulation of Liquids

Another function of the excretory system is to regulate the level of fluids in the body. You might recall that water is an essential nutrient for your body. Some of the water in your body is lost when waste is eliminated. The excretory system controls how much water leaves the body through elimination. This ensures that neither too little nor too much water is lost.

Key Concept Check What does the excretory system do?

FOLDABLES®

Make a four-door book to summarize information about the body systems that make up the excretory system. Label the front *The Excretory System.* Label the inside as shown.

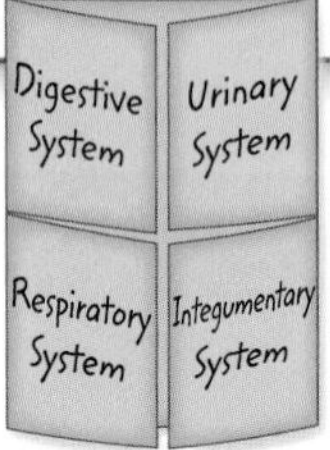

Types of Excretion

Your body excretes, or eliminates, different substances from different body systems. The excretory system is made of four body systems.

- The digestive system collects and removes undigested solids from the foods you eat.
- The urinary system processes, transports, collects, and removes liquid waste from the body.
- The respiratory system removes carbon dioxide and water vapor from the body.
- The integumentary system, which includes the skin, secretes excess salt and water through sweat glands.

Figure 9 illustrates the body systems that make up the excretory system and identifies the substances they excrete. You read previously about how the organs of the digestive system, the respiratory system, and the integumentary system eliminate waste products from the body. In this lesson, you will read about the organs of the urinary system and their roles in eliminating waste from the body.

Reading Check What body systems make up the excretory system?

Figure 9 Several body systems make up the excretory system.

Visual Check What substances are eliminated by the body systems shown below?

The Excretory System

Urinary system
Removes liquid wastes

Integumentary system
Removes excess salt and water

Digestive system
Removes undigested food

Respiratory system
Removes carbon dioxide and water

Organs of the Urinary System

The urinary system produces, stores, and removes liquid waste from the body and helps maintain homeostasis. The organs of the urinary system are shown in **Figure 10.** They include two kidneys, two ureters, the bladder, and the urethra. These organs work together to process, transport, collect, and excrete liquid waste.

Reading Check What is the function of the urinary system?

The Kidneys

The bean-shaped organ that filters, or removes, wastes from blood is the **kidney.** You have two kidneys, one on each side of your body. They are near the back wall of your abdomen, above your waist, and below your rib cage. Each kidney is about the size of your fist. Kidneys are dark red in color because of the large volume of blood that passes through them.

The kidneys have several functions. This lesson will focus on the role of the kidneys in the urinary system. However, the kidneys also produce hormones that stimulate the production of red blood cells. In addition, they control blood pressure and help control calcium levels in the body.

The kidneys contain blood vessels and nephrons (NEH frahnz). **Nephrons** *are networks of capillaries and small tubes, or tubules, where filtration of blood occurs.* Each kidney contains about one million nephrons.

Blood contains waste products, salts, and sometimes toxins from cells that need to be removed from the body. These products are filtered from the blood as it passes through the kidneys. *When blood is filtered, a fluid called* **urine** *is produced.* The kidneys filter the blood and produce urine in two stages. You will read about this two-stage filtration process on the next page.

The Urinary System

Figure 10 Most functions of the urinary system occur in the kidneys. The kidneys connect to the ureters, then the bladder, and finally the urethra.

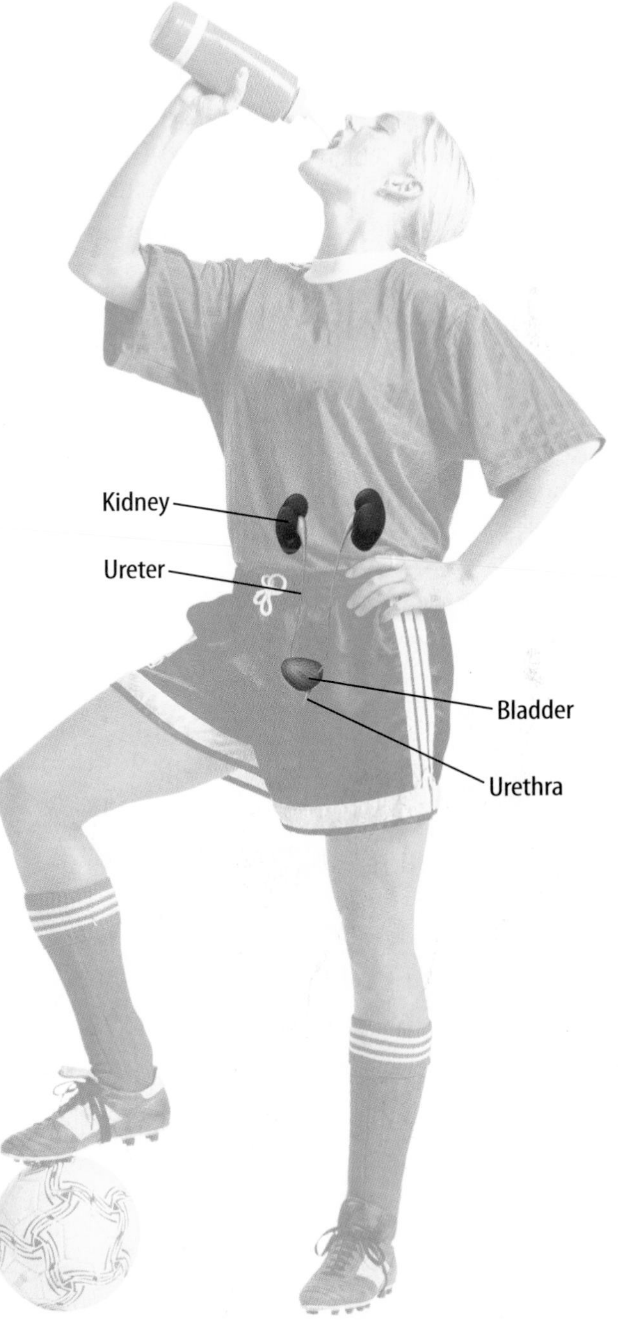

Word Origin

nephron
from Greek *nephros,* means "kidney"

First Filtration Blood is constantly circulating and filtering through the kidneys. In one day, the kidneys filter about 180 L of blood plasma, or the liquid part of blood. That's enough liquid to fill 90 2-L bottles. Your body contains about 3 L of blood plasma. This means your entire blood supply is filtered by your kidneys about 60 times each day. As shown in **Figure 11,** the first filtration occurs in clusters of capillaries in the nephrons. These clusters of capillaries filter water, sugar, salt, and wastes out of the blood.

Second Filtration If all of the liquid from the first filtration were excreted, your body would quickly dehydrate and important nutrients would be lost. To regain some of this water, the kidneys filter the liquid collected in the first filtration again. As shown in **Figure 11,** the second filtration occurs in small tubes in the nephrons. During the second filtration, up to 99 percent of the water and nutrients from the first filtration are separated out and reabsorbed into the blood. The remaining liquid and waste products form urine. On average, an adult excretes about 1.5 L of urine per day.

Figure 11 The kidneys produce urine in two stages.

Visual Check Urine passes through which structure before entering the ureter?

Filtration in the Kidneys

Concepts in Motion Animation

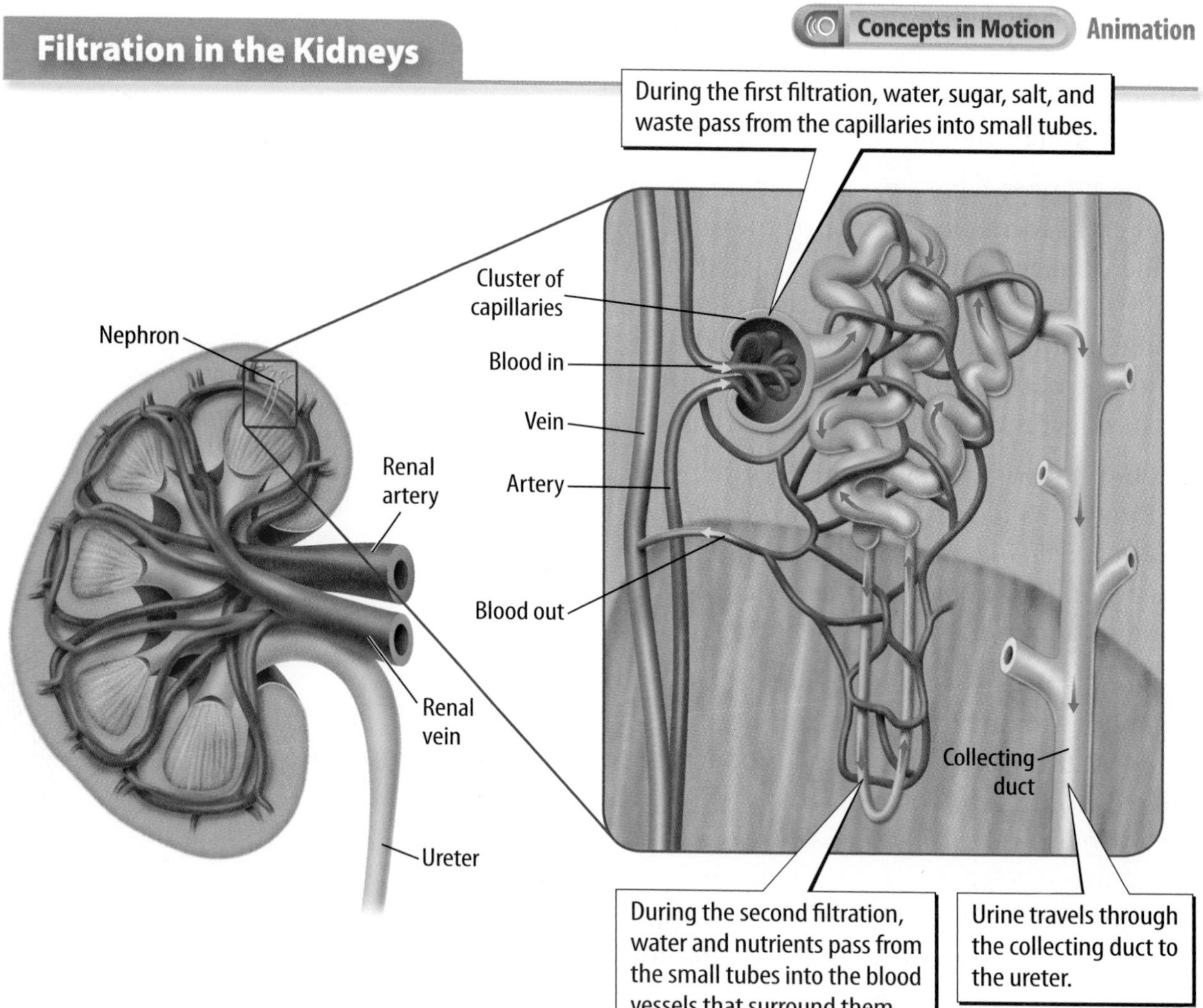

The Ureters, Bladder, and Urethra

Do you remember the trash can you read about earlier in this lesson? What would happen if you put garbage in the trash can but never emptied the trash can? The garbage would pile up. After a while, there would be too much garbage for the trash can to hold. To keep this from happening, you must empty the trash from the trash can. In a similar way, the urine produced by your body cannot stay in the kidney. *Urine leaves each kidney through a tube called the* **ureter** (YOO ruh tur). Refer back to **Figure 10** to see the locations of the ureter and other organs of the urinary system.

Both ureters drain into the bladder. *The* **bladder** *is a muscular sac that holds urine until the urine is excreted.* The bladder expands and contracts like a balloon when filled or emptied. An adult bladder can hold about 0.5 L of urine.

Urine leaves the bladder through a tube called the **urethra** (yoo REE thruh). The urethra contains circular muscles called sphincters (SFINGK turz) that control the release of urine.

Key Concept Check How do the ureters, bladder, and urethra work together to excrete urine?

WORD ORIGIN

ureter
from Greek *ourethra*, means "passage for urine"

Inquiry MiniLab

30 minutes

How can you model the function of a kidney?

The kidneys filter substances from blood plasma. How can you use everyday materials to model the function of the kidneys?

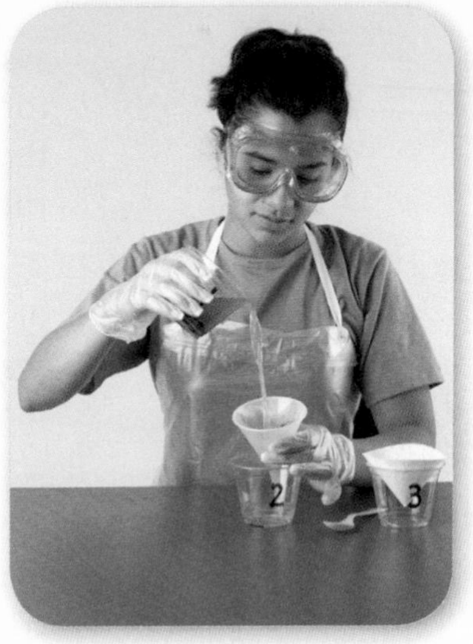

1. Read and complete a lab safety form.
2. Label three **plastic cups** *1, 2,* and *3.*
3. Mix a small amount of **fine gravel** and **sand** with **water** in cup 1.
4. Place a small piece of **wire screen** in a **funnel,** and place the funnel in cup 2.
5. Carefully pour the sand-water-gravel mixture into the funnel. Let it drain. Record your observations in your Science Journal.
6. Remove the screen. Replace it with a piece of **filter paper.** Place the funnel in cup 3.
7. Carefully pour the contents of cup 2 into the funnel. Let it drain. Record your observations.

Analyze and Conclude

1. **Describe** what happened during each filtration.
2. **Key Concept** Summarize how your filtration systems model the function of the kidneys.

Table 3 Urinary Disorders — Concepts in Motion Interactive Table

Urinary Disorder	Description	Possible Causes
Kidney disease	The nephrons are damaged and the ability of the kidneys to filter blood is reduced. However, a person can have the beginning stages of kidney disease and experience no symptoms.	diabetes, high blood pressure, poisons, trauma
Urinary tract infection	Infections usually occur in the bladder or urethra, but infections can also occur in the kidney and ureters. Symptoms can include burning during urination, small and frequent urination, and blood in urine.	bacteria in the urinary system
Kidney stones	Kidney stones are solid substances that form in the kidney. The most common type is made of calcium. Stones that pass through the urinary system can be very painful.	calcium buildup in the kidney
Bladder control problems	The bladder releases urine involuntarily. Occurs in women more often than men.	urinary tract infections, muscle weakness, prostate enlargement

Urinary Disorders

A urinary disorder is an illness that affects one or more organs of the urinary system. Some urinary disorders are described in **Table 3.** Several of these disorders are relatively common. Urinary tract infections, for example, are a leading cause of doctor visits, second only to respiratory infections.

The Excretory System and Homeostasis

You have already read about some of the ways the excretory system helps to maintain homeostasis. For example, the excretory system filters wastes from the blood. The blood is part of the circulatory system. If wastes were allowed to build up in the circulatory system, they would become toxic.

Another example of maintaining homeostasis is the removal of wastes from the digestive system. Similar to the circulatory system, wastes would damage your body if they were not removed from the digestive system by the excretory system.

The excretory system also interacts with the nervous system. The hypothalamus is an **area** of the brain that helps to maintain homeostasis. One function of the hypothalamus is to control the secretion of some hormones. One such hormone causes the tubules in the kidney to absorb more water from the blood. This helps the body to regulate fluid levels. Water is retained in the blood instead of being excreted in the urine.

Academic Vocabulary
area
(noun) a part of something that has a particular function

Key Concept Check How does the excretory system interact with the nervous system?

Lesson 3 Review

Assessment Online Quiz

Visual Summary

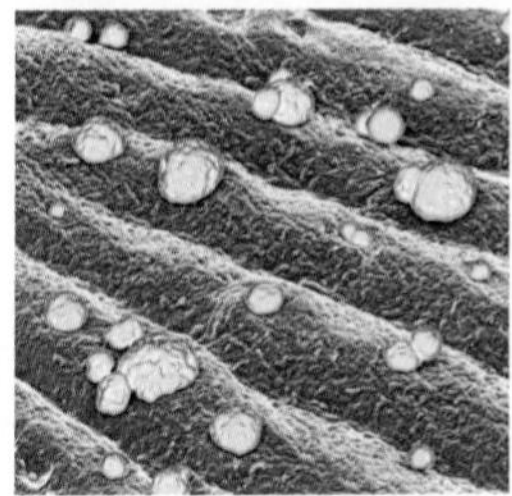

The excretory system collects and eliminates wastes from the body and regulates the level of fluid in the body.

The respiratory system is one of the body systems that make up the excretory system.

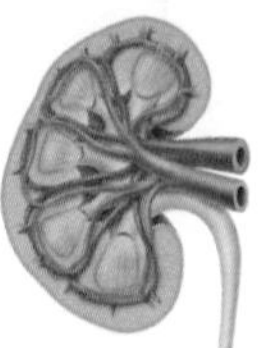

The organs of the urinary system process, transport, collect, and excrete waste.

Use your lesson Foldable to review the lesson. Save your Foldable for the project at the end of the chapter.

What do you think NOW?

You first read the statements below at the beginning of the chapter.

5. Several human body systems work together to eliminate wastes.

6. Blood contains waste products that must be removed from the body.

Did you change your mind about whether you agree or disagree with the statements? Rewrite any false statements to make them true.

Use Vocabulary

1. **Define** the word *nephron* in your own words.
2. **Distinguish** between ureter and urethra.
3. **Use the term** *bladder* in a sentence.

Understand Key Concepts 

4. The kidneys filter wastes from the
   A. blood.
   B. intestine.
   C. lungs.
   D. skin.
5. **Construct** a diagram of the urinary system showing the production and flow of urine.
6. **Distinguish** between the excretory functions of the respiratory system and the integumentary system.

Interpret Graphics

7. **Identify** the function of the highlighted portion of the diagram to the right.

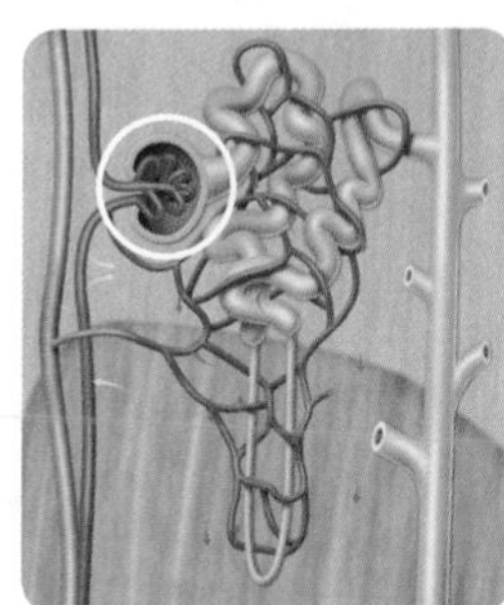

8. **Organize Information** Copy and fill in the table below with details about each organ of the urinary system.

Organ	Structure and Function

Critical Thinking

9. **Hypothesize** What might happen if urine did not go through a second filtration?
10. **Evaluate** the role of the hypothalamus in maintaining the level of fluid in the body.

50 minutes

Model Digestion from Start to Finish

Materials

graham crackers

banana

resealable plastic bag (1 quart size)

nylon hose

nylon netting

funnel

Also needed: scissors, water, paper towel, paper cup, newspaper

Safety

Recall from Lesson 2 that all food goes through four steps: ingestion, digestion, absorption, and elimination. These steps happen in the digestive system. Your task is to model the four steps with the materials provided by your teacher. Before you create your model, think about all the digestive processes and plan each step. Will you model mechanical or chemical digestion or both?

Question

How does food change during the process of digestion? What are the steps in digestion?

Procedure

1. Read and complete a lab safety form.
2. In your Science Journal, make a chart like the one shown here that includes the parts of the digestive system. Record the functions of each part.
3. Using the materials provided by your teacher, design a model to show the steps in digestion. Begin with chewing and end with excretion.
4. Your teacher must approve your design before you test your model.
5. Pass food through your model.
6. Compare the food at the beginning and the end of digestion.
7. Dispose of the materials as directed by your teacher.

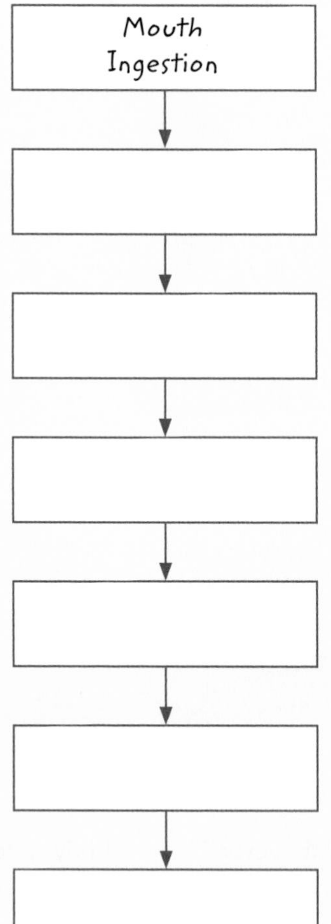

8 Copy and complete the chart at right. Then, compare your model to the four steps as outlined in the text. Does your model include everything? Is there another way to model some of the steps?

9 Make modifications to your model. Record your revisions in your Science Journal.

Part of the Digestive System	Function	Part of the Model	Comparison
Mouth	Ingestion		

Analyze and Conclude

10 **Analyze** Is there a structure or function in digestion that was not included in your model? Did you model mechanical or chemical digestion or both?

11 **Contrast** How did the food change in your model? How does food change in the digestive process?

12 **The Big Idea** How does the digestive system maintain homeostasis in a healthy body?

Lab Tips

- This lab might be messy, so work on several layers of newspaper.
- Be careful not to cut large holes in bags or cups; small holes work better.
- Never eat anything during a lab exercise.

Communicate Your Results

Share your results with the class. Discuss your chart with those of other groups. Demonstrate to the class how you modeled the digestive system.

Extension

How might your model change if you were modeling a disease of the digestive system, such as an inability to produce saliva?

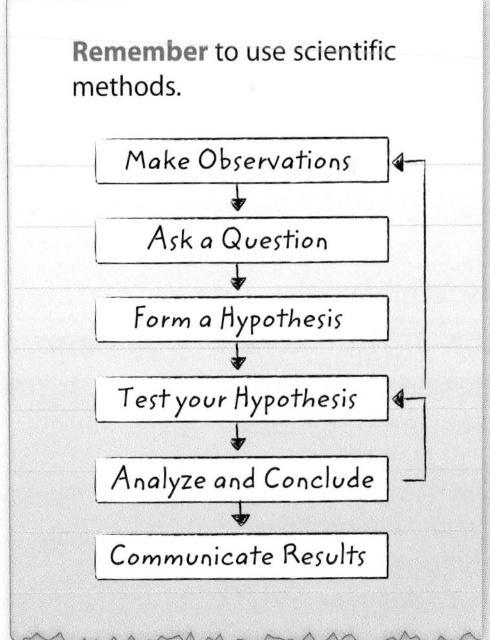

Chapter 15 Study Guide

The digestive and excretory systems move materials through the body and remove waste. The digestive system also absorbs nutrients.

Key Concepts Summary

Lesson 1: Nutrition

- People eat food to obtain the energy their bodies need to function. The amount of energy in food is measured in **Calories.**
- The types and amounts of nutrients a person needs depend on age, gender, and activity level.
- The six groups of nutrients are **proteins, carbohydrates, fats, vitamins, minerals,** and water.
- A balanced diet provides nutrients and energy for a healthful lifestyle.

Vocabulary

Calorie p. 523
protein p. 524
carbohydrate p. 524
fat p. 525
vitamin p. 525
mineral p. 525

Lesson 2: The Digestive System

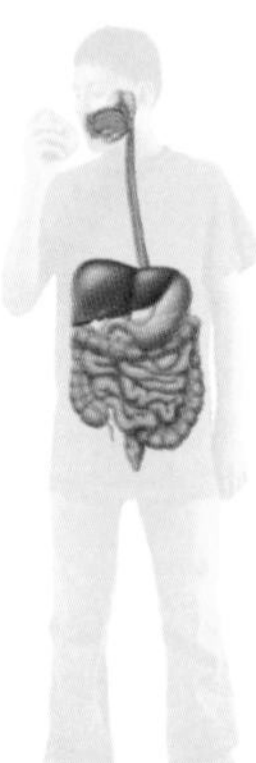

- The function of the digestive system is to break down food and absorb nutrients for the body.
- Organs of the digestive system include the mouth, **esophagus,** stomach, small intestine, and large intestine.
- The digestive system interacts with other body systems to maintain the body's internal balance.

Vocabulary

digestion p. 531
mechanical digestion p. 532
chemical digestion p. 532
enzyme p. 532
esophagus p. 534
peristalsis p. 534
chyme p. 535
villi p. 536

Lesson 3: The Excretory System

- The function of the **excretory system** is to collect and eliminate wastes from the body and regulate the level of fluids in the body.
- The excretory system is made up of the digestive system, respiratory system, urinary system, and the integumentary system.
- The excretory system works with other body systems, including the nervous system, to maintain homeostasis.

Vocabulary

excretory system p. 541
kidney p. 543
nephron p. 543
urine p. 543
ureter p. 545
bladder p. 545
urethra p. 545

Study Guide

- Personal Tutor
- Vocabulary eGames
- Vocabulary eFlashcards

FOLDABLES® Chapter Project

Assemble your lesson Foldables as shown to make a Chapter Project. Use the project to review what you have learned in this chapter.

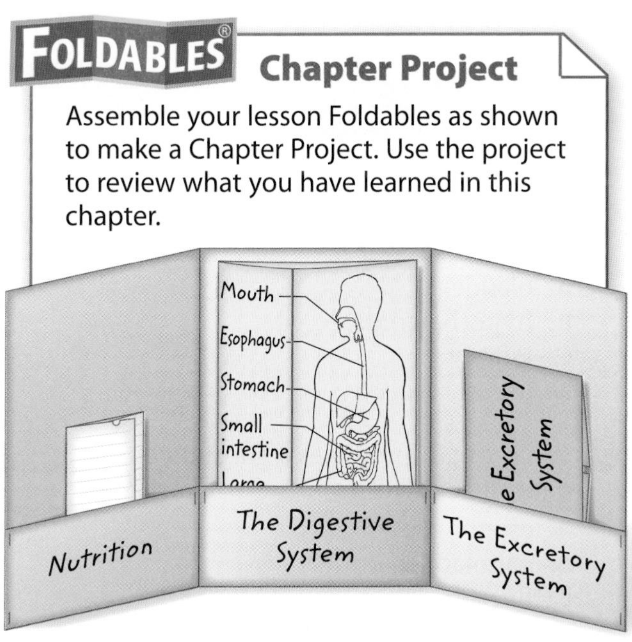

Use Vocabulary

1. About 25–35 percent of your total daily ________ should be from fats.
2. One type of nutrient, ________, is made of long chains of sugars.
3. Food moves down the esophagus by ________.
4. The breakdown of food into small particles and molecules is called ________.
5. A tube that connects a kidney to the bladder is called a(n) ________.
6. Urine is stored in the ________.

Link Vocabulary and Key Concepts

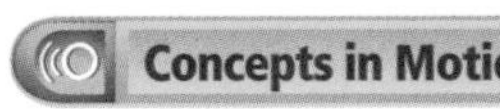

Interactive Concept Map

Copy this concept map, and then use vocabulary terms from the previous page to complete the concept map.

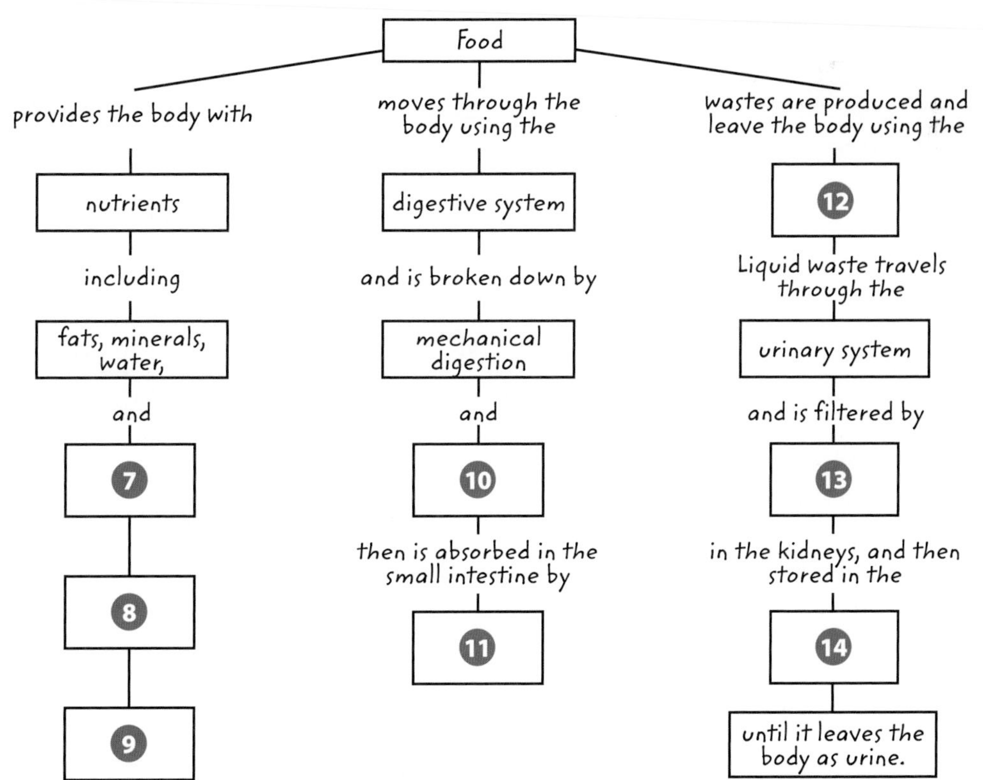

Chapter 15 Review

Understand Key Concepts 

1. What are proteins made of?
   A. amino acids
   B. minerals
   C. sugars
   D. vitamins

2. Which would be considered a grain?
   A. black beans
   B. brown rice
   C. canola oil
   D. lean chicken

3. What is the main source of energy for your body?
   A. carbohydrates
   B. minerals
   C. proteins
   D. water

4. Look at the diagram below. Where does most absorption of nutrients occur?

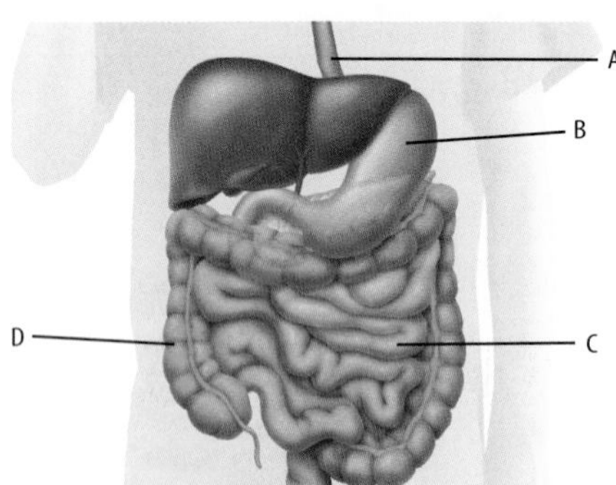

   A. A
   B. B
   C. C
   D. D

5. What is the correct order for how food is processed in the digestive system?
   A. absorption, digestion, ingestion, elimination
   B. elimination, ingestion, absorption, digestion
   C. ingestion, absorption, digestion, elimination
   D. ingestion, digestion, absorption, elimination

6. What organ is shown below?

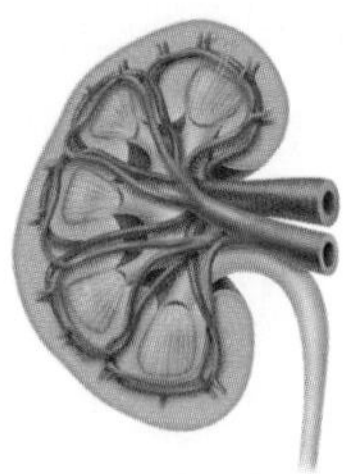

   A. bladder
   B. hypothalamus
   C. kidney
   D. ureter

7. What organ produces a substance that neutralizes acid from the stomach?
   A. esophagus
   B. gallbladder
   C. liver
   D. pancreas

8. What fluid produced in the mouth contains digestive enzymes?
   A. bile
   B. blood
   C. chyme
   D. saliva

9. Carbon dioxide is eliminated by which body system?
   A. digestive system
   B. integumentary system
   C. respiratory system
   D. urinary system

10. What is produced by the urinary system?
   A. blood
   B. feces
   C. perspiration
   D. urine

11. The bladder is most similar to which object?
   A. a balloon
   B. a tube
   C. a folded paper
   D. a rigid container

Assessment
Online Test Practice

Critical Thinking

12 **Distinguish** between minerals and vitamins.

13 **Hypothesize** why a child might have different nutritional needs than an adult over the age of 60.

14 **Select** Study the nutrient information below. Select the snack that would be a better choice as part of a healthful lifestyle. Explain your choice.

Nutrient Information	Tortilla Chips	
	Fried	Baked
Calories	150	110
Calories from fat	60	5
Total fat (g)	7	1
Saturated fat (g)	1	0
Sodium (mg)	135	200
Total carbohydrate (g)	22	24
Sugars	3	0
Protein	3	2

15 **Differentiate** Suppose your teacher showed you a diagram of a small intestine and a diagram of a large intestine. How might you distinguish between them?

16 **Hypothesize** How might digestion be affected if a person swallowed his or her food without first chewing it?

17 **Critique** the following statement: "Bacteria are harmful and should not be in the digestive system."

18 **Compare** the excretions of the urinary system and the digestive system.

Writing in Science

19 **Create** a commercial to encourage people to eat a healthful amount from each food group. Include a setting and dialogue for your commercial.

REVIEW THE BIG IDEA

20 Give examples of how the digestive system and excretory system help to maintain homeostasis.

21 What is the function of the small intestine?

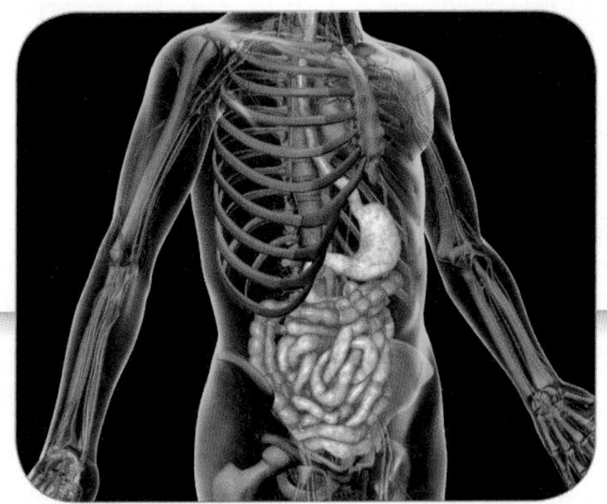

Math Skills

Math Practice

Use Percentages

Use the table below to answer questions 22–24.

Location of food	Time in location (hrs)
Stomach	4
Small intestine	6
Large intestine	24

22 What percentage of the total digestive time does food spend in the stomach?

23 What percentage of the total digestive time does food spend in the large intestine?

24 What percentage of the total digestive time does food spend in the stomach and the small intestine combined?

Standardized Test Practice

Record your answers on the answer sheet provided by your teacher or on a sheet of paper.

Multiple Choice

1 Which process depends on enzymes?

A chemical digestion

B elimination

C mechanical digestion

D respiration

Use the diagram below to answer question 2.

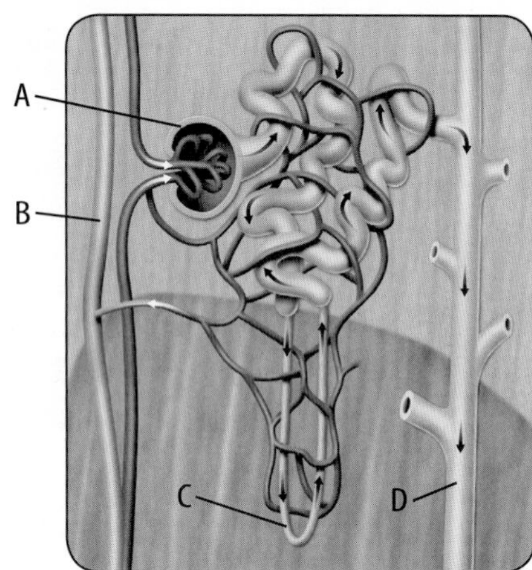

2 Where does the first filtration process occur in the nephron shown above?

A A

B B

C C

D D

3 Which factor does NOT influence how much energy a person needs?

A age

B gender

C height

D weight

Use the diagram below to answer questions 4 and 5.

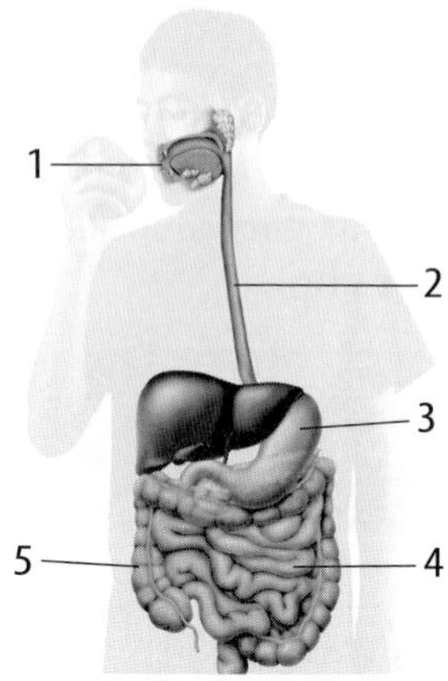

4 In which part of the system pictured above does chemical digestion begin?

A 1

B 2

C 3

D 4

5 In the diagram above, from which organ are nutrients absorbed into the bloodstream?

A 2

B 3

C 4

D 5

6 What is a main function of the excretory system?

A fight diseases

B move limbs

C pump blood

D remove wastes

Assessment
Online Standardized Test Practice

7 Which part of the brain works with the urinary system to help maintain homeostasis?

A cerebellum

B cerebrum

C hypothalamus

D medulla

Use the diagram below to answer question 8.

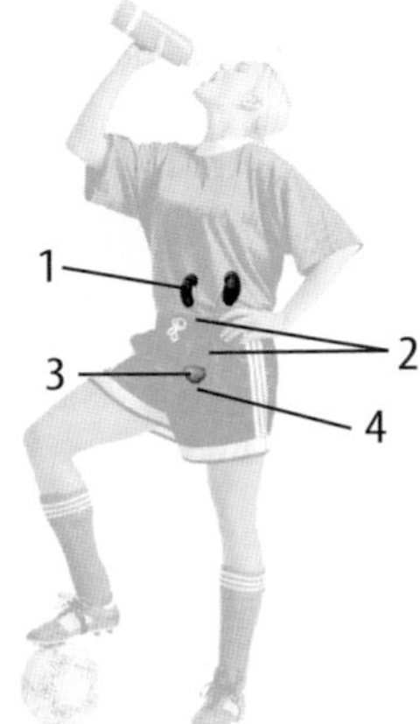

8 In the diagram above, where is urine produced?

A 1

B 2

C 3

D 4

9 Which system works with the digestive system to carry nutrients to the cells of the body?

A circulatory

B excretory

C lymphatic

D respiratory

Constructed Response

Use the table below to answer questions 10 and 11.

Nutrient	Example of Food
Carbohydrates	
Fats	
Minerals	
Proteins	
Vitamins	
Water	

10 In the table above, the six main groups of nutrients are provided. What is an example of a food that contains each nutrient? What is the function of each nutrient in the body?

11 Explain how the nutrients in the table above are related to eating a balanced diet.

Use the table below to answer question 12.

Process	Description
Ingestion	
Digestion	
Absorption	
Elimination	

12 When a person eats food, the food undergoes four processes in the digestive system. Briefly describe each process provided in the table above.

NEED EXTRA HELP?												
If You Missed Question...	1	2	3	4	5	6	7	8	9	10	11	12
Go to Lesson...	2	3	1	2	2	3	3	3	2	1	1	2

Chapter 16

Respiration and Circulation

THE BIG IDEA How do the respiratory and circulatory systems help maintain the body's homeostasis?

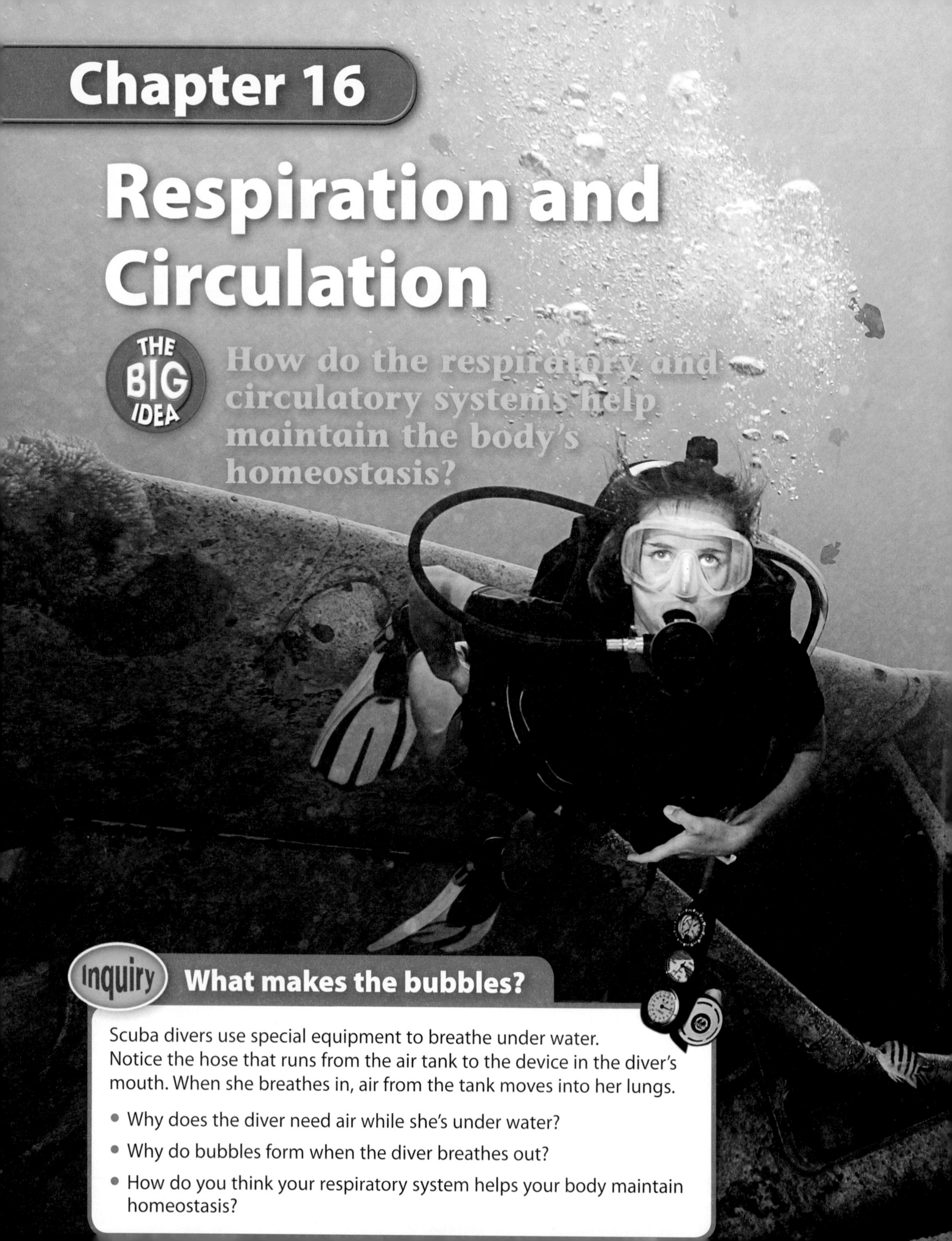

Inquiry What makes the bubbles?

Scuba divers use special equipment to breathe under water. Notice the hose that runs from the air tank to the device in the diver's mouth. When she breathes in, air from the tank moves into her lungs.

- Why does the diver need air while she's under water?
- Why do bubbles form when the diver breathes out?
- How do you think your respiratory system helps your body maintain homeostasis?

Get Ready to Read

What do you think?

Before you read, decide if you agree or disagree with each of these statements. As you read this chapter, see if you change your mind about any of the statements.

1. Breathing and respiration are the same.
2. Lungs are the only parts of the body that use oxygen.
3. There are four chambers in a human heart.
4. Blood travels in both directions in veins.
5. All blood cells are red.
6. Blood plasma is just water.
7. Lymph nodes are only in the neck.
8. The lymphatic system helps fight infections to maintain a healthy body.

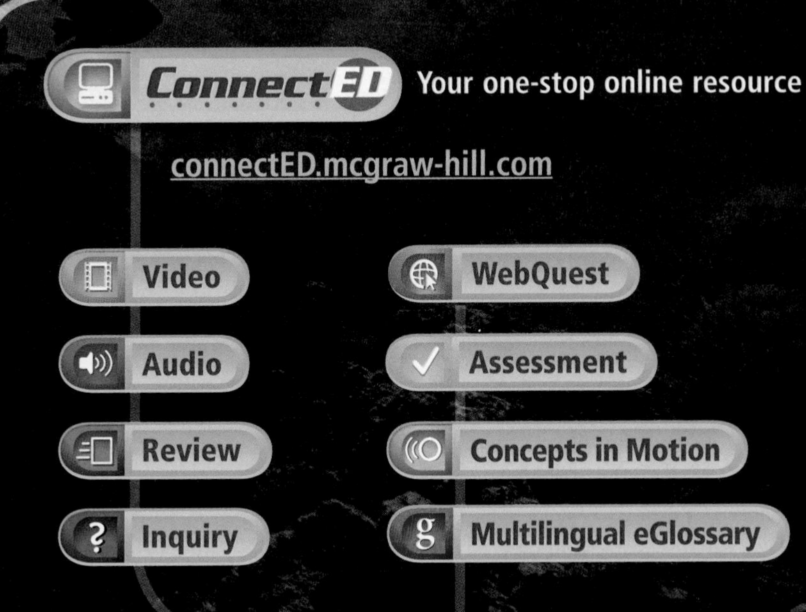

Lesson 1

Reading Guide

Key Concepts
ESSENTIAL QUESTIONS

- What does the respiratory system do?
- How do the parts of the respiratory system work together?
- How does the respiratory system interact with other body systems?

Vocabulary

breathing p. 559
pharynx p. 560
larynx p. 560
trachea p. 560
bronchi p. 561
lungs p. 561
alveoli p. 561
diaphragm p. 562

Multilingual eGlossary

Video BrainPOP®

The Respiratory System

Inquiry Cleaning Up?

The hairlike structures shown here are called cilia (SIH lee uh). They move together in wavelike motions. Cilia line the air passages in your nose, throat, and lungs. The round particles on top of the cilia are bits of dust and other things that can block or irritate airways. What do you think these cilia are doing?

Launch Lab

10 minutes

How much air is in a breath? 

Do your lungs empty completely every time you breathe out? You can use a balloon to find out.

1. Read and complete a lab safety form.
2. Place your hands on your ribs as you breathe in and out. Record your observations in your Science Journal.
3. Breathe in normally. Breathe out normally into a **balloon.** Twist and hold the end of the balloon.
4. Have your partner use a **metric tape measure** to measure around the balloon at its widest point. Record the measurement. Let the air out of the balloon.
5. Breathe in normally again. Breathe out as much air as you can into the balloon. Twist and hold the end. Repeat step 4.
6. Switch roles with your partner, and repeat steps 2–5 using a different balloon.

Think About This

1. Was there a difference in the two measurements? Why do you think this happened?
2.  **Key Concept** How do your lungs interact with the bones and muscles of your chest?

Functions of the Respiratory System

If you've ever held your breath, you probably took deep breaths afterward. That's your body's way of getting the oxygen it needs. **Breathing** *is the movement of air into and out of the lungs.* Breathing enables your respiratory system to take in oxygen and to eliminate carbon dioxide.

Taking in Oxygen

Think about the plumbing pipes that bring water into a house. Your respiratory system is similar. It is a system of organs that brings oxygen into your body. Oxygen is so important for life that your brain will tell your body to breathe even if you try not to. Why is oxygen so important? Every cell in your body needs oxygen for a series of chemical reactions called **cellular respiration.** During cellular respiration, oxygen and sugars react. This reaction releases energy a cell can use.

REVIEW VOCABULARY

cellular respiration
a series of chemical reactions that transform the energy in food molecules to usable energy

Eliminating Carbon Dioxide

The plumbing in a house also includes pipes that take away wastewater. In a similar way, your respiratory system removes carbon dioxide and other waste gases from your body. If waste gases are not removed, cells cannot function.

 **Key Concept Check** What does the respiratory system do?

Concepts in Motion Animation

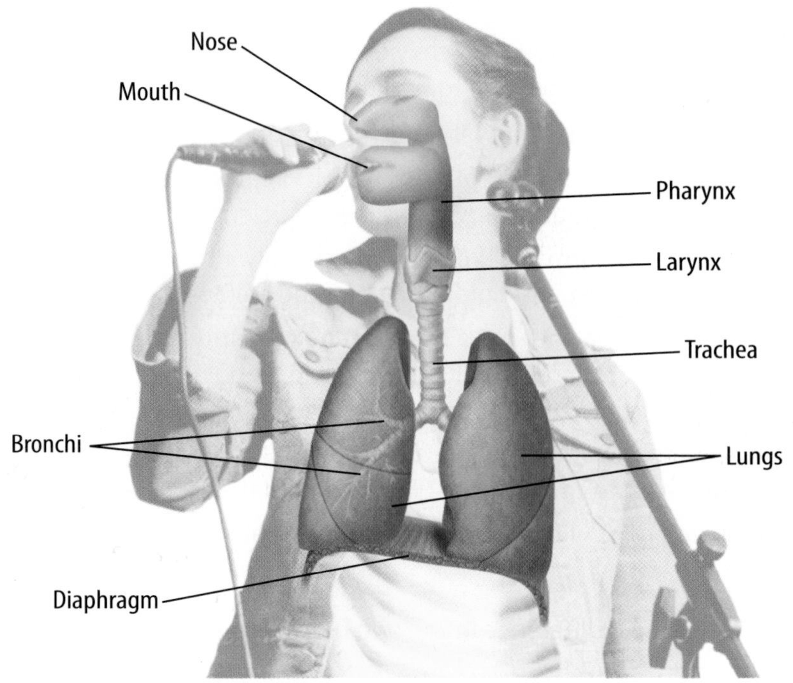

Figure 1 Air moves into and out of the lungs through the respiratory system.

Visual Check Which part of the respiratory system contains bronchi?

FOLDABLES®

Make an eight-tab vocabulary book from a sheet of notebook paper. Use it to organize your notes on the organs of the respiratory system and their functions.

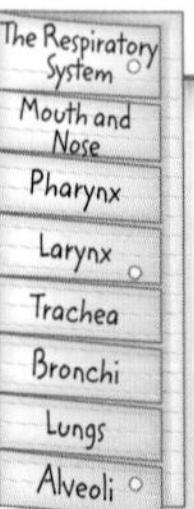

Organs of the Respiratory System

Follow the path of oxygen through the respiratory system in **Figure 1.** Air enters through the mouth and the nose. In the nose, the air is warmed and moistened. Hairs and sticky mucus in the nose help trap dust and dirt from the air. Cilia line the nose and most other airways in the respiratory system. Wavelike motions of the cilia carry trapped particles away from your lungs. The cilia help prevent harmful particles from getting very far into your respiratory system.

Key Concept Check What function do cilia have in the respiratory system?

Pharynx

Air passes from the nose and mouth into the throat. *The* **pharynx** (FER ingks) *is a tubelike passageway at the top of the throat that receives air, food, and liquids from the mouth or nose.* The epiglottis (eh puh GLAH tus) is a flap of tissue at the lower end of the pharynx. It keeps food and liquids from entering the rest of the respiratory system.

Larynx and Trachea

Air passes from the pharynx into a triangle-shaped area called the voice box or **larynx** (LER ingks). Two thick folds of tissue in the larynx—the vocal cords—vibrate and make sounds as air passes over them. Air then enters the **trachea** (TRAY kee uh), *a tube that is held open by C-shaped rings of cartilage.*

Bronchi and Lungs

The trachea branches into two narrower tubes called **bronchi** (BRAHN ki) (singular, bronchus) *that lead into the lungs.* **Lungs** *are the main organs of the respiratory system.* Inside the lungs, the bronchi continue to branch into smaller and narrower tubes called bronchioles.

Alveoli

In the lungs, the bronchioles end in *microscopic sacs, or pouches, called* **alveoli** (al VEE uh li; singular, alveolus), *where gas exchange occurs.* During gas exchange, oxygen from the air you breathe moves into the blood, and carbon dioxide from your blood moves into the alveoli.

WORD ORIGIN
alveoli
from Latin *alveus,* means "cavity"

Alveoli look like bunches of grapes at the ends of the bronchioles. Like tiny balloons, the alveoli fill with air when you breathe in. They contract and expel air when you breathe out. Notice in **Figure 2** how blood vessels surround an alveolus.

The walls of alveoli are only one cell thick. The thin walls and the large surface areas of the alveoli enable a high rate of gas exchange. If you could spread out all the alveoli in your lungs onto a flat surface, they would cover an area bigger than most classrooms. Every time you breathe, your alveoli enable your body to take in billions of molecules of oxygen and get rid of billions of molecules of carbon dioxide.

Reading Check What gases are exchanged in the alveoli?

Figure 2 Red blood cells drop off carbon dioxide and pick up oxygen as they move through the small blood vessels that surround each alveolus.

Visual Check How many layers of cells form the walls of the alveolus shown in this figure?

Gas Exchange

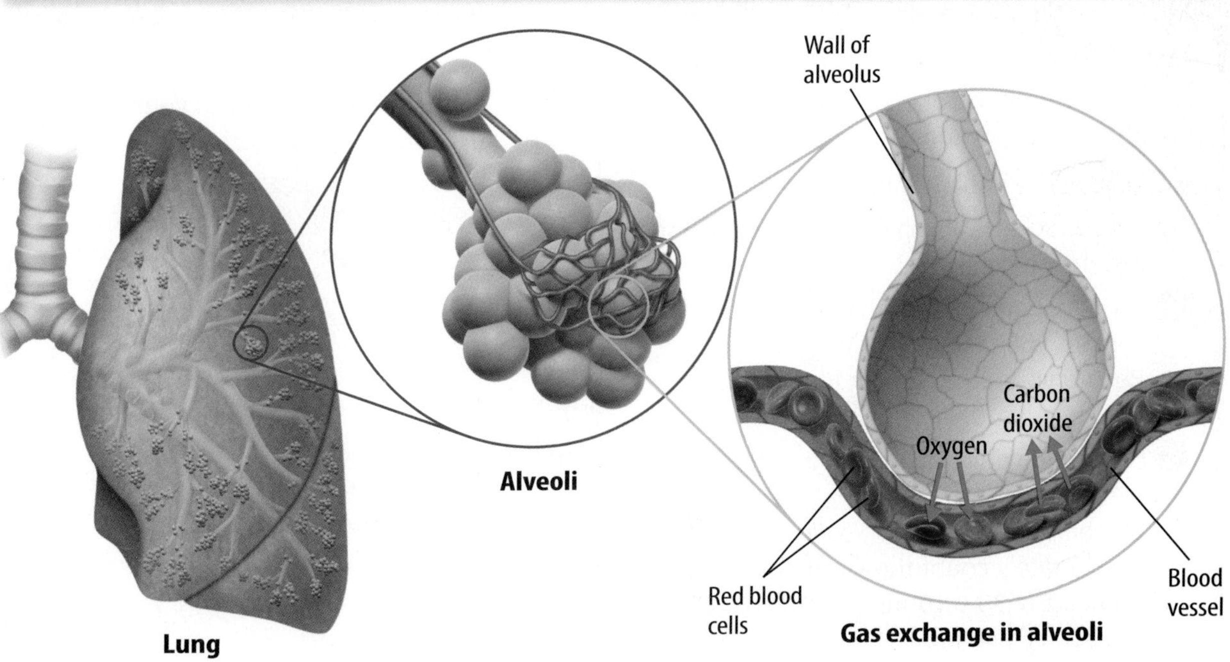

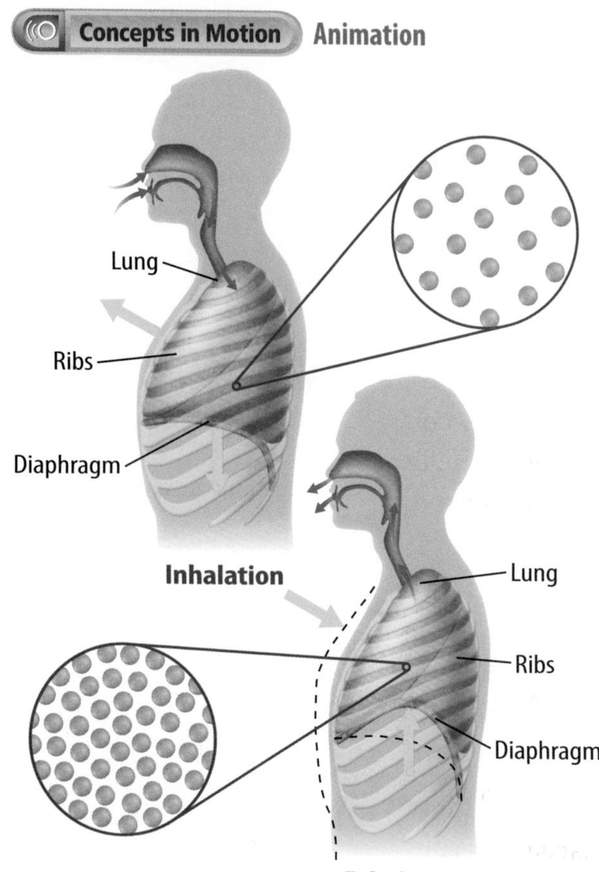

Figure 3 Your diaphragm contracts and moves down when you inhale. The chest cavity gets larger, and air rushes in to equalize the air pressure inside and outside the body. Your diaphragm relaxes and moves up as you exhale. Air rushes out to equalize air pressure.

Breathing and Air Pressure

How does your body know when to breathe? When high levels of carbon dioxide build up in your blood, the nervous system signals your body to breathe out, or exhale. After each exhalation, you breathe in, or inhale. How does this happen?

Below the lungs is a large muscle called the **diaphragm** (DI uh fram) *that contracts and relaxes and moves air in and out of the lungs.* The movement of your diaphragm causes changes in the air pressure inside your chest, as shown in **Figure 3.** Breathing occurs because of these changes in air pressure.

During inhalation, the diaphragm contracts and moves down, enlarging the space around the lungs. The increased space reduces air pressure in the chest. Air rushes into your lungs until the pressure inside your chest equals the air pressure outside it.

During exhalation, the diaphragm relaxes and moves up, reducing the space around the lungs. Air pressure in the chest increases. Waste gases rush out of your lungs.

Inquiry MiniLab

20 minutes

How does exercise affect breathing rate?

If you've ever played or watched a sport, you probably noticed that exercise changes your breathing rate. How does exercise affect the number of breaths you take in 30 seconds?

1. Read and complete a lab safety form.
2. In your Science Journal, create a data table like the one shown.
3. For 30 seconds, count the number of breaths you take while sitting quietly. Record your data. Repeat for two more trials.
4. Following your teacher's instructions, exercise briskly for 1 minute. When your teacher tells you to stop exercising, immediately count the number of breaths you take in 30 seconds. Record your data. Repeat for two more trials.

Activity	Number of Breaths		
	Trial 1	Trial 2	Trial 3
Sitting			
Exercising			

Analyze and Conclude

1. **Calculate** individual and class averages. How does your average breathing rate compare to the class average?
2. **Key Concept** How does the change in breathing rate help your body maintain homeostasis?

Respiratory Health

If you've ever had a cold, allergies, or asthma, you know what it's like to have a respiratory illness. A sore throat or a stuffed-up head makes breathing uncomfortable. Some respiratory illnesses make breathing difficult and can even become life-threatening. Common respiratory illnesses and their causes are listed in **Table 1.**

The best way to maintain good respiratory health is to stay away from irritants and air pollution. Don't smoke, and avoid secondhand smoke. On days when air quality is poor or pollen counts are high, it might be best to spend more time indoors.

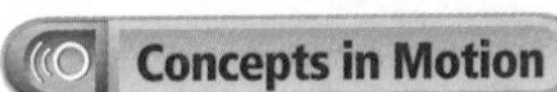

Interactive Table

Table 1 Respiratory Illnesses

Illness	Causes	Symptoms
Colds, flu	viruses	congestion, runny nose, watery eyes, coughing, sneezing
Bronchitis (brahn KI tus)	viruses, bacteria	coughing and fatigue due to mucus blocking the bronchi and bronchioles slows air movement
Pneumonia (noo MOH nyuh)	viruses, bacteria	difficulty breathing due to fluid in the alveoli that slows gas exchange
Asthma (AZ muh)	dust, smoke, pollen, pollution	difficulty breathing due to swollen airways and increased mucus
Emphysema (em fuh SEE muh)	smoking	coughing, fatigue, loss of appetite, and weight loss due to destruction of alveoli
Lung cancer	smoking	coughing, difficulty breathing, and chest pain

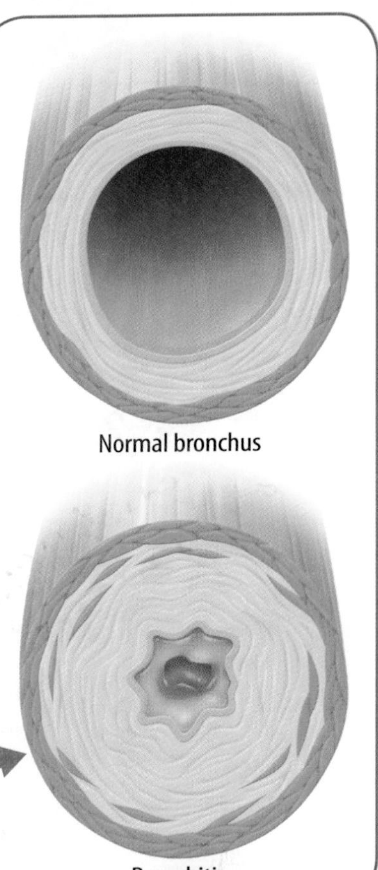
Normal bronchus

Bronchitis

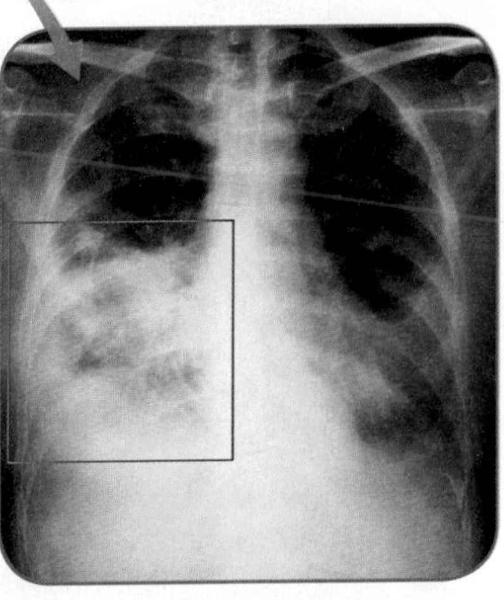
Pneumonia

The Respiratory System and Homeostasis

As you've read in this lesson, the muscular system interacts with the respiratory system so you can breathe. This interaction brings oxygen into your lungs and removes carbon dioxide from your lungs. In the next lesson, you'll read how the circulatory and respiratory systems work together to bring oxygen to body cells and remove carbon dioxide. All these systems help maintain homeostasis.

Key Concept Check How do the respiratory and muscular systems work together to maintain your body's homeostasis?

Lesson 1 Review

Assessment Online Quiz

Visual Summary

Air enters the body through the nose and mouth. It passes through the pharynx, larynx, and trachea on its way into the lungs.

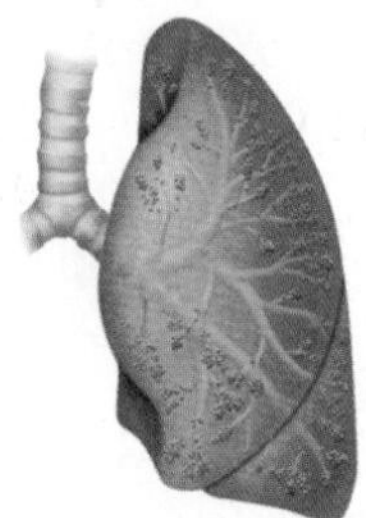

Inside the lungs, air moves through bronchi and bronchioles to the alveoli, where gas exchange takes place.

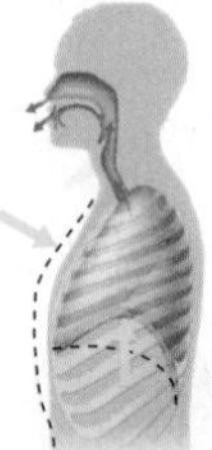

Breathing results from air pressure changes inside the chest that are created by the movement of the diaphragm muscle.

FOLDABLES®

Use your lesson Foldable to review the lesson. Save your Foldable for the project at the end of the chapter.

What do you think NOW?

You first read the statements below at the beginning of the chapter.

1. Breathing and respiration are the same.
2. Lungs are the only parts of the body that use oxygen.

Did you change your mind about whether you agree or disagree with the statements? Rewrite any false statements to make them true.

Use Vocabulary

1. The trachea branches into two narrower airways called ________.
2. Capillaries surround the ________, where gas exchange occurs.
3. **Distinguish** between breathing and respiration.

Understand Key Concepts 

4. **Explain** how the nose helps to clean air as the air enters the respiratory system.
5. **Describe** the functions of the respiratory system.
6. Which body system helps the respiratory system bring oxygen into the body?
   - **A.** circulatory
   - **B.** digestive
   - **C.** excretory
   - **D.** muscular

Interpret Graphics

7. **Explain** how oxygen moves into and out of the structures shown to the right.

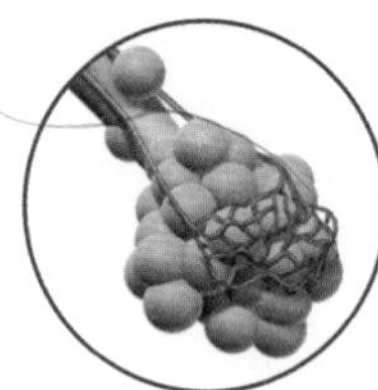

8. **Compare** Copy and fill in the Venn diagram below to explain the similarities and differences between the trachea and the bronchi.

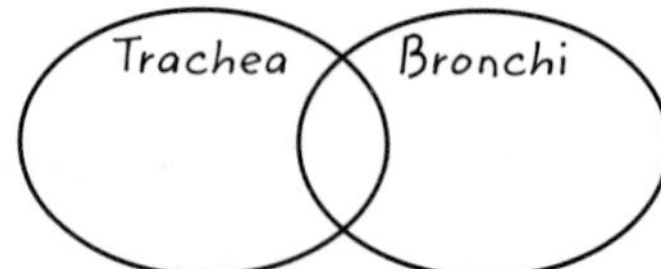

Critical Thinking

9. **Compose** a letter explaining why a friend of the family should stop smoking. Focus on the health reasons.
10. **Justify** Imagine that you answered a question in class by saying the contraction of the diaphragm causes a person to inhale. Another student disagrees. Justify your answer using words and a drawing.

20 minutes

How can a model show the physics of breathing?

Materials

1-L plastic drink bottle

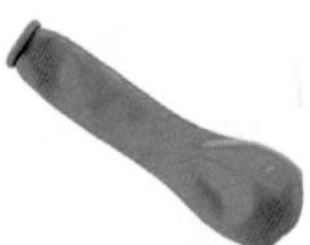

1 small balloon

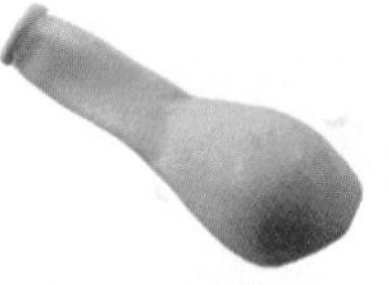

1 large balloon

duct tape

scissors

Safety

Air flows from areas of higher pressure to areas of lower pressure. This physics principle explains how air gets into and out of your lungs. Can this principle be observed in a model of the lungs?

Learn It

In science, a **model** is a representation of how something in the natural world works. A model can be used to explain or predict a natural process.

Try It

1. Read and complete a lab safety form.
2. Cut off the bottom one-third of a 1-liter clear, plastic drink bottle.
3. Blow into a small balloon two or three times to stretch it. Have your partner hold on to the bottle. Place the balloon inside the top of the bottle. Stretch the opened end of the balloon over the opening of the bottle.

4. Blow into a large balloon two or three times to stretch it. Tie a knot in the open end of the balloon. Cut off the tip of the opposite end of the balloon. Stretch the cut end of the balloon over the cut end of the bottle. Secure the balloon to the bottle with duct tape.
5. Pull down on the knotted end of the large balloon and then slowly release it. Observe what happens as you do this several times. Record your observations in your Science Journal.

Apply It

6. **Make a diagram** of your model. Label the parts representing the chest cavity, mouth and nose, diaphragm, and lungs.
7. **Describe** what happens to the volume inside the bottle when the large balloon is pulled downward.
8. On your diagram, label the areas of higher and lower pressure when you pull down on the large balloon. Label the areas of higher and lower pressure when the large balloon is released.
9. **Key Concept** Use your model and air pressure diagram to explain how air gets into your lungs when you inhale.

Lesson 2

Reading Guide

Key Concepts
ESSENTIAL QUESTIONS

- What does the circulatory system do?
- How do parts of the circulatory system work together?
- How does the circulatory system interact with other body systems?

Vocabulary

atrium p. 568
ventricle p. 568
artery p. 570
capillary p. 570
vein p. 570
systemic circulation p. 571
coronary circulation p. 571
pulmonary circulation p. 571
atherosclerosis p. 572

Multilingual eGlossary

Video BrainPOP®

The Circulatory System

Inquiry Where To?

How does food get from where it's grown to your dinner table? Food and most other products that people need are transported on roads and highways. Believe it or not, the vessels that carry blood through your body share similarities with roads and highways.

Launch Lab

10 minutes

How fast does your heart beat?

Have you ever felt your heartbeat speed up when you're exercising or when you're watching a scary movie? You can take your own pulse to find out how many times your heart beats every minute.

1. Read and complete a lab safety form.
2. Sit quietly for 1 minute.
3. Feel your pulse by placing the middle and index fingers of one hand on an artery in your neck or an artery in your wrist.
4. While sitting quietly, count the number of heartbeats you feel in 30 seconds. Multiply this number by two to calculate your pulse. Record your data in your Science Journal.
5. Jog in place for 1 minute.
6. Immediately repeat step 4.

Think About This

1. How did your pulse after exercising compare to your resting pulse?
2. **Key Concept** Why do you think your pulse changed when you exercised?

Functions of the Circulatory System

Have you ever looked at a road map of the United States? A complex network of highways and roads crisscrosses the country. This road network is important for transporting people and materials from place to place. In a similar way, your circulatory system is important for transporting materials from one part of your body to another.

Transportation

Trucks haul food, fuel, and other products from factories and farms to markets and businesses around the country. Your circulatory system is like the network of roads, and your blood cells are like the vehicles that travel on those roads. Blood carries food, water, oxygen, and other materials through your circulatory system to your body's cells and tissues.

Elimination

Blood also carries away waste materials, just as garbage trucks haul away trash. As blood travels through the circulatory system, it picks up carbon dioxide produced during cellular respiration. It also picks up wastes produced by all the other chemical reactions that take place inside cells.

 **Key Concept Check** What does the circulatory system do?

FOLDABLES®

Make a horizontal two-tab book from a sheet of notebook paper. Label the front *The Circulatory System,* and label the inside as shown. Use it to organize your notes on the functions of the circulatory system and the organs associated with those functions.

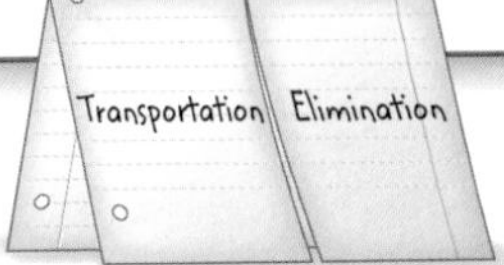

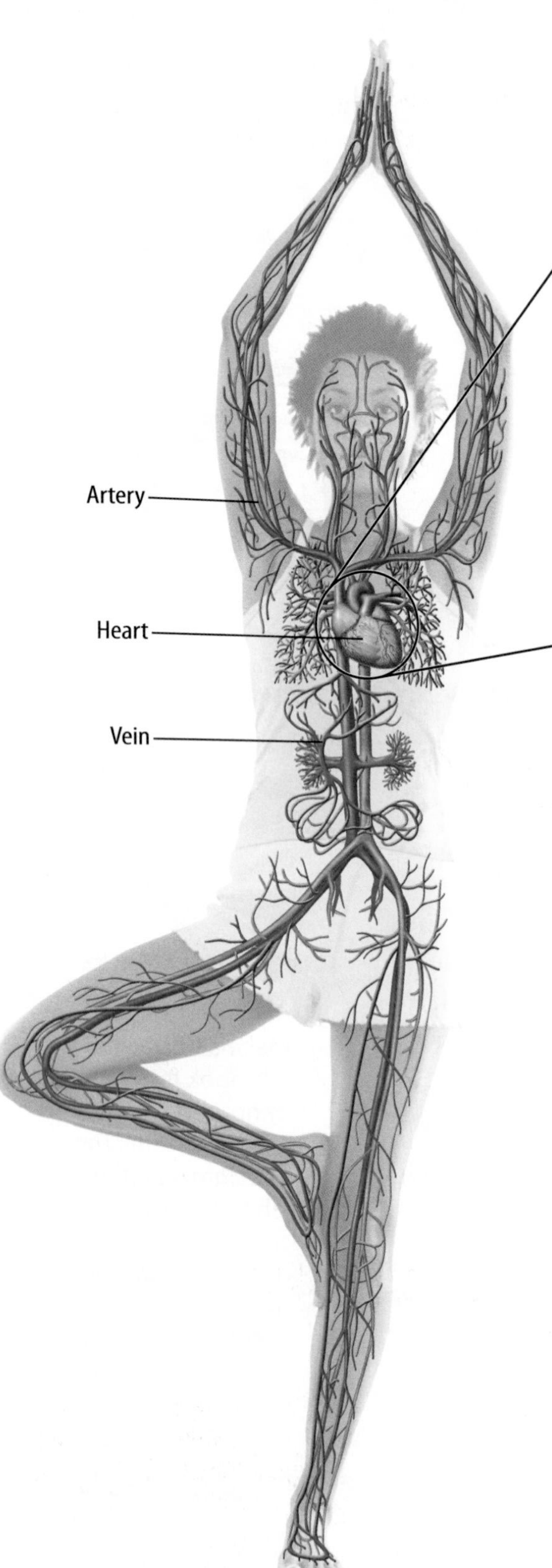

Figure 4 Your heart muscle is about the size of your fist. It acts as a pump that pushes blood through your circulatory system.

Circulatory System Organs

Highways connect and intersect and provide routes for traffic. **Figure 4** illustrates how your circulatory (SUR kyuh luh tor ee) system is similar. It provides routes for blood to flow through your body. Just as every vehicle on a highway is powered by its engine, your heart powers the flow of blood through your circulatory system.

The Heart

Your heart is always at work. The heart is a muscle that pushes blood through the circulatory system, as shown in **Figure 5.** On average, a human heart beats 70 to 75 times per minute, every minute of your life. It slows when you sleep. It speeds up when you exercise or are frightened.

 **Reading Check** What does the heart do?

Look again at **Figure 4.** Notice that your heart has four chambers, two upper and two lower. *Blood enters the upper two chambers of the heart, called the* **atria** (AY tree uh; singular, atrium). *Blood leaves through the lower two chambers of the heart, called the* **ventricles** (VEN trih kulz).

The Path of Blood Through the Heart

Concepts in Motion Animation

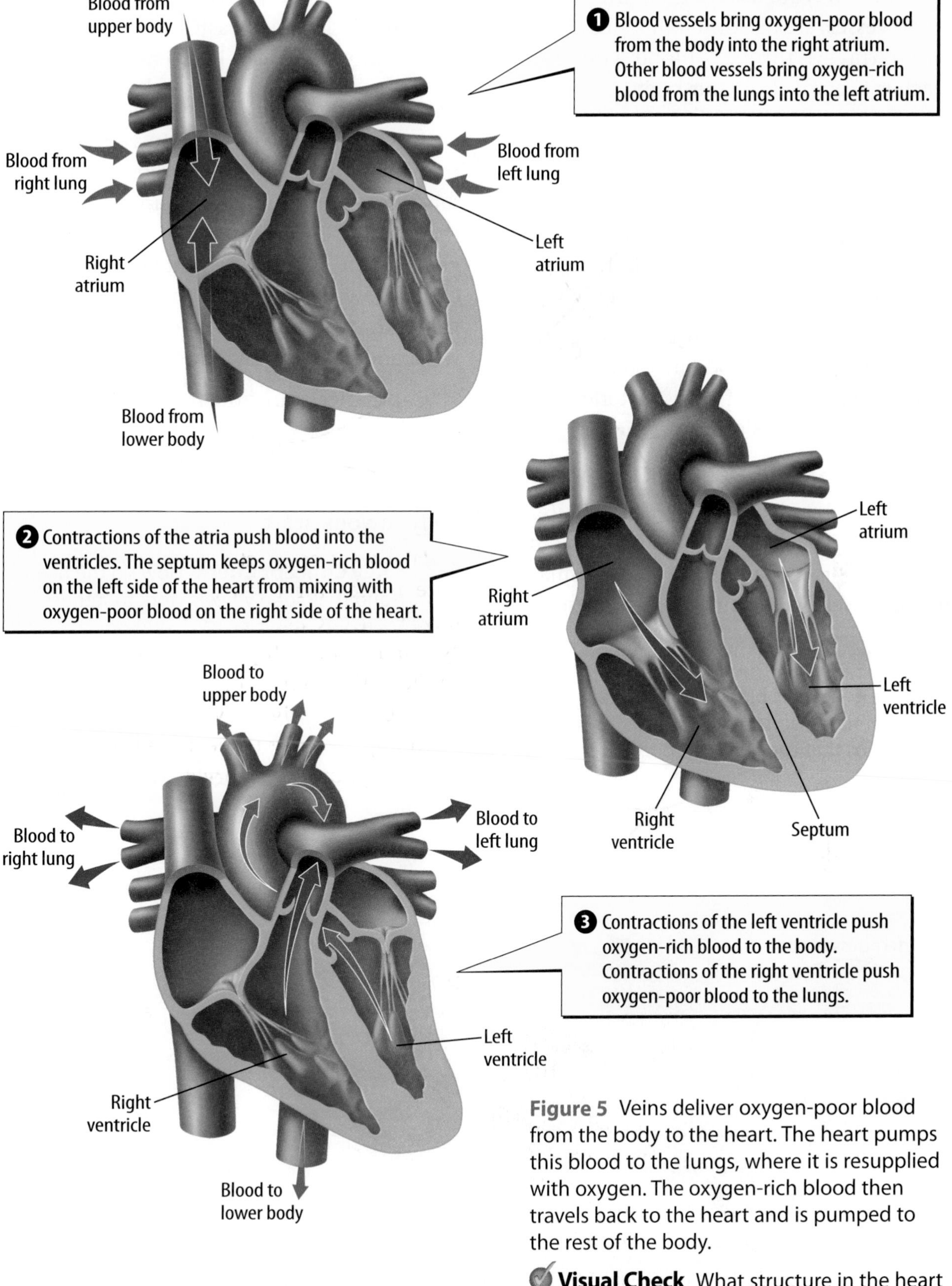

Figure 5 Veins deliver oxygen-poor blood from the body to the heart. The heart pumps this blood to the lungs, where it is resupplied with oxygen. The oxygen-rich blood then travels back to the heart and is pumped to the rest of the body.

Visual Check What structure in the heart separates oxygen-poor blood from oxygen-rich blood?

Blood Vessels

If the circulatory system is like a network of roads for your body, then the different blood vessels are like different kinds of roads. Blood travels through blood vessels and reaches every cell in your body.

Arteries As shown in **Figure 6,** *a vessel that takes blood away from the heart is an* **artery.** Blood pressure in arteries is high because arteries are near the pumping action of the heart. Artery walls are thick and can withstand the high pressure of flowing blood.

The aorta is the largest artery. It carries a large volume of blood, just like freeways carry a high volume of traffic. Arteries branch into smaller vessels called arterioles.

Capillaries Notice in Figure 6 that arterioles branch into **capillaries,** *tiny blood vessels that deliver supplies to individual cells and take away waste materials.* Capillaries are the smallest blood vessels in the circulatory system.

Many capillary walls are only one cell thick. This makes it possible for molecules of oxygen, food, water, wastes, and other materials to move between blood and body cells.

Veins *A vessel that brings blood toward the heart is a* **vein.** The pressure in veins is lower than in arteries. This is because capillaries separate veins from the pumping action of the heart. Because there is less pressure in veins, there is a greater chance that blood could flow backward. Veins have one-way valves that prevent blood from moving backward and keep it moving toward the heart.

Capillaries join and form larger vessels called venules. Venules join and form veins. The inferior vena cava is the largest vein. It carries blood from the lower half of your body to your heart.

Key Concept Check How do the heart and the blood vessels work together?

Figure 6 Arteries branch into arterioles and capillaries and bring oxygen-rich blood to cells. Capillaries combine to form venules and veins that carry oxygen-poor blood back to the heart. ▼

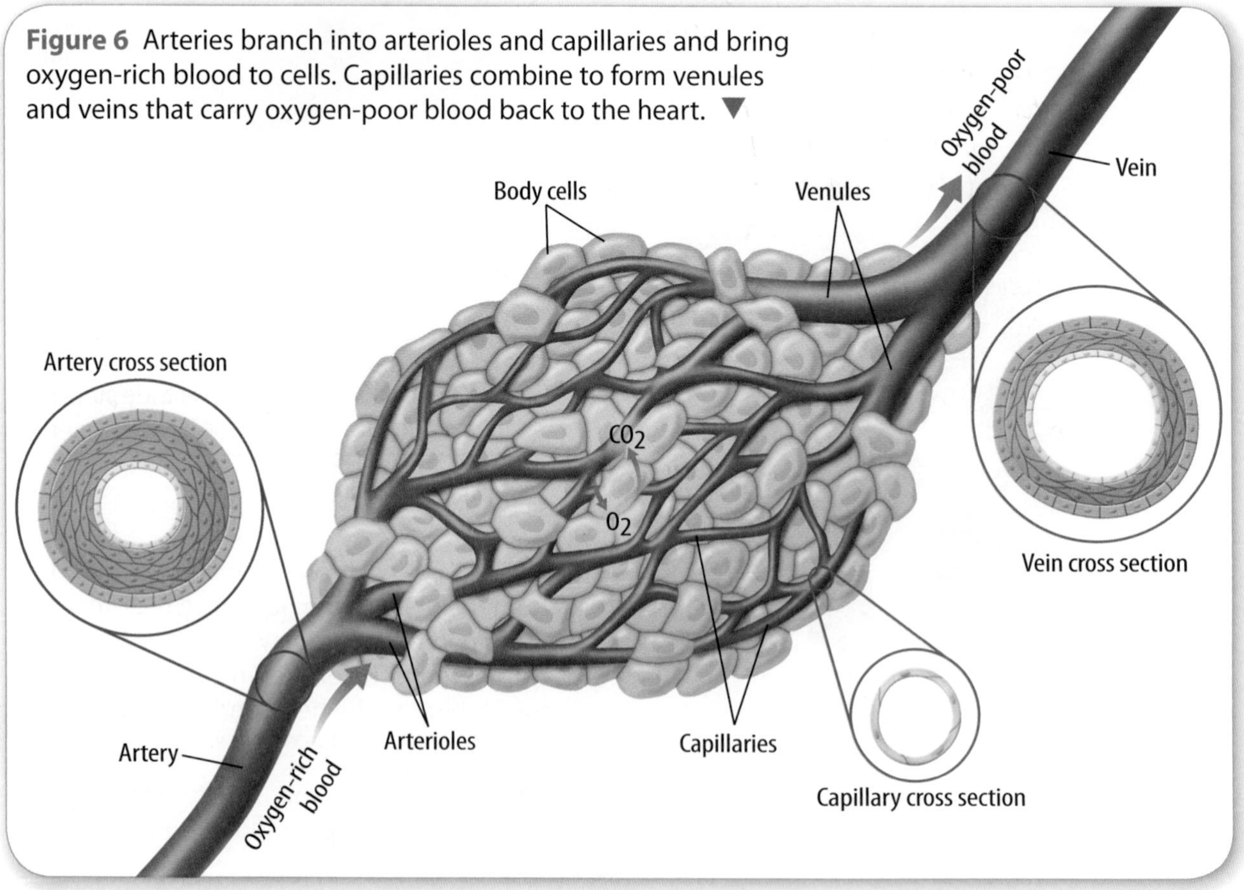

Types of Circulation

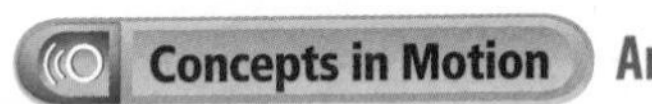

Your circulatory system is one large system that circulates blood throughout your entire body. However, when scientists and medical professionals discuss the circulatory system, they name three different types of circulation. One type supplies blood to the body. A second type supplies blood to the heart. A third type carries blood to and from the lungs.

Systemic Circulation

Blood leaves your heart through the aorta and travels to your arms, your toes, your head, and all other parts of your body, as shown by the orange vessels in **Figure 7.** **Systemic circulation** *is the network of vessels that carry blood from the heart to the body and from the body back to the heart.*

Coronary Circulation

You might think the cells of the heart get oxygen and nutrients from the blood that travels through the heart. However, the heart is a thick organ made up of many layers of cells. As a result, most heart cells don't come into contact with the blood inside the heart. *A network of arteries and veins called the* **coronary circulation** *supplies blood to all the cells of the heart.* Some of these vessels are on the outside of the heart, as shown in **Figure 7.**

Reading Check What does coronary circulation do?

Pulmonary Circulation

The purple part of the circulation shown in **Figure 7** illustrates how blood moves back and forth between the heart and the lungs. *The network of vessels that carries blood to and from the lungs is called* **pulmonary circulation.** Pulmonary circulation carries oxygen-poor blood from the heart to the lungs. It also carries oxygen-rich blood from the lungs back to the heart. Blood that enters the heart from the lungs is then pushed to the rest of the body.

Figure 7 Coronary circulation, shown in green, provides oxygen to heart cells. Pulmonary circulation, shown in purple, supplies blood with oxygen and removes carbon dioxide. Systemic circulation, shown in orange, supplies the rest of the body with oxygen and nutrients and removes wastes.

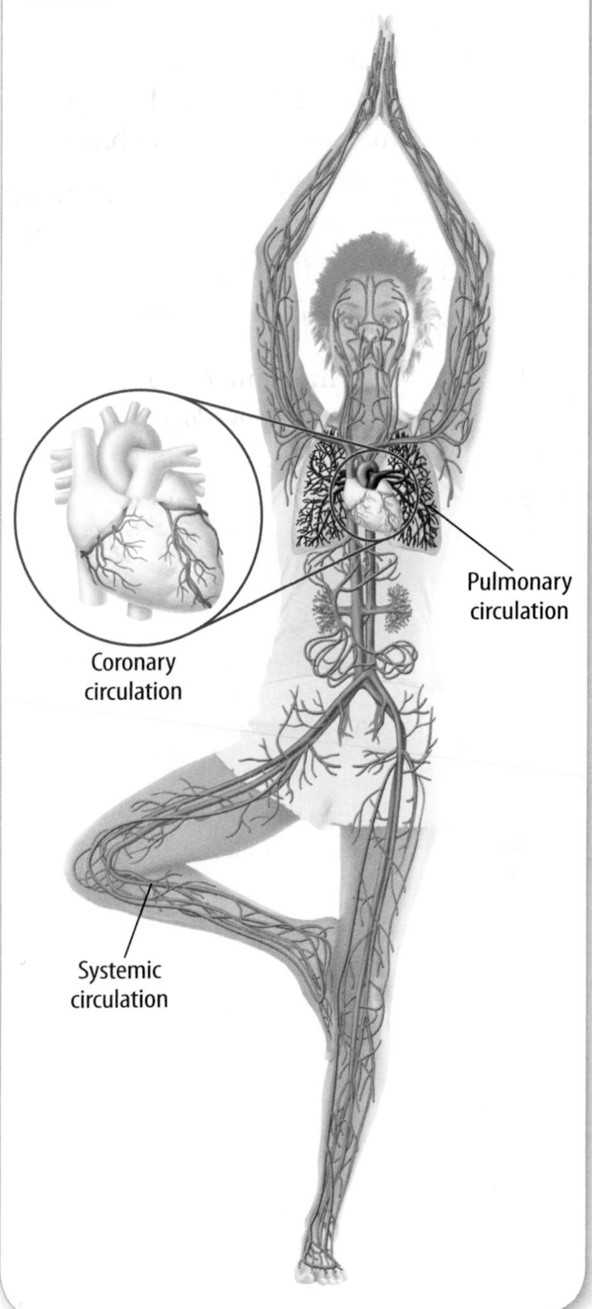

WORD ORIGIN

pulmonary
from Latin *pulmonarius,* means "of the lungs"

MiniLab

20 minutes

How can you model atherosclerosis?

Atherosclerosis is the buildup of fatty deposits in arteries. You can model atherosclerosis to see how it affects blood flow.

1. Read and complete a lab safety form.
2. Fit the end of a **funnel** into a piece of **plastic tubing.** Place the end of the tubing into a **beaker.**
3. Have your partner measure 10 mL of **water** into a **graduated cylinder.**
4. Pour the water into the funnel while your partner uses a **stopwatch** to time how long it takes the water to move through the tubing. Record the data in your Science Journal.

5. Fill the open end of the tubing with a small amount of **modeling clay.** Use a **toothpick** to make a hole in the center of the clay.
6. Repeat steps 3 and 4.

Analyze and Conclude

1. **Describe** any differences in the flow rate of water with and without clay in the tubes.
2. **Identify** which parts of your model represent blood, blood vessels, and fatty deposits of atherosclerosis.
3. **Key Concept** Explain how the presence of fatty deposits could affect the function of the circulatory system.

Circulatory System Health

Good health depends on a healthy circulatory system. Your heart muscle must be strong enough to push blood through all the blood vessels in your body. Your blood vessels must be flexible so that the volume of blood flowing through them can change. The valves in your heart and veins must work properly to keep blood from flowing in the wrong direction.

Circulatory diseases are illnesses that occur when some part of the circulatory system stops working properly. About one-third of all adults in the United States have some form of circulatory disease. Nearly 2,400 people die from it every day.

Hypertension

When the ventricles of the heart contract, they push blood into the arteries. When this happens, the arteries bulge a little. The bulging of an artery is what you feel when you check your pulse. The bulge happens because blood presses against the sides of the artery. That pressure is called blood pressure.

Have you ever had your blood pressure measured? Normal blood pressure is considered to be 120 mm Hg (millimeters of mercury) or less during the contraction of the ventricles. It is 80 mm Hg or less after the contraction. Normal blood pressure can be written as 120/80 mm Hg. Blood pressure higher than 140/90 mm Hg is known as hypertension, or high blood pressure. Hypertension can lead to weakened and less flexible artery walls.

Atherosclerosis

The buildup of fatty material within the walls of arteries is called **atherosclerosis** (a thuh roh skluh ROH sus). Fat deposits can interfere with the artery's blood flow. If a deposit breaks loose, it can flow to and block a narrower artery. A blockage in the heart can cause a heart attack. A blockage in a blood vessel in the brain can cause a stroke.

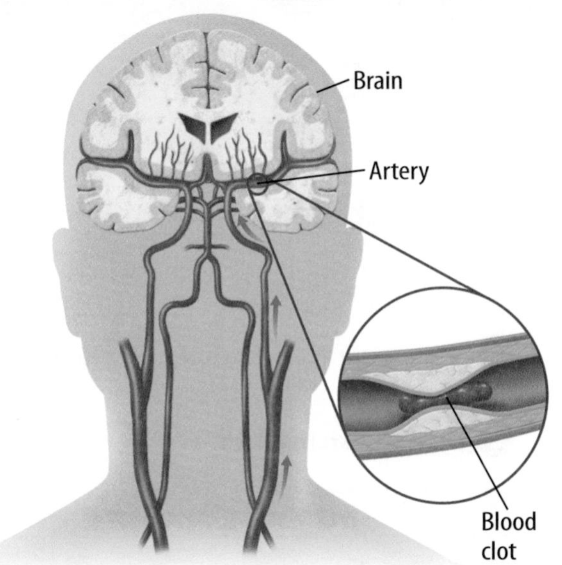

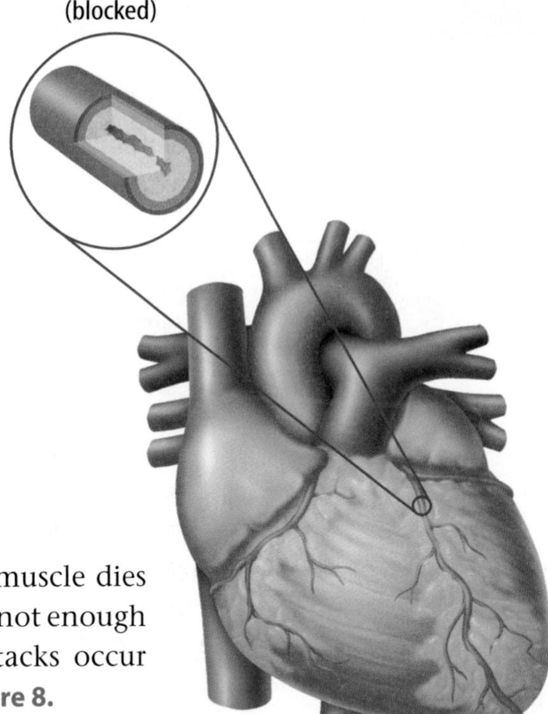

Heart Attacks, Strokes, and Heart Failure

A heart attack happens when part of the heart muscle dies or is damaged. A heart attack is usually caused when not enough oxygen reaches cells in the heart. Most heart attacks occur when a coronary vessel is blocked, as shown in **Figure 8.**

A stroke happens when part of the brain dies or is damaged. Most strokes are caused when not enough oxygen reaches cells in the brain. A stroke might occur if a blood clot blocks a blood vessel in the brain, also shown in **Figure 8.**

Heart failure occurs when the heart is not working efficiently. It can result from a previous heart attack, a problem with heart valves, or diseases that damage the heart.

Figure 8 Most heart attacks occur when a vessel of the coronary circulation is blocked. Most strokes occur when a blood clot blocks a vessel in the brain.

Preventing Circulatory System Disorders

Some risk factors for circulatory system diseases cannot be avoided. For example, if one of your parents has a circulatory disease, you might have inherited a slightly higher risk of developing a similar disease. However, most risk **factors** can be controlled by making good life choices, like eating a healthful diet, controlling weight, exercising, and not smoking.

Academic Vocabulary
factor
(noun) something that helps produce a result

The Circulatory System and Homeostasis

The circulatory system is closely connected with other body systems. Once oxygen enters your body, the respiratory system interacts with the circulatory system and transports oxygen to your body's cells. It also transports nutrients from the digestive system and hormones from the endocrine system. The nervous system regulates your heartbeat. Later in this chapter, you'll read how the circulatory and skeletal systems work together.

Key Concept Check How do the circulatory system and the respiratory system work together to maintain homeostasis?

Lesson 2 Review

Assessment Online Quiz
Inquiry Virtual Lab

Visual Summary

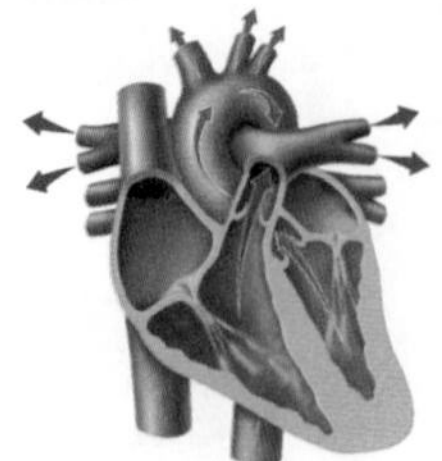

The contractions of the heart push blood through the circulatory system.

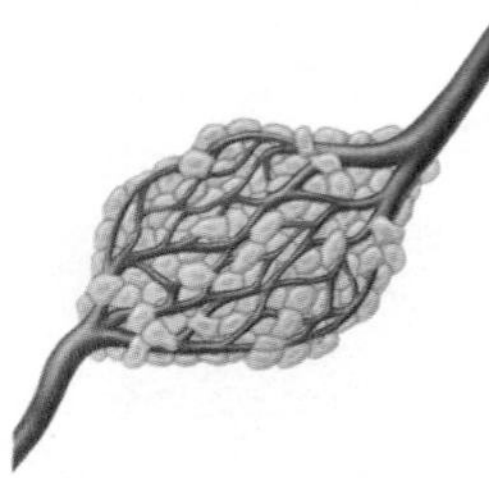

Arteries and veins carry blood throughout the body. Materials move between blood and cells through capillary walls.

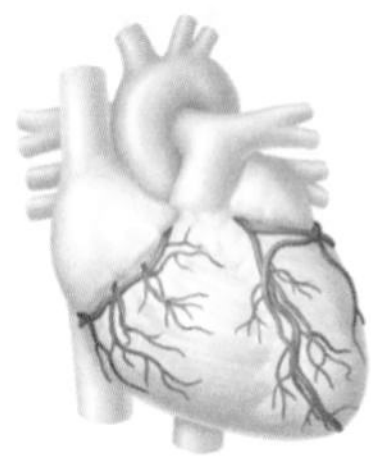

Coronary circulation supplies blood to heart cells.

FOLDABLES®

Use your lesson Foldable to review the lesson. Save your Foldable for the project at the end of the chapter.

Use Vocabulary

1 The narrow blood vessels where gas exchange occurs are ________.

2 The two lower chambers of the heart are ________.

3 **Distinguish** between veins and arteries.

Understand Key Concepts 

4 **Explain** how blood keeps flowing continuously through the body.

5 **Illustrate** at least five parts of the heart and the main blood vessels that enter and leave the heart.

6 A blockage of blood vessels in the brain can cause

A. a heart attack.
B. a stroke.
C. heart failure.
D. hypertension.

Interpret Graphics

7 **Identify** the artery and the vein in the figure below. Explain your answer.

8 **Summarize** Copy and fill in the graphic organizer below to identify the three types of circulation.

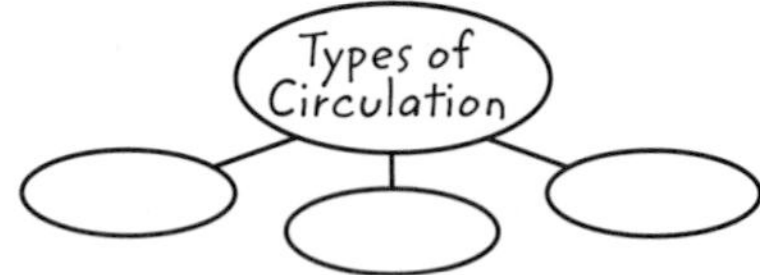

Critical Thinking

9 **Design** a daily schedule that includes at least three lifestyle choices that can help you avoid circulatory system disorders.

10 **Justify** A physician has a patient, age 42, whose blood pressure averages 141/89 mm Hg. Why might the physician recommend healthful life choices to the patient?

What do you think NOW?

You first read the statements below at the beginning of the chapter.

3. There are four chambers in a human heart.

4. Blood travels in both directions in veins.

Did you change your mind about whether you agree or disagree with the statements? Rewrite any false statements to make them true.

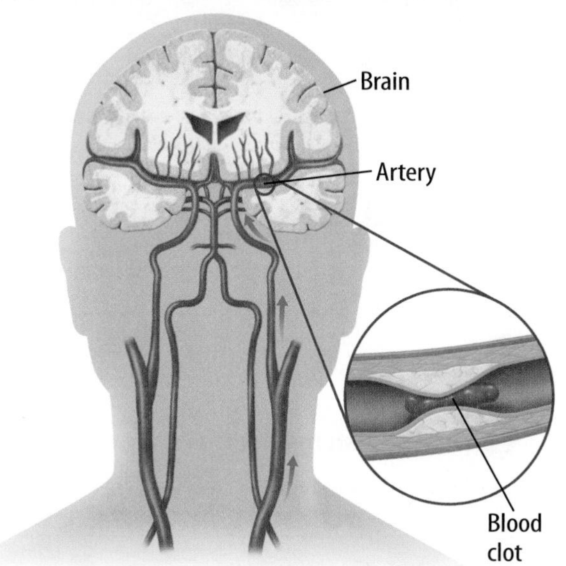

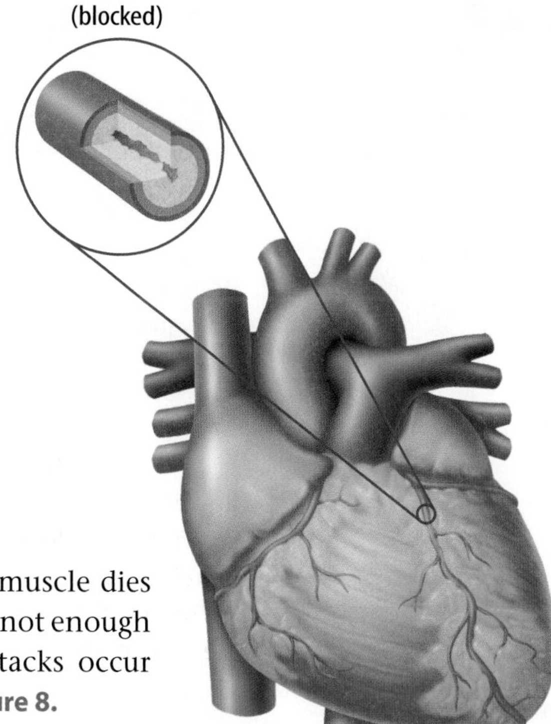

Heart Attacks, Strokes, and Heart Failure

A heart attack happens when part of the heart muscle dies or is damaged. A heart attack is usually caused when not enough oxygen reaches cells in the heart. Most heart attacks occur when a coronary vessel is blocked, as shown in **Figure 8.**

A stroke happens when part of the brain dies or is damaged. Most strokes are caused when not enough oxygen reaches cells in the brain. A stroke might occur if a blood clot blocks a blood vessel in the brain, also shown in **Figure 8.**

Heart failure occurs when the heart is not working efficiently. It can result from a previous heart attack, a problem with heart valves, or diseases that damage the heart.

Figure 8 Most heart attacks occur when a vessel of the coronary circulation is blocked. Most strokes occur when a blood clot blocks a vessel in the brain.

Preventing Circulatory System Disorders

Some risk factors for circulatory system diseases cannot be avoided. For example, if one of your parents has a circulatory disease, you might have inherited a slightly higher risk of developing a similar disease. However, most risk **factors** can be controlled by making good life choices, like eating a healthful diet, controlling weight, exercising, and not smoking.

ACADEMIC VOCABULARY
factor
(noun) something that helps produce a result

The Circulatory System and Homeostasis

The circulatory system is closely connected with other body systems. Once oxygen enters your body, the respiratory system interacts with the circulatory system and transports oxygen to your body's cells. It also transports nutrients from the digestive system and hormones from the endocrine system. The nervous system regulates your heartbeat. Later in this chapter, you'll read how the circulatory and skeletal systems work together.

 **Key Concept Check** How do the circulatory system and the respiratory system work together to maintain homeostasis?

Lesson 2 Review

Assessment Online Quiz
Inquiry Virtual Lab

Visual Summary

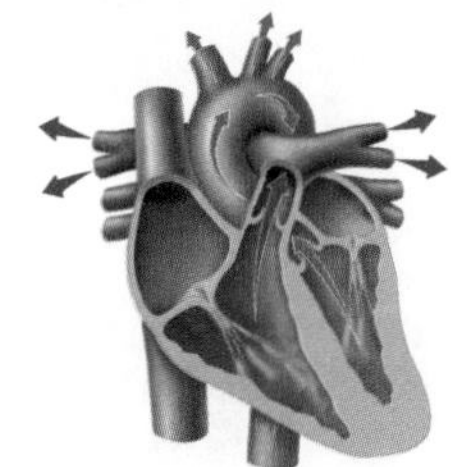

The contractions of the heart push blood through the circulatory system.

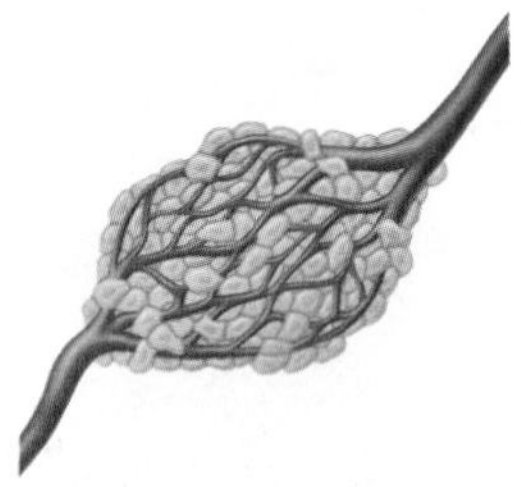

Arteries and veins carry blood throughout the body. Materials move between blood and cells through capillary walls.

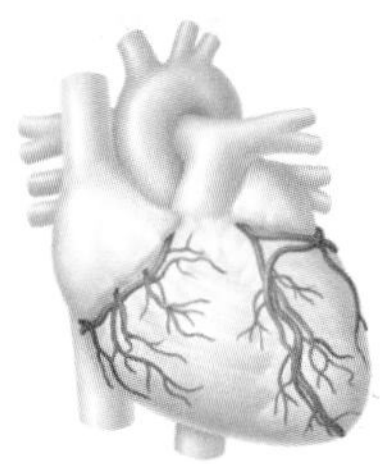

Coronary circulation supplies blood to heart cells.

FOLDABLES®

Use your lesson Foldable to review the lesson. Save your Foldable for the project at the end of the chapter.

What do you think NOW?

You first read the statements below at the beginning of the chapter.

3. There are four chambers in a human heart.

4. Blood travels in both directions in veins.

Did you change your mind about whether you agree or disagree with the statements? Rewrite any false statements to make them true.

Use Vocabulary

1. The narrow blood vessels where gas exchange occurs are ________.
2. The two lower chambers of the heart are ________.
3. **Distinguish** between veins and arteries.

Understand Key Concepts 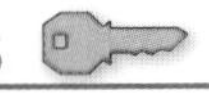

4. **Explain** how blood keeps flowing continuously through the body.
5. **Illustrate** at least five parts of the heart and the main blood vessels that enter and leave the heart.
6. A blockage of blood vessels in the brain can cause
   A. a heart attack. **C.** heart failure.
   B. a stroke. **D.** hypertension.

Interpret Graphics

7. **Identify** the artery and the vein in the figure below. Explain your answer.

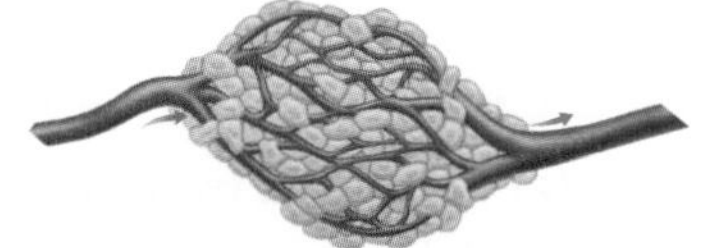

8. **Summarize** Copy and fill in the graphic organizer below to identify the three types of circulation.

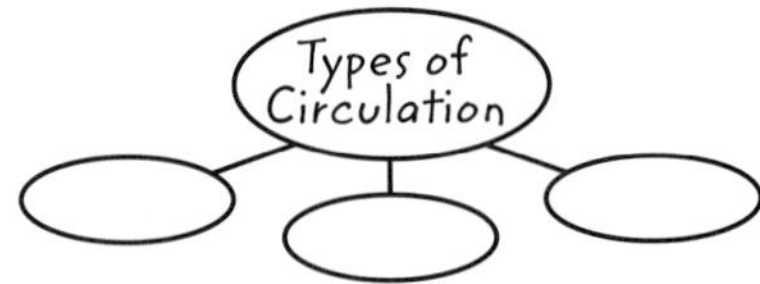

Critical Thinking

9. **Design** a daily schedule that includes at least three lifestyle choices that can help you avoid circulatory system disorders.
10. **Justify** A physician has a patient, age 42, whose blood pressure averages 141/89 mm Hg. Why might the physician recommend healthful life choices to the patient?

How strong is your heart muscle?

Materials

plastic basin

turkey baster bulb

stopwatch

Safety

Your heart is a muscle that beats about 75 times every minute. With each beat, it pushes about 60 mL of blood into your arteries. Heart muscle cells differ from skeletal muscle cells. Heart muscle cells have about 35 percent more mitochondria than skeletal muscle cells do. More mitochondria means that a heart cell can obtain more energy from the nutrients it receives. Can you use the skeletal muscles in your hand to pump liquid as fast as your heart pumps blood—without getting tired?

Learn It

To **infer** means to reach a conclusion based on facts. Your heart beats continuously to keep blood flowing through your body. Modeling the function of the heart can help you infer how hard your heart works.

Try It

1. Read and complete a lab safety form.
2. Fill a plastic basin with enough water to cover your fist.
3. Remove the bulb from a turkey baster. Hold the bulb in your fist and submerge it completely in the water. Move slowly to avoid spilling water.
4. Aim the opening of the bulb toward the side of the basin. Let the bulb fill with water.
5. Squeeze the bulb repeatedly as quickly as you can to simulate the beating of your heart.

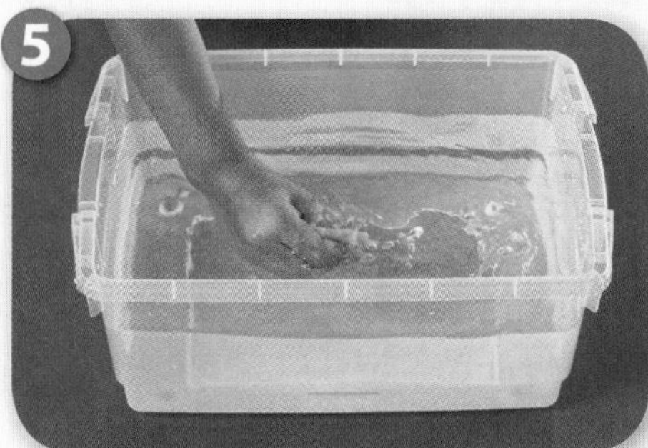
5

6. Have a partner count the number of times you squeeze the bulb. Have another partner time the experiment.
7. Continue the squeezing action for as long as you can, but not more than 3 minutes. Record the number of squeezes and the time in your Science Journal.
8. Repeat steps 3–7 for all the students in your group.

Apply It

9. **Describe** how your hand felt at the end of the experiment.
10. **Compare** how tired your hand muscles got with how tired you think your heart muscle is after the same number of beats.
11. **Calculate** Suppose a man who lives to the age of 80 has an average heart rate of 75 beats per minute. About how many times would his heart beat in his lifetime?
12. **Key Concept** Infer which type of muscle, heart or skeletal, is able to perform more consistently without tiring.

Lesson 3

Blood

Reading Guide

Key Concepts 
ESSENTIAL QUESTIONS

- What does the blood do?
- How do the parts of the blood differ?

Vocabulary

platelet p. 579
plasma p. 579
Rh factor p. 581

g Multilingual eGlossary

Working Hard?

You might have noticed that the cheeks of some people turn bright red during vigorous exercise. Why does this happen? The red color comes from blood flowing near the surface of the skin. It helps release excess thermal energy from the body. What else does your blood do?

Launch Lab

10 minutes

What do blood cells look like? 

Like every tissue in your body, blood is a tissue made of different cells. Take a look in the microscope to see the different types of blood cells.

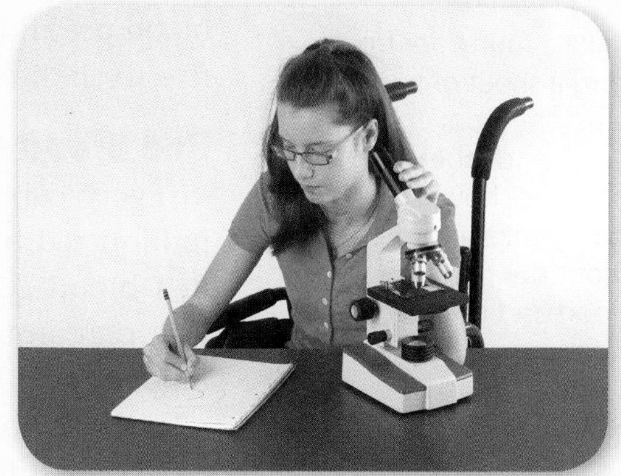

1. Read and complete a lab safety form.
2. Observe a **prepared blood smear slide** under low power on a **microscope.**
3. Switch to high power, and observe the different cell types on the slide. In your Science Journal, draw one example of each type of cell you see.
4. Return to low power, and remove the slide.

Think About This

1. How many kinds of cells did you observe? How did their appearances differ?
2. **Key Concept** Why do you think there are different kinds of blood cells?

Functions of Blood

Have you ever had an injury that caused bleeding? Blood is a red liquid, slightly thicker than water. At the end of Lesson 2 you read that your circulatory system works closely with all your other body systems to maintain homeostasis. Blood is the link that connects the circulatory system with all other body systems. Blood transports substances throughout your body, helps protect your body against infection, and helps regulate your body's temperature.

Transportation

Blood transports many different substances throughout your body. You've read that blood carries oxygen to and carbon dioxide from the lungs. Blood also picks up nutrients in the small intestine and carries them to all body cells. It transports hormones produced by the endocrine system. Blood also carries waste products to the excretory system. Most of the substances carried in blood are dissolved in the liquid part of blood.

Protection

Some blood cells fight infection. They help protect you from harmful organisms, such as bacteria, viruses, fungi, and parasites. Blood also contains materials that help repair torn blood vessels and heal wounds. These materials help protect the body from losing too much blood.

Temperature Regulation

Blood helps your body maintain a steady temperature of about 37°C. When your body temperature rises, blood vessels near the surface of your skin widen. This increases blood flow to your skin's surface and releases more thermal energy to the air. Your body cools down. When your body temperature lowers, the vessels at your skin's surface get narrower. This decreases blood flow to the surface of your skin and reduces the amount of thermal energy that is lost to the air. Your body warms up.

Key Concept Check What does the blood do?

Parts of Blood

Blood is considered a tissue because it is made up of different kinds of cells that work together. As shown in **Figure 9,** blood consists of four main parts: red blood cells, white blood cells, platelets, and plasma. Most adults have about 70 mL of blood per kilogram of body weight. An average adult has about five to six liters of blood.

FOLDABLES®

Make a four-door book from a sheet of paper. Label the front *The Blood,* and label the inside as shown. Use it to organize information about the parts of blood and their functions.

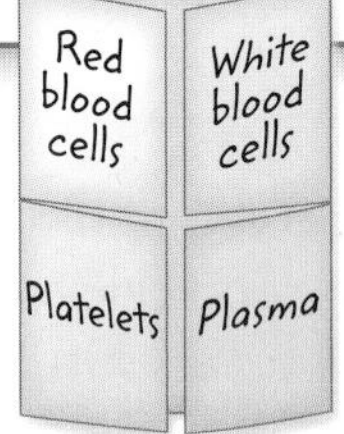

Red Blood Cells

Every cubic millimeter of your blood contains four to six million red blood cells, or erythrocytes (ih RIHTH ruh sites). Red blood cells contain hemoglobin (HEE muh gloh bun), iron-rich protein molecules. In the alveoli of the lungs, oxygen binds to the hemoglobin in red blood cells. The hemoglobin releases the oxygen when red blood cells enter the capillaries and come into close contact with body cells.

How would you describe the shape of the red blood cells in **Figure 9?** Some people describe them as a doughnut without a hole. This flattened disk shape gives them more surface area. This means red blood cells can carry more oxygen than if they were round like a ball. Your body produces new red blood cells all the time because they wear out in a few months.

White Blood Cells

Your blood also contains several kinds of white blood cells, or leukocytes (LEW kuh sites), shown in **Figure 9.** All white blood cells protect your body from illness and infection. Some attack viruses, bacteria, fungi, and parasites that might invade your body. Most white blood cells last only a few days and are constantly replaced. You have far fewer white blood cells—5,000 to 10,000 per cubic millimeter—than red blood cells.

Figure 9 Blood flows through blood vessels. It is made of liquid plasma, red blood cells, white blood cells, and platelets.

Parts of Blood

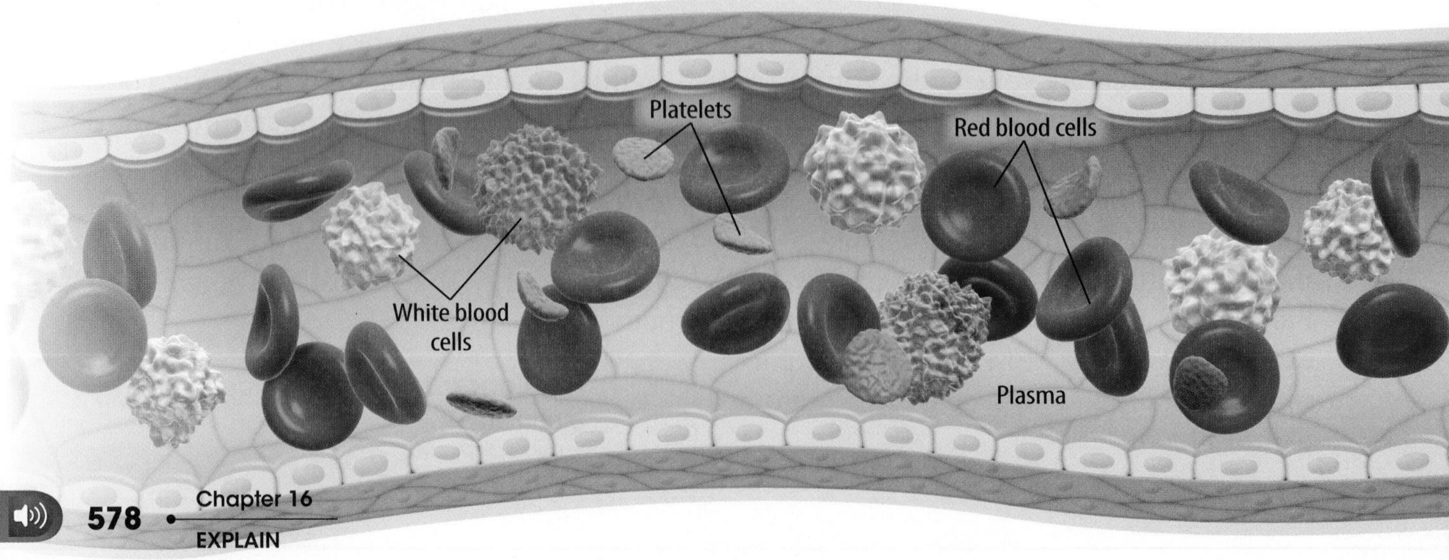

Platelets

What happens if you get a cut? The cut bleeds for a short time, and then the blood clots, as shown in **Figure 10.** **Platelets** *are small, irregularly shaped pieces of cells that plug wounds and stop bleeding.* Platelets produce proteins that help strengthen the plug. Without platelets, blood would not stop flowing. Your blood contains 150,000 to 440,000 platelets per cubic millimeter.

Plasma

The yellowish, liquid part of blood, called **plasma**, *transports blood cells.* Plasma is 90 percent water, which helps thin the blood. This enables it to travel through small blood vessels. Many molecules are dissolved in plasma. They include salts, vitamins, sugars, minerals, proteins, and cellular wastes.

Plasma also plays an important role in regulating the activities of cells in your body. For example, plasma carries chemical messengers that control the amounts of salts and glucose that enter cells.

 **Key Concept Check** How do the parts of blood differ?

Figure 10 When a blood vessel breaks, platelets rush to the wound. They cause the formation of a threadlike protein that makes a net. A blood clot forms as blood cells are trapped in the net.

Inquiry MiniLab — 20 minutes

How does a cut heal?

How does a scab help a cut stop bleeding? Use gauze and liquid bandage to find out.

1. Read and complete a lab safety form.
2. Place two 5-cm square pieces of **gauze** side by side on a piece of **waxed paper.**
3. Make a 2-cm circle of **liquid bandage** in the center of one piece of gauze. Allow it to dry for 5 minutes.
4. Use a **plastic dropper** to place a drop of **water** in the center of each piece of gauze.

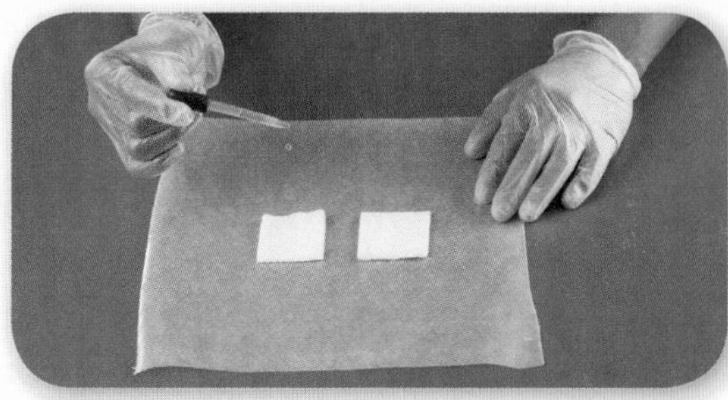

Analyze and Conclude

1. **Compare** what happened to the drop of water on each piece of gauze.
2. **Identify** which piece of gauze best models a scab. Explain why you think so.
3. **Key Concept** Explain how blood helps heal cuts in the skin.

How a Blood Clot Forms

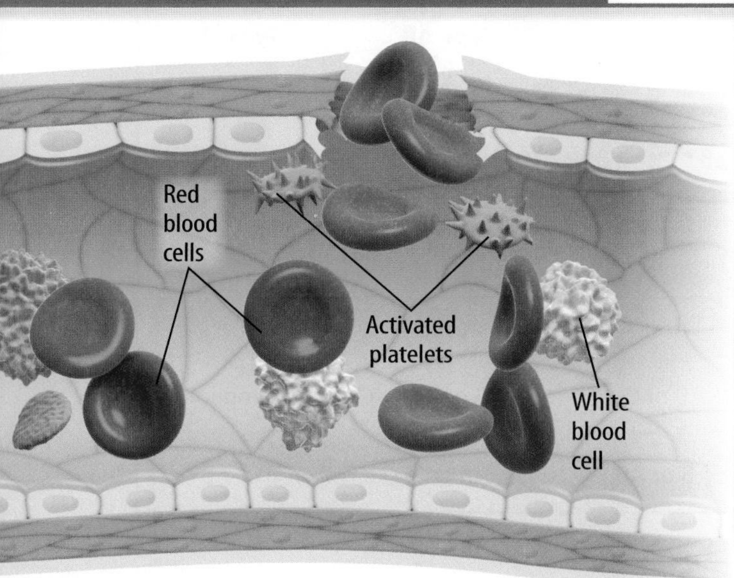

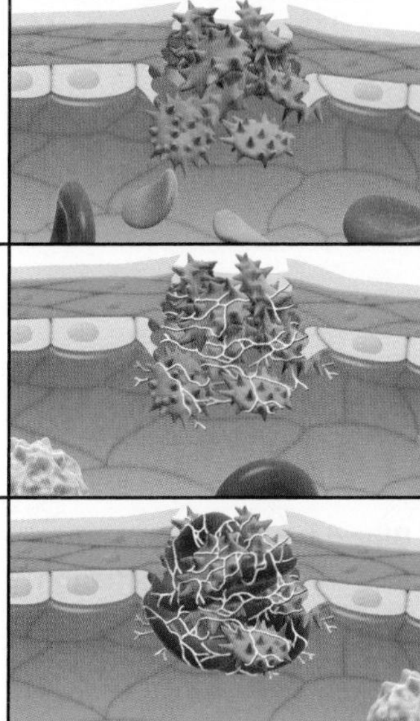

Step 1 Platelets rush to the tear and form a plug to stop the bleeding.	
Step 2 A web of fibrin forms around the platelets to hold them in place.	
Step 3 More platelets and red blood cells are caught in the fibrin web, forming a blood clot.	

Blood Types

Do you know anyone who donates blood? Donated blood is used to help people who have lost too much blood from an injury or surgery and need a transfusion. A blood transfusion is the transfer of blood from one person to another. Even though your blood has the same four parts as everyone else's—red cells, white cells, platelets, and plasma—you can't receive blood from just anyone. Why? Because different people have different blood types.

WORD ORIGIN

plasma
from Greek *plassein,* means "to mold"

The ABO System

You inherited your blood type from your parents. Blood type refers to the type of proteins, or antigens, on red blood cells. The four human blood types are A, B, AB, and O. Type A blood cells have the A antigen. Type B cells have the B antigen. Type AB has both A and B antigens. Type O has no antigens.

Why is blood type important? Any time a blood transfusion brings foreign antigens to someone's blood, the red blood cells will clump together and lose their ability to function. Clumps form because of clumping proteins in blood plasma. As shown in **Table 2,** people with types A, B, and O blood have clumping proteins in their plasma. For example, a person with type A blood has anti-B clumping proteins. If given type B blood, his or her anti-B proteins would attack the type B antigens and cause the type B red blood cells to clump together.

Type AB blood has no clumping proteins. People with AB blood are known as "universal recipients" because they can receive transfusions of any blood type. Type O blood has clumping proteins that attack both A and B antigens. People with type O blood are known as "universal donors" because they can donate blood to anyone.

Table 2 People with blood types A, B, and O have clumping proteins in their plasma. These proteins determine what blood type a person can safely receive in a blood transfusion.

Table 2 Human Blood Types Concepts in Motion Interactive Table

Blood Type	Type A	Type B	Type AB	Type O
Antigens on red blood cells	A A A A A A	B B B B B B	B A A B B A	
Percentage of US population with this blood type	42	10	4	44
Clumping proteins in plasma	Anti-B	Anti-A	None	Anti-A and anti-B
Blood type(s) that can be RECEIVED in a transfusion	A or O	B or O	A or B or AB or O	O only
This blood type can DONATE TO these blood types	A or AB	B or AB	AB only	A or B or AB or O

The Rh Factor

Another protein found on red blood cells is a chemical marker called the **Rh factor.** People with blood cells that have this protein are Rh positive. People who do not have it are Rh negative. If Rh positive blood is mixed with Rh negative blood, clumping can result. Blood types usually have a plus (+) sign or a minus (–) sign to indicate whether a person is Rh positive or negative. For example, a person with an A+ blood type has red cells with A antigens and the Rh factor. Someone with O– blood has no antigens and no Rh factor.

 **Reading Check** What kinds of antigens are found in AB+ blood?

Blood Disorders

Some medical conditions disrupt the normal functions of blood. People with hemophilia lack a protein needed to clot blood. A person who has hemophilia bleeds at the same rate as other people, but the bleeding does not stop as quickly.

People suffering from anemia have low numbers of red blood cells or have red blood cells that do not contain enough hemoglobin. As a result, the blood might not carry as much oxygen as the body needs.

Bone marrow, the soft tissue in the center of bones, produces red blood cells. Cancer of the bone marrow is called leukemia. This kind of cancer can slow or prevent blood cell formation. Leukemia can lead to anemia and a damaged immune system.

People who inherit sickle-cell disease have red blood cells shaped like crescents, as shown in **Figure 11.** Sickle-shaped cells do not move through blood vessels as easily as normal disk-shaped cells. Sickle cells can prevent oxygen from reaching tissues and cause sickle-cell anemia.

Math Skills

Use Percentages

If percentages refer to the same factor, they can be added or subtracted. For example, you could add the percentages of people with each of the four blood types:

42% + 10% + 4% + 44% = 100%

You could also subtract to find what percentage of people do not have type O blood:

100% – 44% = 56%

Practice

Forty-four percent of people have type O blood. If 7 percent of people have type O blood and are Rh negative, what percent has type O Rh positive blood?

- Math Practice
- Personal Tutor

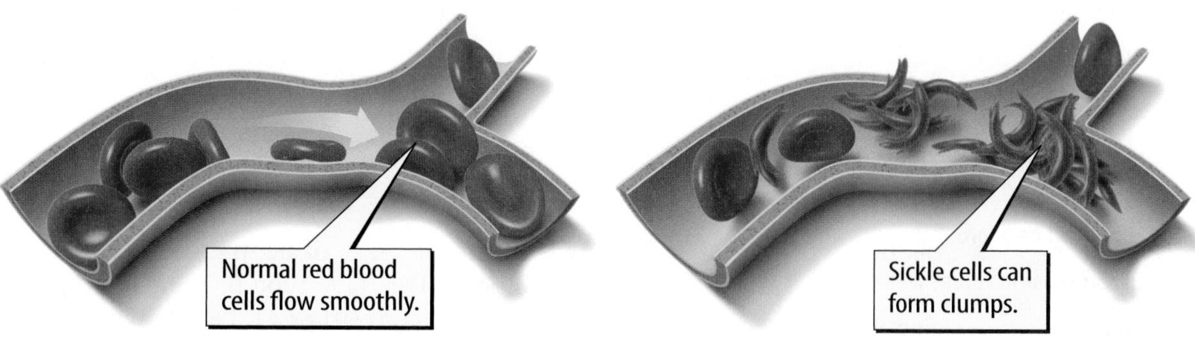

Figure 11 The crescent-shaped red blood cells in sickle-cell disorder form clumps that can block blood vessels.

Lesson 3 Review

Visual Summary

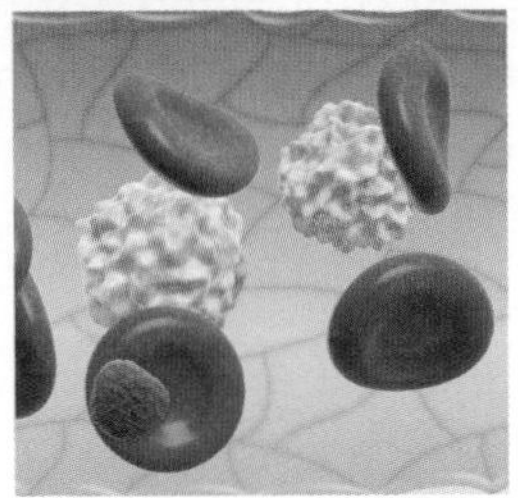

Red blood cells contain hemoglobin and carry oxygen. White blood cells help fight disease.

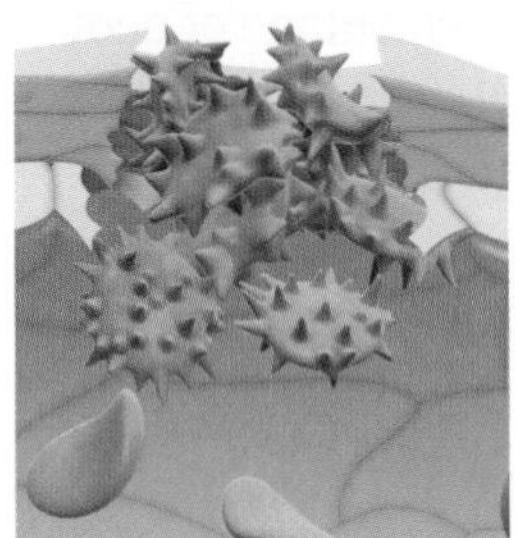

Platelets are pieces of cells that aid in blood clotting.

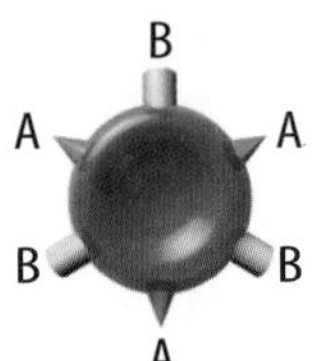

Blood type is determined by antigens on the surface of red blood cells and clumping proteins in blood plasma.

FOLDABLES®

Use your lesson Foldable to review the lesson. Save your Foldable for the project at the end of the chapter.

Use Vocabulary

1. **Identify** the yellowish liquid part of blood.
2. **Distinguish** between plasma and platelets.
3. **Define** the term *Rh factor* using your own words.

Understand Key Concepts

4. **Give an example** of a blood disorder and explain how it can affect health.
5. Which part of the blood carries dissolved molecules such as glucose and salt?
   A. plasma
   B. platelets
   C. red blood cells
   D. white blood cells

Interpret Graphics

6. What blood type is shown below? What blood type(s) can a person with this blood type receive in a transfusion?

7. **Organize Information** Copy the table below and list the parts of blood and their functions.

Part of Blood	Functions

Critical Thinking

8. **Design** a model to show how blood can help regulate body temperature.

Math Skills

Math Practice

9. In the US, 42% of people have type A blood, 10% have type B, 4% have type AB, and 44% have type O. People with type B and type AB blood can receive type B blood. What percentage of people can receive transfusions of type B blood?

What do you think NOW?

You first read the statements below at the beginning of the chapter.

5. All blood cells are red.

6. Blood plasma is just water.

Did you change your mind about whether you agree or disagree with the statements? Rewrite any false statements to make them true.

SCIENCE & SOCIETY

Very Special Blood Cells

▲ **Technicians remove only a small portion of the crabs' blood. After this procedure, the crabs are returned to the ocean. Their blood cell levels return to normal in a couple of weeks.**

Horseshoe crabs, living relatives of extinct trilobites, have been gathering on beaches for 350 million years. They usually become food for fish and birds. Yet someday your life might depend on horseshoe crabs—or at least on their blood. Unlike human blood, horseshoe crab blood contains only one type of blood cell. If bacteria enter the crab's bloodstream from an open wound, blood cells secrete a clotting factor. This secretion closes the wound, and the blood cells engulf the bacteria. When scientists saw that horseshoe crab blood turned to a gel in the presence of harmful bacteria, they realized its value. Today, medical professionals use an extract made from horseshoe crab blood to screen all intravenous medicines for bacteria. A quart of this special blood costs about $15,000!

The horseshoe crab blood can do even more. Another component of the blood can stop the human immunodeficiency virus (HIV) from replicating, or making copies of itself. Part of horseshoe crab blood can act as an antibiotic. Scientists also are using horseshoe crab blood in the development of a hand-held instrument that helps to diagnose human illnesses. The instrument uses enzymes from the blood as illness detectors.

It's Your Turn

REPORT Medical professionals use certain types of snake venom to treat strokes. Research to find other unusual animal products that have medical uses.

Lesson 4

The Lymphatic System

Reading Guide

Key Concepts 
ESSENTIAL QUESTIONS

- What does the lymphatic system do?
- How do the parts of the lymphatic system work together?
- How does the lymphatic system interact with other body systems?

Vocabulary

lymphatic system p. 585
lymph p. 586
lymph node p. 586
thymus p. 587
spleen p. 587

g Multilingual eGlossary

Inquiry Healthy or Not?

Do you know anyone who has had his or her tonsils removed? Tonsils are clusters of lymph tissue that help the body fight off disease. Why do tonsils sometimes get swollen and inflamed, like the ones shown here? Knowing how the lymphatic system works will help you answer that question.

Launch Lab

10 minutes

How can you model a lymph node?

Fluid surrounds your body cells. Body cells absorb materials from and release materials into this fluid. Some of the fluid drains into vessels and then drains into spongy structures called lymph nodes. What happens in the lymph nodes?

1. Read and complete a lab safety form.
2. Observe a **liquid** provided by your teacher. Record the observations in your Science Journal.
3. Use a **rubber band** to attach a square of **cheesecloth** to a **plastic drinking straw.** Hold the straw upright over a **paper plate.**
4. Use a **plastic dropper** to squeeze about 1 mL of the liquid into the open end of the straw.
5. Allow the liquid to drain from the cheesecloth and onto the plate. Observe the liquid. Record the observations in your Science Journal.

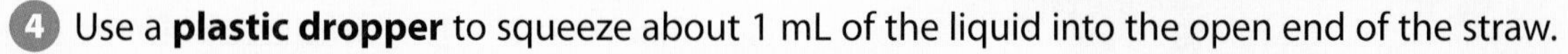

Think About This

1. What differences did you observe in the liquid after it passed through the cheesecloth?
2.  **Key Concept** What do you think the function of the lymph nodes might be?

Functions of the Lymphatic System

At times when you were sick, you might have noticed small, swollen structures under your jaw on each side of your neck. These structures can become swollen when they're working to fight off an infection in your body.

The **lymphatic system** *is part of the immune system and helps destroy microorganisms that enter the body.* The lymphatic system works closely with the circulatory system. Both systems move liquids through the body, and both contain white blood cells. However, their functions are different. There are four main functions of the lymphatic system.

- It absorbs some of the tissue fluid that collects around cells.
- It absorbs fats from the digestive system and transports them to the circulatory system.
- It filters dead cells, viruses, bacteria, and other unneeded particles from tissue fluid and then returns the tissue fluid to the circulatory system.
- It helps fight off illness and infections and includes structures in which white blood cells develop.

 **Key Concept Check** What does the lymphatic system do?

FOLDABLES®

Fold a sheet of paper into an eight-page book. Use it to organize your notes about the parts of the lymphatic system and their functions.

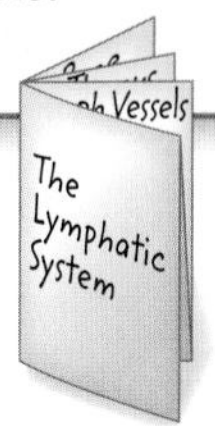

The Lymphatic System 

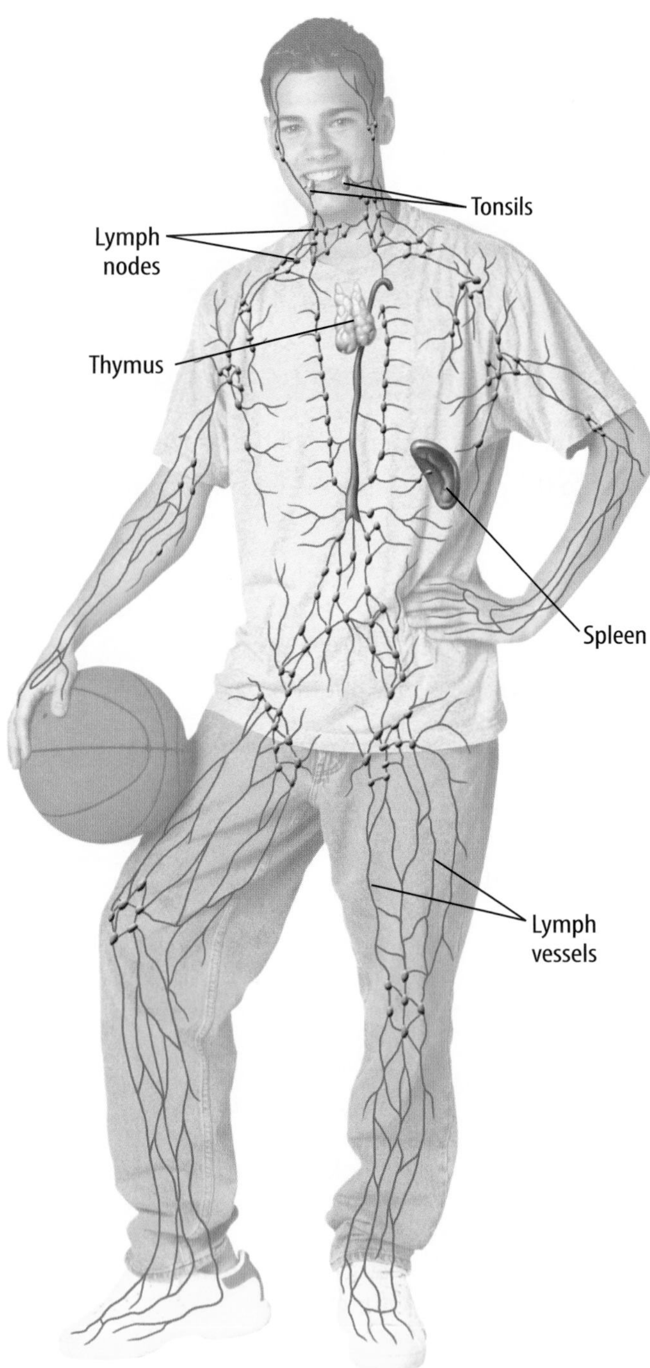

Figure 12 The lymphatic system is a network of vessels and organs. Vessels transport lymph. When it reaches the area beneath the collarbone, it re-enters the circulatory system.

Visual Check What organs of the lymphatic system are in the throat?

Parts of the Lymphatic System

The lymphatic system, shown in **Figure 12,** includes lymph vessels and the fluid they carry. It also includes several other structures.

Lymph

Water, white blood cells, and dissolved materials such as salts and glucose leak out of capillary walls and into the spaces that surround tissue cells. This fluid is called tissue fluid. Cells absorb the materials they need from tissue fluid and release wastes into it. About 90 percent of the tissue fluid is reabsorbed by the capillaries. *About 10 percent of the tissue fluid is absorbed by the lymph vessels and is called* **lymph.**

 **Reading Check** What is lymph?

Lymph Vessels

The lymphatic system forms a network of lymph vessels that look similar to the circulatory system's network of blood vessels. Lymph vessels absorb and transport lymph. The lymph is pushed through the lymph vessels by contractions of the muscles you use to move your body. Lymph is not pumped through the lymph vessels by the heart.

Lymph Nodes

Lymph vessels include *clusters of small, spongy structures called* **lymph nodes** *that filter particles from lymph.* Bacteria, viruses, fungi, and pieces of dead cells are trapped and removed from the lymph as it flows through a lymph node. Lymph nodes also store white blood cells that attack and destroy the trapped particles.

Large groups of lymph nodes are in the neck, the groin, and the armpits. When you have an infection, your body increases its production of white blood cells that fight the infection. Many of these white blood cells gather in your lymph nodes and cause the nodes to swell. The swelling disappears when the infection is gone.

Bone Marrow and Thymus

Lymphocytes (LIHM fuh sites) are white blood cells that destroy pathogens—infection-causing microorganisms such as viruses and bacteria. Bone marrow is the spongy center of bones where red and white blood cells, including lymphocytes, form. Lymphocytes include B cells and T cells. As shown in **Figure 13,** B cells mature in the bone marrow, and T cells mature in the thymus gland.

The **thymus** *is the organ of the lymphatic system in which T cells complete their development.* After immature T cells move from the bone marrow to the thymus, they develop the ability to recognize and destroy body cells that have been infected by microorganisms. Mature B cells and T cells move into the lymph and blood to help fight infection.

Key Concept Check How do bone marrow and the thymus work together?

Spleen

You read earlier that the life of a red blood cell is only a few months. *The* **spleen** *is an organ of the lymphatic system that recycles worn-out red blood cells and produces and stores lymphocytes.* The spleen also stores blood and platelets. If a person is injured and loses a lot of blood, the spleen can release stored blood and platelets into the circulatory system.

WORD ORIGIN

lymph
from Latin *lympha,* means "water"

Figure 13 Lymphocytes attack and destroy disease-causing microorganisms. B cells mature in the bone marrow. T cells mature in the thymus.

Where Lymphocytes Mature

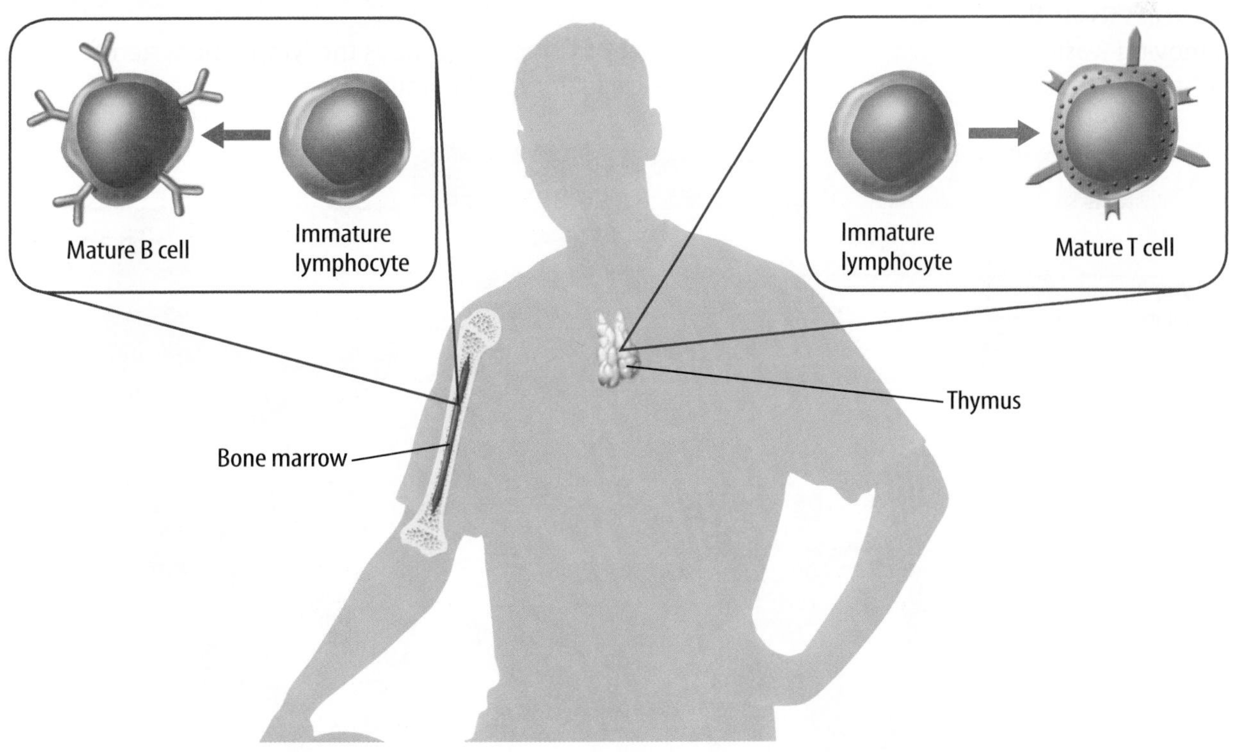

Tonsils

Your tonsils are clusters of lymph tissue on the sides of your throat. They help protect your body from infection by trapping and destroying bacteria and other pathogens that enter your nose and mouth. However, you can live without tonsils.

Lymph Diseases and Disorders

Science Use v. Common Use
vessel
Science Use a tube through which a body fluid travels
Common Use a container for holding something

Damage to the lymphatic system from injury or surgery can prevent tissue fluid from draining into lymph **vessels.** As a result, tissue fluid can build up around cells and cause swelling. Recall that the action of your body muscles pushes lymph through the lymph vessels. Inactivity can also cause lymph buildup and swelling.

Do you recall the swollen tonsils shown at the beginning of this lesson? If the cells of your tonsils become infected, you have tonsillitis—an inflammation of the tonsils.

The uncontrolled production of white blood cells is a type of cancer called lymphoma. Cancer of the lymph nodes is a related disease called Hodgkin's lymphoma.

The Lymphatic System and Homeostasis

The lymphatic system helps maintain your body's homeostasis by regulating fluid buildup around cells, as shown in **Figure 14.** It supports the circulatory system by cleaning fluids and replacing them in the bloodstream. It also supports overall health by helping to fight infection throughout the body.

Key Concept Check How does the lymphatic system interact with the circulatory and immune systems?

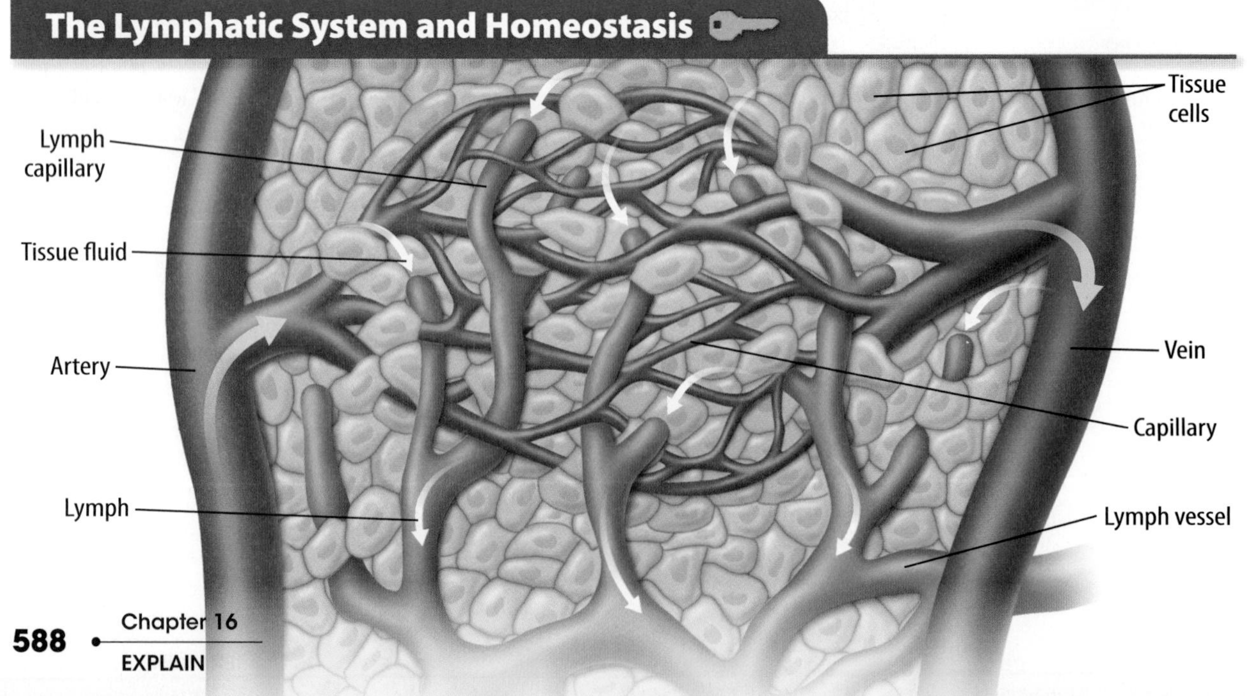

Figure 14 The lymphatic system helps maintain the body's homeostasis by preventing the buildup of excess tissue fluid, removing wastes, and fighting infection.

Lesson 4 Review

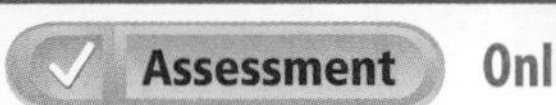

Visual Summary

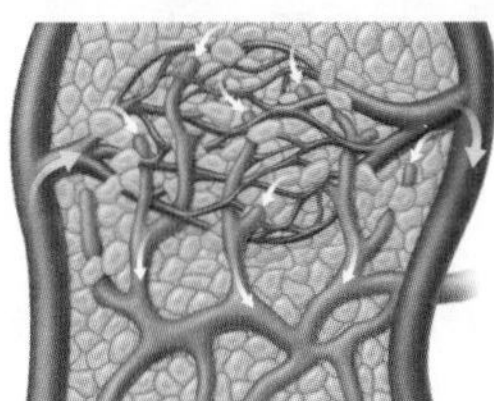

Tissue fluid that drains into the lymph vessels becomes lymph.

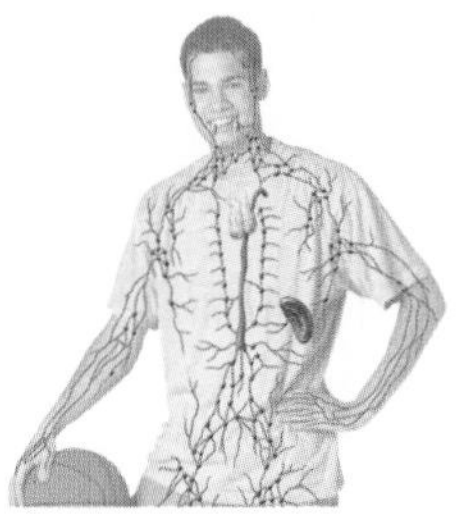

The lymphatic system consists of lymph nodes, lymph vessels, lymph, and several other organs.

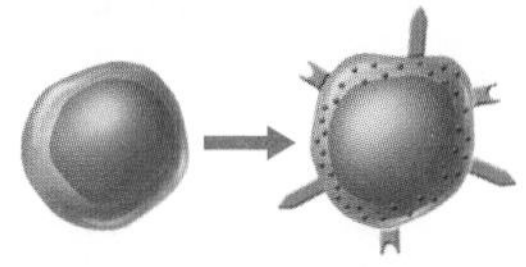

The lymphatic system cleans lymph, fights infection, and includes structures in which white blood cells develop.

FOLDABLES®

Use your lesson Foldable to review the lesson. Save your Foldable for the project at the end of the chapter.

What do you think NOW?

You first read the statements below at the beginning of the chapter.

7. Lymph nodes are only in the neck.

8. The lymphatic system helps fight infections to maintain a healthy body.

Did you change your mind about whether you agree or disagree with the statements? Rewrite any false statements to make them true.

Use Vocabulary

1. **Define** the term *lymph.*
2. **Distinguish** between the spleen and the thymus.
3. Clusters of small, spongy structures that filter particles from lymph are called ________ ________.

Understand Key Concepts 

4. **Describe** the function of the lymph nodes.
5. **Distinguish** between lymph and tissue fluid.
6. The lymphatic system cleans fluid for which system?

   A. circulatory
   B. digestive
   C. immune
   D. respiratory

Interpret Graphics

7. **Identify** the structures of the lymphatic system shown in the figure below.

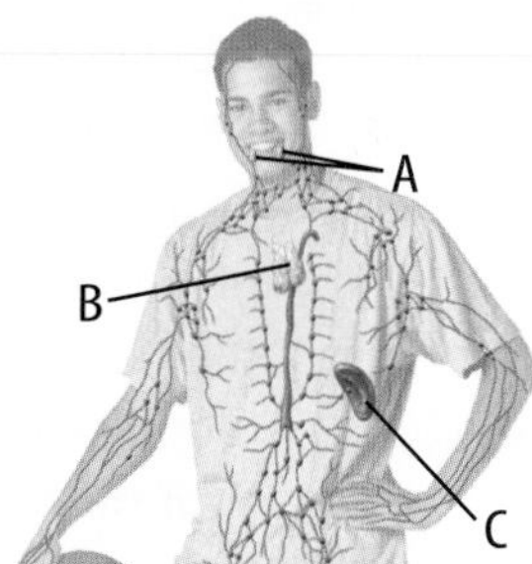

8. **Summarize** Copy and fill in the graphic organizer below to identify the functions of the lymphatic system.

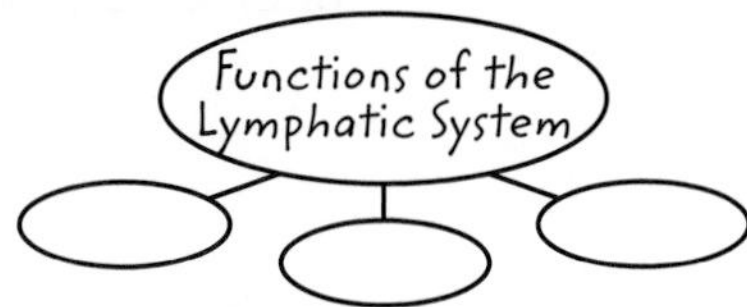

Critical Thinking

9. **Synthesis** Design a three-dimensional model of the lymphatic system.
10. **Evaluate** In what ways are the circulatory and lymphatic systems similar? In what ways are they different?

Lab — 40 minutes

Using a Blood Count to Evaluate Health

Materials

Model of blood

Paper plates

Safety

You are a physician who uses blood test results to help evaluate the health of your patients. One of the most common blood tests is a complete blood count, or CBC. A CBC measures the number of red blood cells, white blood cells, and platelets in a small sample of a person's blood. As a physician, you can use the results of a CBC to tell whether your patient is healthy or might have health problems.

Ask a Question

What do the results of a CBC tell you about your patient's health status?

Make Observations

1. Read and complete a lab safety form.
2. Obtain a bag of beans representing a blood sample from a healthy patient. This is your control sample. Red beans represent red blood cells, white beans represent white blood cells, and lentils represent platelets.

   ⚠ *Do not put beans or lentils in your mouth.*
3. In your Science Journal, make a data table like the one below to record your data.
4. Count the number of each type of blood cell in the sample. Calculate the percentage of each type of blood cell in your sample. Record your data.

Form a Hypothesis

5. After looking at the data you collected, form a hypothesis about the relationship between the percentages of blood cell types and the health of a patient.

Patient Number	Red Blood Cells		White Blood Cells		Platelets	
	Number	Percent	Number	Percent	Number	Percent
Control						

Test Your Hypothesis

6. Select blood samples from three other patients.
7. Repeat step 4 for each sample.

Analyze and Conclude

8. **Analyze** the CBC results for each patient. Based on the data from your healthy control patient, explain whether each patient is healthy.
9. **Conclude** Use information from the table below to conclude whether each of your patients is probably healthy or might have a health problem.

Type of Blood Cell	Normal Levels	Illnesses Associated with High Levels	Illnesses Associated with Low Levels
Red (RBCs)	94%	Smoking Lung disease	Anemia
White (WBCs)	1%	Infection Leukemia	Infection Leukemia
Platelets	5%	Chronic inflammation Bone marrow problems	Weak immune system

10. **The Big Idea** How do changes in the circulatory system affect homeostasis?

Communicate Your Results

Write a medical report explaining the results of each patient's CBC test and the conclusions you reached as their physician.

Extension

Think of some other blood-related diseases you could model with the materials used in this lab. Are there some diseases that would not be easy to model with these materials?

Lab Tips

- ☑ Calculate percentages by taking the number of one type of blood cell and dividing it by the total number of all blood cell types. Then multiply by 100.
- ☑ Some of the white beans might be different sizes.

Remember to use scientific methods.

Make Observations → Ask a Question → Form a Hypothesis → Test your Hypothesis → Analyze and Conclude → Communicate Results

Chapter 16 Study Guide

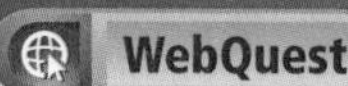

The respiratory and circulatory systems move materials through the body and remove wastes.

Key Concepts Summary	Vocabulary
Lesson 1: The Respiratory System • The respiratory system provides the body with oxygen and removes carbon dioxide. • In the **lungs,** oxygen is carried by the **bronchi** and the bronchioles to the **alveoli.** • The respiratory system works with the circulatory and muscular systems to maintain homeostasis. 	**breathing** p. 559 **pharynx** p. 560 **larynx** p. 560 **trachea** p. 560 **bronchi** p. 561 **lungs** p. 561 **alveoli** p. 561 **diaphragm** p. 562
Lesson 2: The Circulatory System 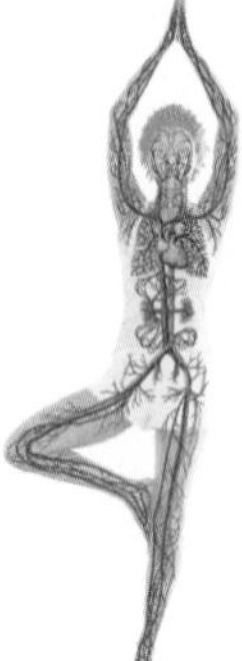 • The circulatory system moves materials throughout the body. • **Arteries** carry blood away from the heart. **Capillaries** allow the exchange of materials between blood and body cells. **Veins** return blood to the heart. • The circulatory system works with the respiratory, digestive, nervous, and endocrine systems to maintain homeostasis.	**atrium** p. 568 **ventricle** p. 568 **artery** p. 570 **capillary** p. 570 **vein** p. 570 **systemic circulation** p. 571 **coronary circulation** p. 571 **pulmonary circulation** p. 571 **atherosclerosis** p. 572
Lesson 3: Blood • Blood transports oxygen, nutrients, and wastes; protects against illness and injury; and regulates body temperature. • Red blood cells contain hemoglobin and carry oxygen. White blood cells fight infection. **Platelets** help stop bleeding. **Plasma** is the liquid portion of blood. 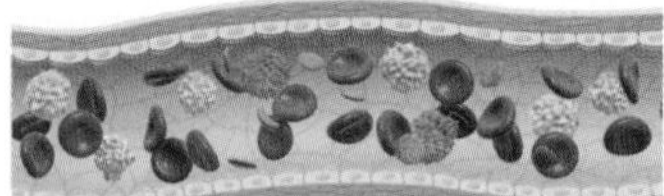	**platelet** p. 579 **plasma** p. 579 **Rh factor** p. 581
Lesson 4: The Lymphatic System • The **lymphatic system** drains away excess tissue fluid and produces white blood cells that fight infection. • **Lymph nodes** filter **lymph.** The **spleen** recycles worn-out red blood cells. B cells and T cells produced in the bone marrow fight disease-causing organisms. T cells mature in the **thymus.** • The lymphatic system works together with the circulatory system to regulate the amount of fluid between cells. 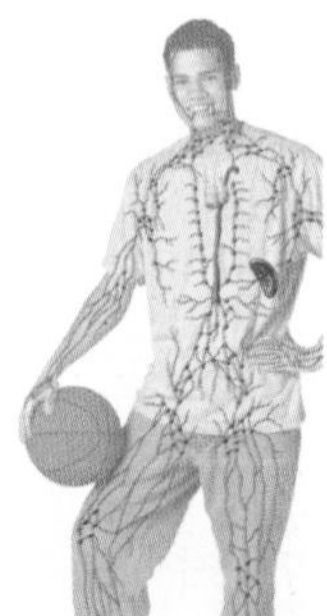	**lymphatic system** p. 585 **lymph** p. 586 **lymph node** p. 586 **thymus** p. 587 **spleen** p. 587

Study Guide

- Personal Tutor
- Vocabulary eGames
- Vocabulary eFlashcards

FOLDABLES® Chapter Project

Assemble your lesson Foldables as shown to make a Chapter Project. Use the project to review what you have learned in this chapter.

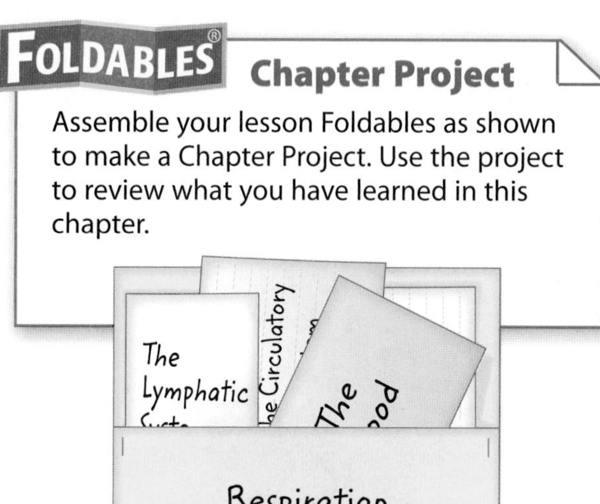

Use Vocabulary

1. The large muscle that contracts and relaxes to move gases into and out of the lungs is the ________.
2. A respiratory infection in which the bronchi swell is ________.
3. The smallest blood vessels are ________.
4. The two lower chambers of the heart are called ________.
5. Small, irregularly shaped pieces of cells in blood are ________.
6. The organ that holds a reserve supply of blood and produces white blood cells is the ________.

Link Vocabulary and Key Concepts

Concepts in Motion Interactive Concept Map

Copy this concept map, and then use vocabulary terms from the previous page to complete the concept map.

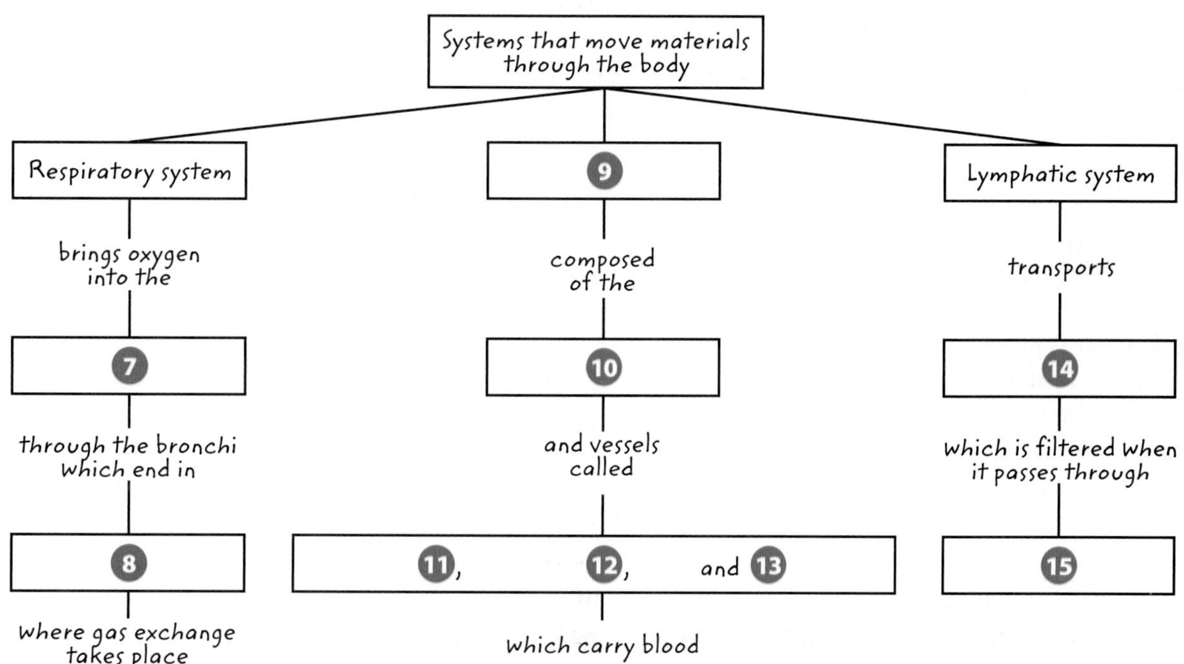

Chapter 16 Review

Understand Key Concepts

1. Which process takes place in the structure shown below?

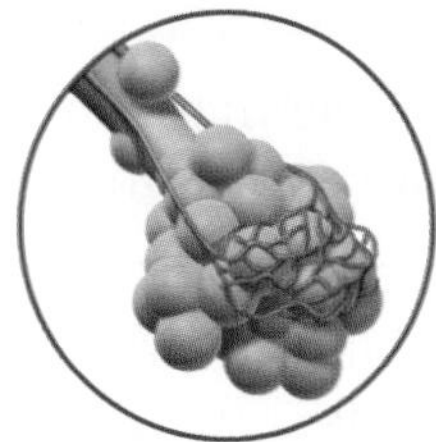

A. breathing
B. immunity
C. blood clotting
D. gas exchange

2. Which structure is held open by rings of cartilage?
A. alveolus
B. larynx
C. pharynx
D. trachea

3. What happens to the diaphragm during inhalation?
A. It contracts and moves down.
B. It contracts and moves up.
C. It relaxes and moves down.
D. It relaxes and moves up.

4. Which term describes the main function of pulmonary circulation?
A. fight infection
B. oxygenate blood
C. produce T cells
D. stop bleeding

5. Which type of circulation supplies oxygen to the cells of the heart?
A. coronary
B. lymphatic
C. pulmonary
D. systemic

6. Which organ keeps blood flowing through the body?
A. heart
B. lungs
C. spleen
D. thymus

7. The arrow in the diagram below points to which structure?

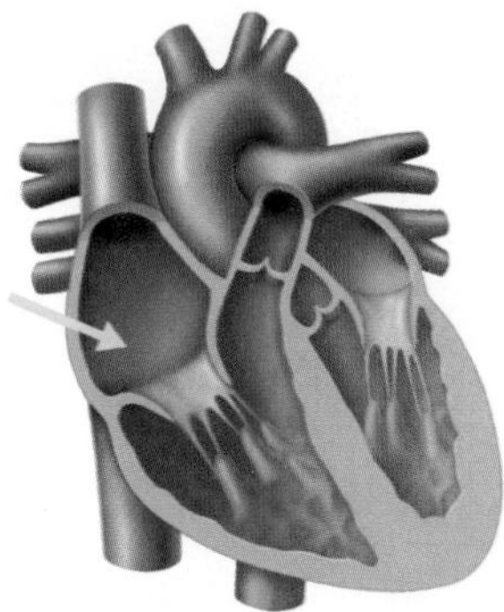

A. aorta
B. atrium
C. vein
D. ventricle

8. Which part of the blood helps defend the body from a virus infection?
A. plasma
B. platelets
C. red blood cells
D. white blood cells

9. What is the primary role of hemoglobin in blood?
A. attract platelets
B. blood typing
C. carry oxygen
D. fight parasites

10. Which body system filters infection-causing organisms from tissue fluids?
A. circulatory
B. immune
C. lymphatic
D. respiratory

Critical Thinking

11 **Illustrate** the path of air from the nose into the lungs.

12 **Compare** the structures and functions of the circulatory and lymphatic systems.

13 **Give an example** of a life choice that can harm the health of both the respiratory and circulatory systems.

14 **Interpret Graphics** The arrow below points to one of the chambers of the heart. Where does blood entering this chamber come from? Where does it go when it leaves this chamber? Is the blood oxygen-rich or oxygen-poor?

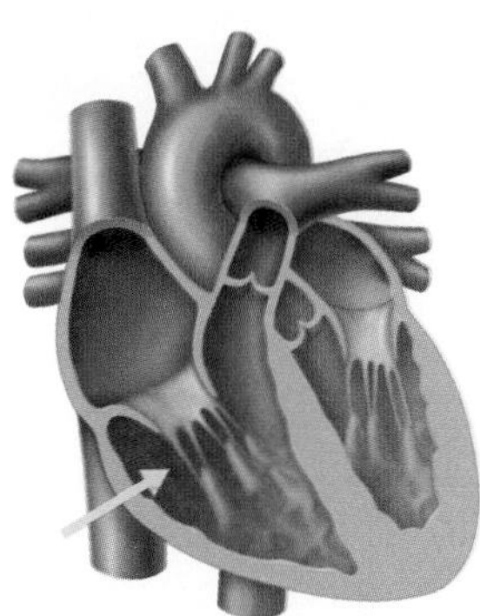

15 **Determine** A person with type AB blood regularly donates blood as a community service. Determine the blood type(s) that can receive this blood in a transfusion.

16 **Synthesis** Design a brochure for a tour through the lymphatic system. Include all the structures of the lymphatic system.

Writing in Science

17 **Write** a paragraph comparing the functions of the alveoli, the capillaries, and the lymph nodes. Your paragraph should have a topic sentence, supporting details, and a concluding sentence.

REVIEW THE BIG IDEA

18 How does oxygen reach the cells of the body? Explain how the respiratory and circulatory systems work together to supply cells with the materials they need.

19 The photo below shows bubbles of gas exhaled by a diver who is breathing from an air tank. Why is it important that the bubbles be released into the water rather than going back into the air tank?

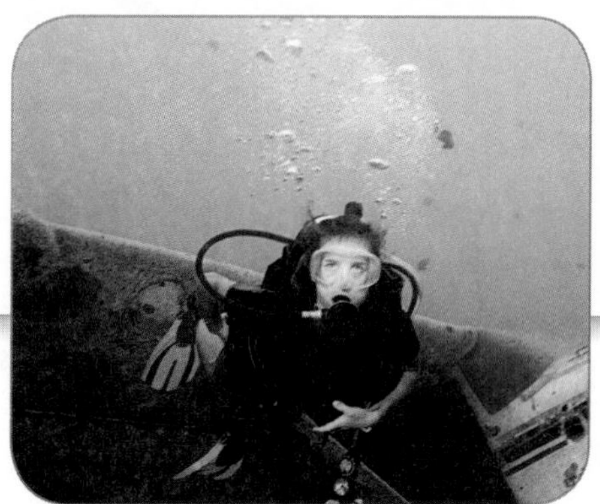

Math Skills

Review
Math Practice

Use Percentages

The table below shows the percentages of the total population in the US with different blood types and with Rh– blood. Use the table to answer questions 20–22.

Blood type	A	B	AB	O
Percent with blood type	42	10	4	44
Percent who are Rh–	6	2	1	7

20 What percentage of people has Rh+ blood?

21 What percentage of the total population has AB+ blood?

22 What percentage of people could donate blood to a person with O+ blood?

Standardized Test Practice

Record your answers on the answer sheet provided by your teacher or on a sheet of paper.

Multiple Choice

1 Where in the human body does gas exchange occur?

A alveoli

B bronchi

C pharynx

D trachea

Use the diagram below to answer question 2.

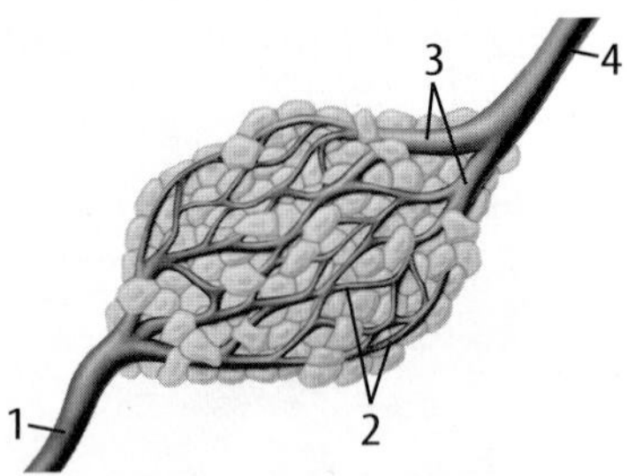

2 Which numbered blood vessel in the diagram above could be the aorta?

A 1

B 2

C 3

D 4

3 Which blood component stops the bleeding after a cut?

A plasma

B platelets

C red blood cells

D white blood cells

4 Which shows the general path of blood from the time it leaves the heart until it returns?

A arteries → capillaries → veins

B arteries → veins → capillaries

C capillaries → arteries → veins

D veins → capillaries → arteries

Use the diagram below to answer question 5.

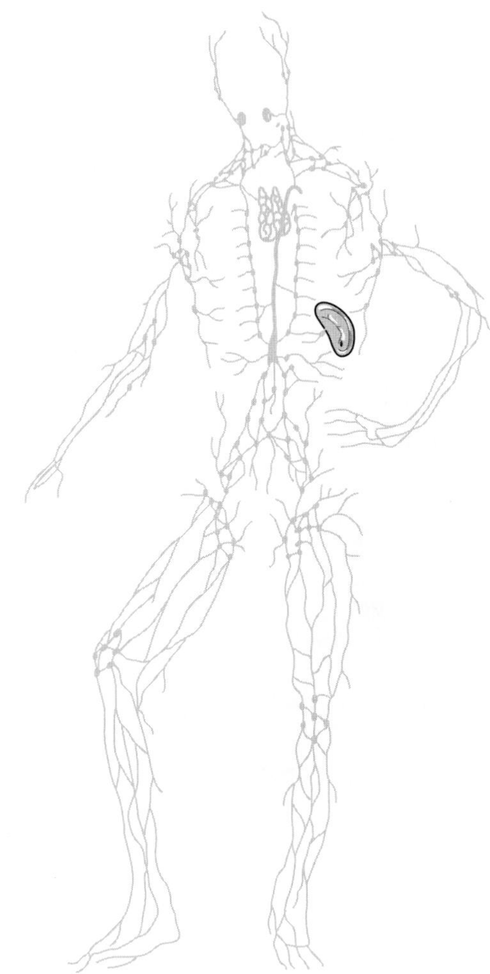

5 Which organ is highlighted in the diagram above?

A heart

B lung

C spleen

D stomach

6 Which blood type can be donated to all humans?

A type A

B type AB

C type B

D type O

7 Which is a function of the lymphatic system?

A circulate blood

B digest food

C fight infection

D transport gas

Use the diagram below to answer question 8.

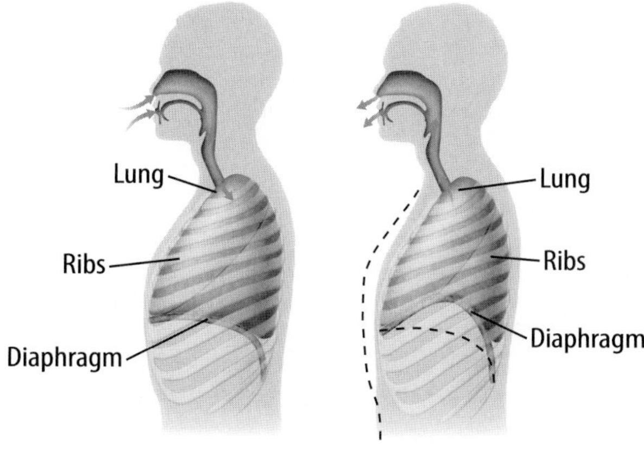

8 What process do the dotted lines in the diagram represent?

A gas exchange

B diaphragm contraction

C muscle expansion

D weight increase

9 Which contracts to move lymph through the lymphatic system?

A heart

B stomach

C body muscle

D heart muscle

Constructed Response

Use the diagram below to answer questions 10 and 11.

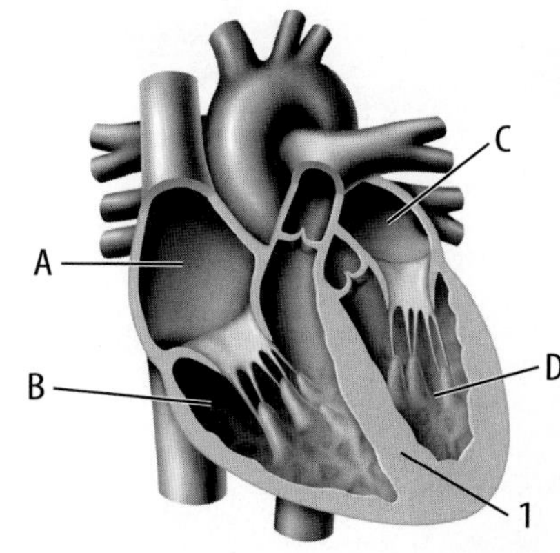

10 The four main chambers of the heart are represented by letters in the diagram above. Identify each chamber. What is the structure in the heart represented by the number 1? Describe its function.

11 Use the labels in the diagram to describe in a paragraph the path of blood through the heart.

12 Describe the three types of circulation in the human body.

13 Which organs of the respiratory system help keep dirt and food from entering the lungs? Describe how each functions to protect the lungs.

14 How does your body know when to breathe? How and why does air pressure change in your body as you breathe?

NEED EXTRA HELP?														
If You Missed Question...	1	2	3	4	5	6	7	8	9	10	11	12	13	14
Go to Lesson...	1	2	3	2	4	3	4	1	4	2	2	2	1	1

Chapter 17

Immunity and Disease

THE BIG IDEA **How does the immune system help maintain the body's homeostasis?**

Inquiry What is going on?

Your immune system protects your body against invaders. Notice the two small brown-yellow cells on the large green cell.

- Why might the small cells be attacking the large cell?
- How does your immune system help your body maintain homeostasis?

Get Ready to Read

What do you think?

Before you read, decide if you agree or disagree with each of these statements. As you read this chapter, see if you change your mind about any of the statements.

1. Some diseases are infectious, and others are noninfectious.
2. Cancer is an infectious disease.
3. The immune system helps keep the body healthy.
4. All immune responses are specific to the invading germs.
5. Exercise and sleep can help keep you healthy.
6. Chemicals make you sick and should not be used.

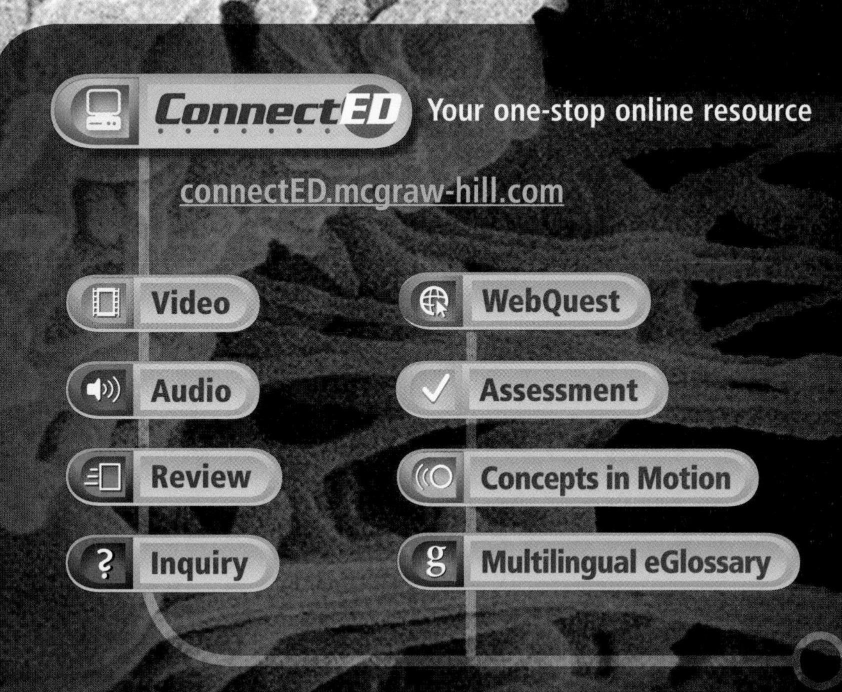

Lesson 1

Reading Guide

Key Concepts
ESSENTIAL QUESTIONS

- Why do we get diseases?
- How do the two types of diseases differ?

Vocabulary

pathogen p. 601
pasteurization p. 603
infectious disease p. 605
vector p. 605
noninfectious disease p. 606
cancer p. 607

Multilingual eGlossary

Video BrainPOP®

Diseases

Inquiry A Bull's-Eye?

Have you ever seen a bull's-eye on someone's skin? It is a rash caused by Lyme disease. The disease is spread by a tick like the one shown here.

Launch Lab

15 minutes

Which well is contaminated?

Imagine that you live in a town with four wells. You get your water regularly from one well, but sometimes you also drink water from another well. People are getting sick. Some suspect the water in one well is causing the sickness. Which well is contaminated?

1. Take an **envelope** from your assigned well. Do not look inside the envelope.
2. Write your name on the envelope, and then pass it to another person from any well. You should also receive a different envelope from another person.
3. Repeat step 2.
4. Write your name on the third envelope and open it. If there is an "X" on the card inside, the three people who signed the envelope drank from the contaminated well and are sick. As a class, compile the results in a table.

Think About This

1. Which well was contaminated? How could you tell?
2.  **Key Concept** What do you think you might do to stop the sickness from spreading?

Disease Through History

Imagine you are sick, and your doctor suggests scraping your skull with a rock until a hole is created, as shown in **Figure 1.** The doctor tells you that the hole in your skull will allow the cause of your illness to escape. Today you might think this was strange. But thousands of years ago, this was an accepted treatment for disease.

Today we know that many diseases are caused by bacteria and viruses. *Disease-causing agents, such as bacteria and viruses, are called* **pathogens.** Pathogens have always caused illnesses, but only in the last few hundred years has the relationship between pathogens and diseases been understood.

Before then little was known about disease and immunity, and superstitions were common. Today we know the hole-in-the-head treatment would be painful and would create an opportunity for even more pathogens to enter the body.

WORD ORIGIN
pathogen
from Greek *pathos,* means "disease"

Figure 1 Archaeologists have found skulls with smooth holes made while the patient was alive. Bone growth around the hole in this skull shows that the patient lived after the procedure.

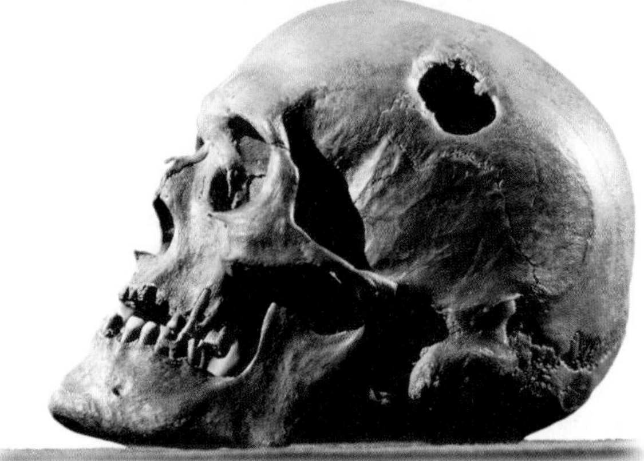

Early Research on Diseases

Despite limited technology and equipment, doctors in the eighteenth and nineteenth centuries learned a lot about the causes and treatments of diseases. The research and experiments performed by a few scientists saved many lives.

First Vaccination

In 1796 a doctor in England named Edward Jenner developed the first vaccination—a procedure that helps the body defend itself against disease. Jenner knew that women who milked cows often developed a mild disease called cowpox. However, these women were resistant to the deadly disease smallpox. He made a cut in the arm of a young boy and inserted pus from a cowpox sore. Two weeks later, he infected the boy with smallpox, but the boy did not develop smallpox. Although the smallpox vaccination saved many lives, scientists did not understand why or how it worked.

Connecting Disease with a Source

In the mid-1800s people realized there was a connection between pathogens and disease. During this period, many people in London were dying from cholera, a bacterial disease of the intestinal tract. Dr. John Snow mapped outbreaks of the disease, as shown in **Figure 2.** He tracked the origin of one outbreak to a water pump. He had the pump closed, and new cases of cholera decreased immediately. John Snow thought a microscopic organism that he saw in the water—the cholera bacteria shown in **Figure 2**—caused the disease. Not everyone agreed, but people were beginning to think pathogens existed.

Figure 2 John Snow mapped the outbreak of cholera and realized the origin was the water from a specific pump.

Visual Check How did Snow use his map to identify the source of the cholera outbreak?

John Snow

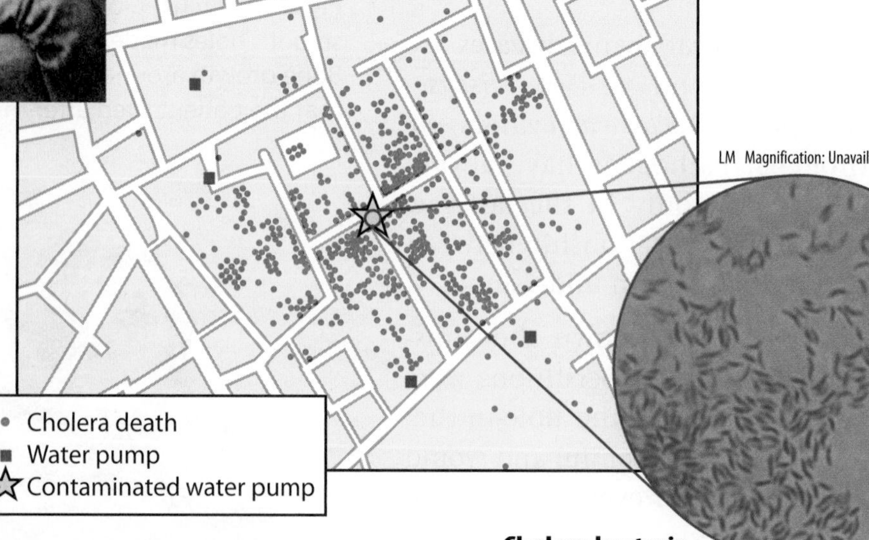

Cholera bacteria

Figure 3 Anton van Leeuwenhoek made many scientific discoveries using simple microscopes he designed. He named the moving organisms he saw in pond water *animalcules.* Today they are called bacteria and protozoa.

The Development of Microscopes

One of the reasons people were slow to accept the idea of pathogens was because they could not see them. The development of microscopes changed that. In the late 1600s, Dutch merchant Anton van Leeuwenhoek (LAY vun hook) made one of the first microscopes. He discovered bacteria in pond water, as illustrated in **Figure 3.** However, van Leeuwenhoek did not share how he made the lenses, so bacteria were not observed again until the nineteenth century.

Connecting Bacteria to Infections

When scientists first realized bacteria were present in wounds, they thought the wounds caused the bacteria to appear. When Louis Pasteur began doing experiments in the mid-1800s, he realized that this idea was backward. Instead, bacteria from outside the body caused the tissue in the wound to decay. Pasteur discovered that he could kill bacteria in boiling liquids. **Pasteurization** *is the process in which a food is heated to a temperature that kills most harmful bacteria.* It is based on the work of Pasteur.

Joseph Lister used Pasteur's discoveries to make surgery safer for patients. He found that carbolic acid killed bacteria. He developed a misting system to spray carbolic acid throughout an operating room during surgery. Infection and death from surgeries decreased greatly. In the late 1800s, doctors improved on Lister's idea. They used carbolic acid to sterilize tools before surgery and steam to sterilize the linens and clothes.

Reading Check How did Lister make surgery safer?

Koch's Rules

1. The bacterium must be found in all organisms suffering from the disease, but not in healthy organisms.

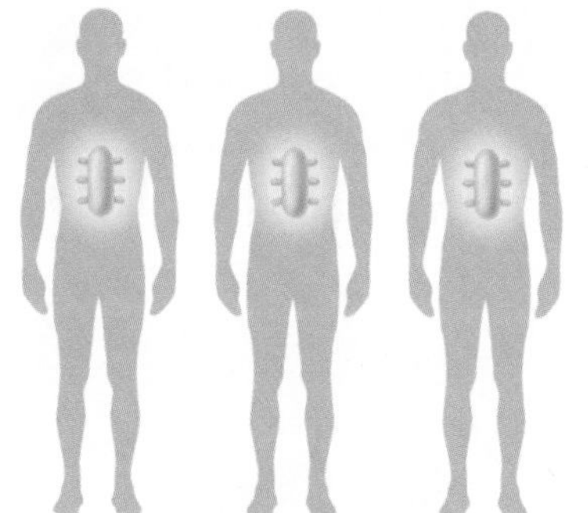

2. The bacterium must reproduce in the lab.

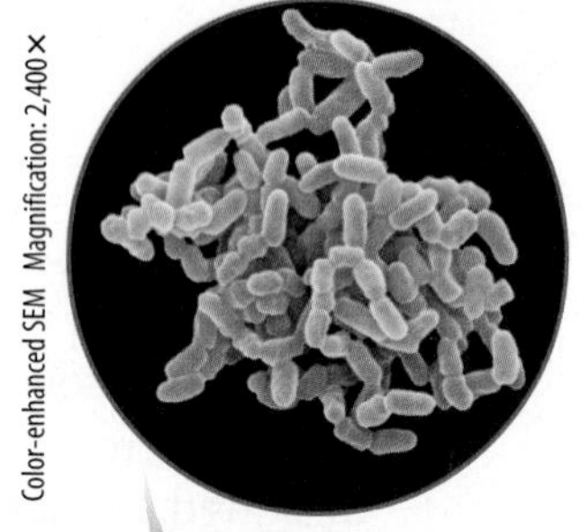

3. A sample of the newly grown pathogen must cause the illness when injected into a healthy animal.

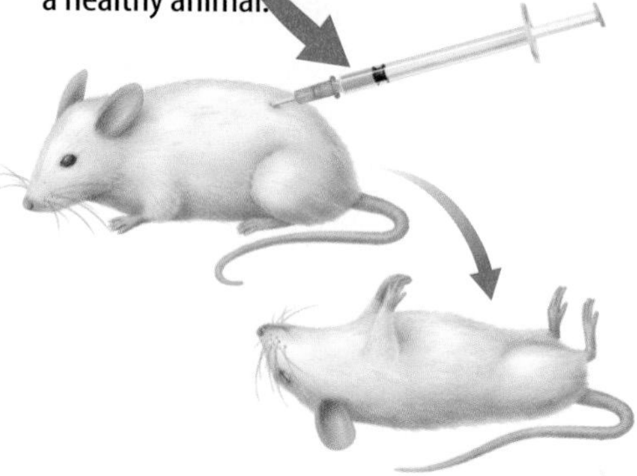

4. When the suspected pathogen is removed from the infected animal and grown in the lab, it must be identical to the original pathogen.

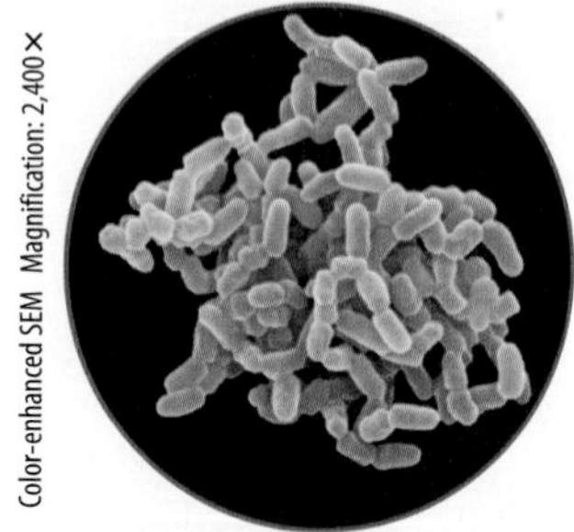

Figure 4 Koch developed a procedure to determine if a bacterium caused an illness.

Discovering Disease Organisms

Despite the research on bacteria in wounds, most people did not think bacteria could make a healthy person sick. In 1867, Robert Koch was one of the first scientists to argue that bacteria could cause illness in an animal as large as a cow. He developed a set of rules to determine if specific bacteria caused an illness. Koch's rules are illustrated in **Figure 4.** The research based on these rules convinced most scientists that some bacteria were disease-causing pathogens. Although the roles of pathogens in disease are not as simple as Koch thought, current understandings are based on his findings.

Reading Check What are Koch's rules?

Bacteria are not the only pathogens that cause disease—viruses are others. However, they are so small that many years passed before scientists understood that viruses could be pathogens too. Some fungi and protists can also cause diseases. Some of the diseases in humans caused by different pathogens include the following:

- Viruses cause the flu, colds, chickenpox, and AIDS.
- Bacteria cause ear infections, strep throat, pneumonia, meningitis, whooping cough, and syphilis, a sexually transmitted disease.
- Fungi cause athlete's foot, ringworm, and yeast infections.
- Protists cause malaria, African sleeping sickness, and dysentery.

Pathogens can be transmitted through food and water and carried by insects. They also can be passed directly among people by physical contact, sneezing, coughing, or exchange of bodily fluids. Some pathogens, such as the bacterium that causes syphilis, require a host to reproduce.

Types of Diseases

Have you ever heard anyone say they "caught" a cold? The common cold is contagious. This means that the pathogens that cause the common cold can be passed from person to person. Not all diseases are caused by pathogens. Your inherited traits are responsible for some diseases. Others can be caused by external factors, such as your environment and the choices you make about diet, exercise, and sleep.

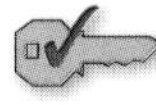 **Key Concept Check** Why do we get diseases?

Infectious Diseases

Diseases caused by pathogens that can be transmitted from one person to another are **infectious diseases.** The way this happens can vary depending on the pathogen.

Flu and cold viruses can pass to others through direct contact, such as shaking hands. The human immunodeficiency virus (HIV) can pass through the exchange of blood or bodily fluids. HIV causes acquired immunodeficiency syndrome (AIDS), a disease that attacks the body system that fights pathogens.

The protist that causes malaria is transferred by a **vector,** *a disease-carrying organism that does not develop the disease.* The vector for malaria is a certain type of mosquito. The mosquito bites an animal that has the protist in its bloodstream. Then the pathogen enters the saliva of the mosquito but the mosquito does not develop malaria. When the mosquito bites another animal, the pathogen moves into that animal's blood.

Inquiry MiniLab

20 minutes

How does an infectious disease spread through a population?

Imagine that one of the students in your class has a new disease, similar to a mild case of the flu. You can model the spread of this disease to indicate infection.

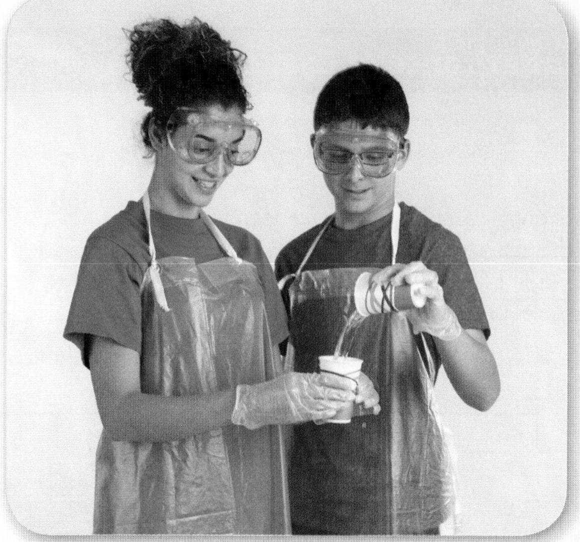

1. Read and complete a lab safety form.
2. Obtain a **cup** of **clear liquid** from your teacher.
3. Partner with another student in round 1. Pour the contents of your cup into his or her cup. Your partner should then pour the contents of his or her cup into your empty cup. Pour half the contents of your cup into your partner's cup.
4. Repeat step 3 three more times with a different partner in each round.
5. Your teacher will add an **indicator** to your cup. If your liquid changes color, you are infected.

Analyze and Conclude

1. **Determine** how many students were infected in each round.
2. **Key Concept** Describe the pattern of how the number of infected persons changes as an infection progresses.

FOLDABLES®

Make a horizontal two-tab book and label it as shown. Use your book to compare infectious diseases and noninfectious diseases.

Infectious Diseases | Noninfectious Diseases

Noninfectious Diseases

A disease that cannot pass from person to person is a **noninfectious disease.** For example, you cannot catch lung cancer from another person. Pathogens do not directly cause noninfectious diseases. Two common causes of noninfectious diseases include:

- genetics, or traits inherited in your DNA from your biological parents, and
- environmental conditions, including lifestyle choices.

In many cases of noninfectious disease, a person has a genetic trait for a disease that environmental conditions make worse. It is the combination of genetics and environment that causes the disease to develop.

Childhood Diseases Noninfectious diseases that affect children are primarily due to genetics. One genetic disease is cystic fibrosis. It causes the body to produce mucus thicker than normal. This affects breathing and other body functions. Children with cystic fibrosis inherit a form of the gene that causes this disorder. It is a recessive trait, which means a person must inherit the gene from each parent, as shown in **Figure 5.** The parents might not have the disease, but they each must carry at least one gene form, or allele (uh LEEL), for cystic fibrosis. Like many genetic disorders in children, environmental conditions can make the disease worse. A poor diet, air pollution, and lack of exercise can make the symptoms of cystic fibrosis worse.

Figure 5 Diseases caused by genetic disorders are inherited.

Visual Check How many children inherited a gene for cystic fibrosis?

Inheritance of Cystic Fibrosis

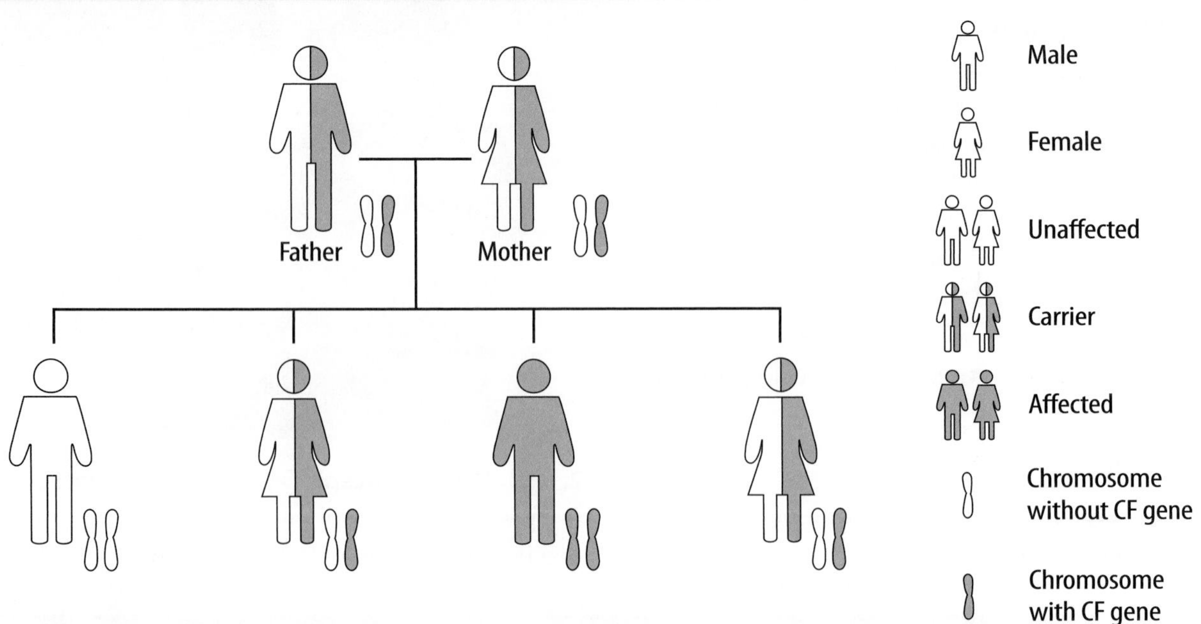

Other Diseases Many noninfectious diseases that affect adults are due primarily to environmental causes and life choices. For example, an unhealthful diet, obesity, a lack of regular exercise, and smoking cause most cases of heart disease. Osteoporosis is a disease in which bones become weak and less dense. People inherit a tendency to develop osteoporosis. However, years of poor lifestyle choices such as an unhealthful diet, lack of calcium and **vitamin** D, smoking, and a lack of exercise can all lead to weakened bones. There is also a type of diabetes that develops in adults that is strongly linked to environmental conditions, although there might also be a genetic link.

Review Vocabulary

vitamin
nutrient needed for growth, regulation of body functions, and prevention of some diseases

Reading Check List some causes of noninfectious diseases.

Cancer Tumors form when cells reproduce uncontrollably. **Cancer** *is a disease in which cells reproduce uncontrollably without the usual signals to stop.* For example, lung-cancer tumors form in the lungs and interfere with normal lung function. In **Figure 6,** notice the color difference of the lung that has not been functioning properly due to cancer. People can inherit forms of genes that make them more likely to develop lung cancer. However, if they are not exposed to such environmental conditions as poor air quality, or they do not smoke, they might not develop lung cancer.

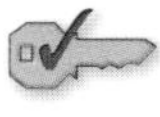

Key Concept Check How do infectious and noninfectious diseases differ?

Figure 6 Cancer cells in the lung form tumors and interfere with normal functioning.

Visual Check Identify the differences between the healthy lungs and the diseased lungs.

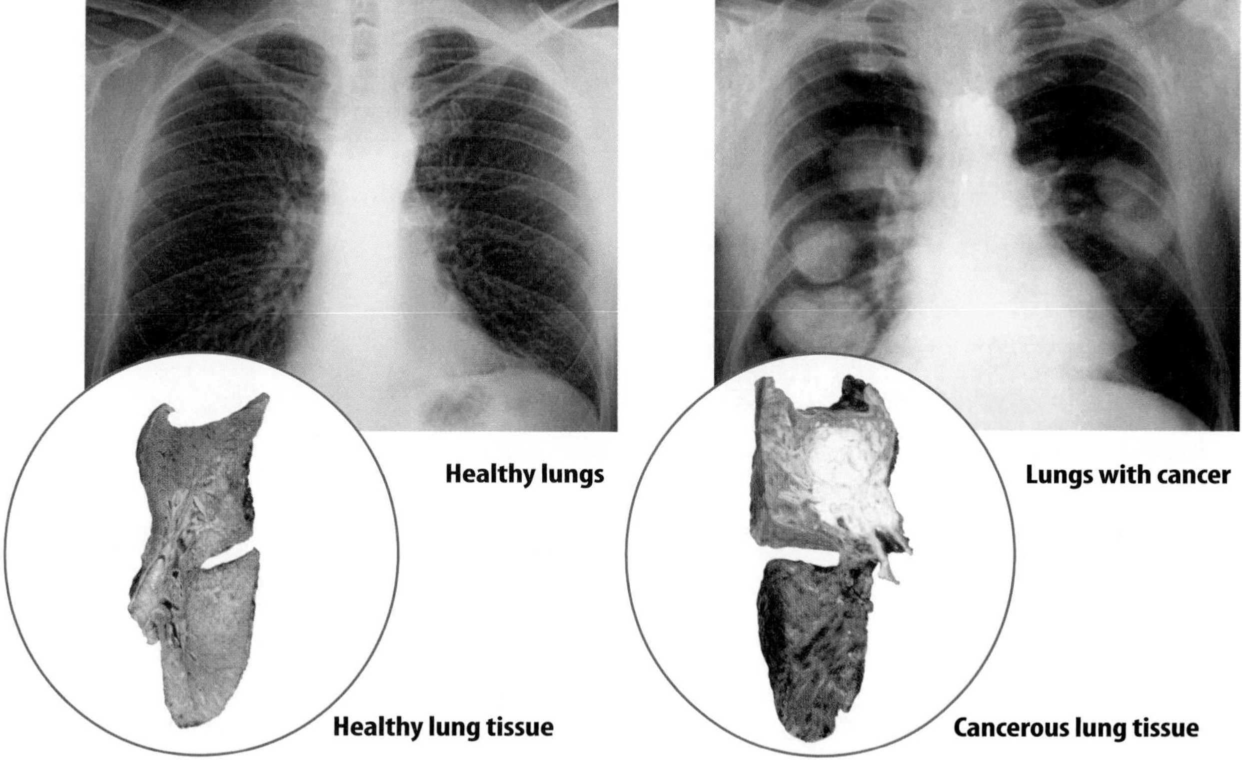

Lesson 1 Review

Assessment Online Quiz

Visual Summary

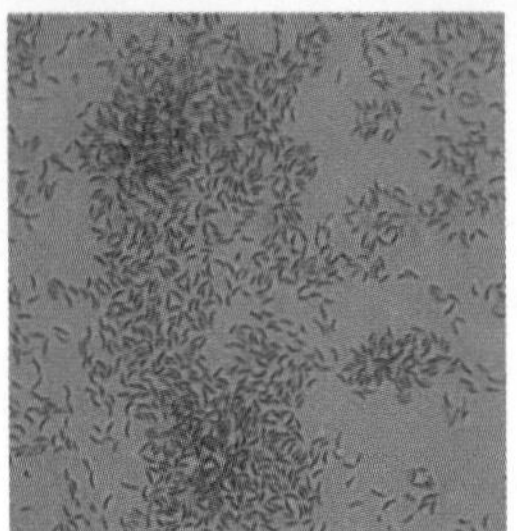

How a disease spreads depends on the pathogen. Some pathogens can be transmitted by a vector, such as a tick or a mosquito.

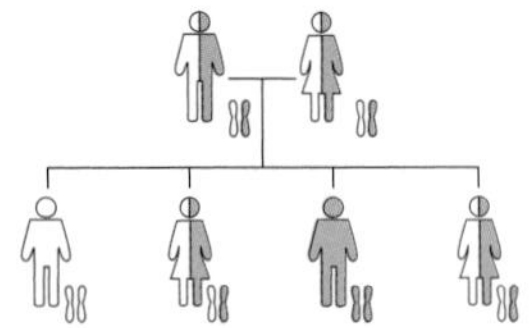

The two common causes of noninfectious diseases are environmental conditions and genetics.

People might inherit forms of genes that make them more likely to develop cancer.

FOLDABLES®

Use your lesson Foldable to review the lesson. Save your Foldable for the project at the end of the chapter.

What do you think NOW?

You first read the statements below at the beginning of the chapter.

1. Some diseases are infectious, and others are noninfectious.
2. Cancer is an infectious disease.

Did you change your mind about whether you agree or disagree with the statements? Rewrite any false statements to make them true.

Use Vocabulary

1. **Define** the term *pathogen* in your own words.
2. The process of boiling a liquid to kill bacteria and sealing it so bacteria cannot enter is called ________.
3. A host that transmits a pathogen but does not develop the disease is called a(n) ________.

Understand Key Concepts 

4. **List** two main causes of disease.
5. Which is NOT a pathogen?
   A. bacterium
   B. fungus
   C. vector
   D. virus

Interpret Graphics

6. **Summarize** Copy and fill in the table below using these terms: *viruses, bacteria, unhealthful diet, smoking, protists, gene forms, fungus.*

Heredity	Environmental Conditions	Pathogens

7. **Explain** the illustration below using one of Koch's rules.

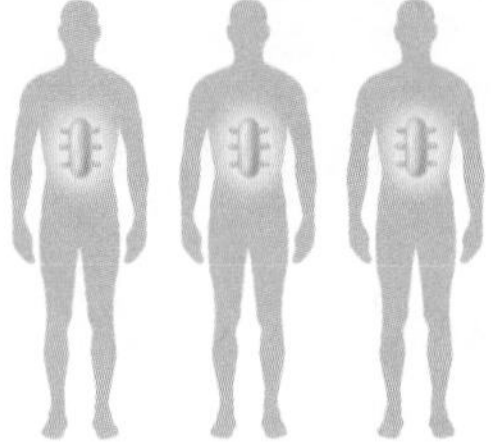

Critical Thinking

8. **Design** a plan to determine the source of a pathogen found in food.
9. **Support** the claim that genetics and environmental conditions can both contribute to a disease.

Use Lab Techniques

25 minutes

How would you prepare a work area for procedures that require aseptic techniques?

Aseptic techniques are methods used to prevent microbial contamination. Hospitals use aseptic techniques to reduce the risk of infection by microorganisms during surgery. Although their procedures might be less stringent than in an operating room, microbiologists also use aseptic techniques to prevent their work from becoming contaminated by unwanted microorganisms.

Materials

bleach

spray bottle

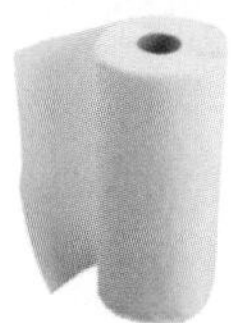

paper towels

black light

Safety

Learn It

You can practice aseptic **techniques** by disinfecting a bench top or desk. You will use a smooth, nonporous surface because rough surfaces are hard to clean and they provide many surfaces for microorganisms to grow.

Try It

1. Read and complete a lab safety form.
2. Thoroughly wash your hands. Then, wearing disposable gloves, prepare a bleach solution according to your teacher's directions. Place the solution in a spray bottle.
3. Spray the entire work area and wipe it with a paper towel. Instead of drying the area with another paper towel, let it air dry. After disinfecting, do not recontaminate the area by touching or leaning over it.
4. Test the effectiveness of your cleaning procedure by using a black light. The invisible test solution your teacher applied will glow under the black light. In your Science Journal, draw a diagram to indicate the areas you missed.
5. Reclean the work area and check the effectiveness of your cleaning procedure again.
6. Thoroughly wash your hands after the activity.

Apply It

7. Some people think polished granite countertops are better for food preparation than rough tiles. Explain why this might be so.
8. **Propose** an argument to defend this statement: Hand washing is an essential part of aseptic techniques.
9. **Key Concept** How do aseptic techniques reduce the risk of infection and disease?

Preparing a Work Area for Aseptic Procedures
Wash your hands.
Choose an area with a smooth, dry surface.
Thoroughly disinfect the entire area.
Do not recontaminate.

Lesson 2

Reading Guide

Key Concepts
ESSENTIAL QUESTIONS

- What does the immune system do?
- How do the parts of the immune system work together?
- How does the immune system interact with other body systems?

Vocabulary

inflammation p. 615
antigen p. 616
antibody p. 616
B cell p. 616
T cell p. 616
allergy p. 616
immunity p. 617
active immunity p. 617
vaccination p. 617
passive immunity p. 617

 Multilingual eGlossary

 Video

- BrainPOP®
- What's Science Got to do With It?

The Immune System

Inquiry Mysterious Blobs?

The large yellow blobs you see are bacteria. The bacteria grew after a human hand touched the red agar plate. With all that bacteria on your hand, what keeps you from getting sick? How does the body protect itself?

Launch Lab

10 minutes

Can you escape the pox?

The loffpox disease is an imaginary disease. How might it affect you?

1. Your teacher will give you one of **three cards**: *healthy, in poor health,* or *pox.* Only one person will be given the *pox* card. Do not tell anyone which card you receive.
2. As you stand in a circle looking at each other, the person with the *pox* card will wink at the other students. If he or she winks at you and you have an *in poor health* card, you have caught the disease and you must sit down. If you have a *healthy* card, you do not catch the disease and you remain standing. However, if the person with the *pox* card winks at you a second time, you must sit down.

Think About This

1. Who is left standing?
2. **Key Concept** How does a person's state of health affect the pox disease? Why do you think it took the pox more than one wink to infect a healthy person?

Functions of the Immune System

Your body is constantly exposed to different pathogens. In Lesson 1 you read that disease-causing agents, such as bacteria and viruses, are pathogens. Pathogens also include fungi and protists. Pathogens are in the air, on objects, and in water. Like a spacesuit protects an astronaut, your immune system works to protect your body. There are many barriers to keep pathogens from entering your body.

Sometimes pathogens get past your body's initial barriers. When this happens, your immune system also has defenses to stop any pathogens that get past the barriers. For example, there are cells in your body that can destroy the pathogens. The immune system interacts with other body systems and helps keep you healthy, even as the environment outside your body changes.

 **Key Concept Check** What does the immune system do?

You can improve the effectiveness of these prevention methods by making healthful choices every day. Choices such as eating healthful food, getting enough sleep, exercising regularly, and using sunscreen support your immune system. As you read about the parts of the immune system, consider how the choices you make every day could affect how well your immune system functions.

FOLDABLES®

Fold two sheets of paper into a layered book. Label it as shown. Use it to organize information about the immune system's lines of defense.

Parts of the Immune System
First-Line Defenses
Second-Line Defenses
Third-Line Defenses

MiniLab

20 minutes

How do different layers of your skin protect your body?

Your skin has three layers. The top layer, called the epidermis, is thin but tough. It helps prevent harmful microorganisms from getting into the tissues and provides physical protection. The middle layer, called the dermis, is the thickest layer. It provides strength and elasticity. The bottom layer, called the subcutaneous layer, insulates against heat and cold and helps cushion the skin. In this activity, you will build a model of skin.

1. Read and complete a lab safety form.
2. Start with a sheet of **cardboard** or **cardstock paper** as the base. **Glue** a piece of **cotton batting** to the base.
3. Glue a **sponge** onto the cotton batting.
4. Glue a piece of **vinyl** to the sponge.
5. Test your skin model with a drop of **water** containing **food coloring.** Record your observations in your Science Journal.
6. Use **scissors** to make a small cut in the top layer of your model. Repeat step 5.

Analyze and Conclude

1. **Identify** Which skin layer does the vinyl represent? The sponge? The cotton batting?
2. **Key Concept** Relate the results of the water and food coloring test to what would happen with real skin.

Parts of the Immune System

Different parts of your body work together to keep pathogens from making you sick. The integumentary system (skin), the respiratory system, the circulatory system, the digestive system, and the nervous system all work with the immune system and protect you against disease.

First-Line Defenses

Keeping germs from reaching the parts of your body where they can make you sick is the function of first-line defenses. Skin, hair, mucus, and acids are first-line defenses. They are effective against many types of pathogens. An immune defense that protects against more than one type of pathogen is a nonspecific defense.

Skin Often, the first nonspecific defense that protects you from pathogens is your skin, as shown in **Figure 7.** Your skin keeps dirt and germs from entering your body. Sweat and acids from skin cells kill some bacteria. Natural oils make skin waterproof so you can easily wash it.

You encounter pathogens every day, but your skin stops most of them from entering your body. Pathogens, such as cold and flu viruses, can survive for short periods on objects such as doorknobs or telephones. When you touch these objects, the pathogens can be transferred to your hand. If they reach your mouth, nose, eyes, or a cut, they can enter your body. Washing your hands often with soap and water easily removes most pathogens from your skin.

Your skin protects you from other dangers. It forms a chemical called melanin that protects you from the Sun's ultraviolet (UV) rays. Nerve endings in your skin can help you sense the warmth of a stove or the sharpness of a pin to protect you from injury.

Reading Check Why is your skin considered a first-line defense?

Respiratory System You can inhale pathogens from the air through your nose or mouth. The hairs in your nose help protect you by trapping dirt and pathogens. This keeps them from reaching the rest of your respiratory system. Small hairlike structures called cilia, shown in **Figure 7,** also trap pathogens and move them up and out of the upper respiratory system. If pathogens get past the cilia, they might encounter mucus. Mucus traps pathogens and enables your respiratory system to remove them by coughing, sneezing, or swallowing.

Digestive System Pathogens can enter your digestive system on or in the food you eat. The digestive system is effective at stopping bacteria from making you seriously ill. The stomach, also shown in **Figure 7,** contains strong acids. Stomach acids destroy many pathogens. Like the mucus in the respiratory system, mucus in the digestive system traps disease-causing bacteria and viruses, too.

Sometimes when you feel nauseated, it is actually your immune system clearing your body of pathogens. When disease-causing bacteria are not destroyed by stomach acids, your digestive system can reverse the usual direction of muscle contractions, and you vomit. Other times, muscle contractions speed up, and pathogens are removed through diarrhea.

Reading Check List ways the digestive system helps defend against pathogens.

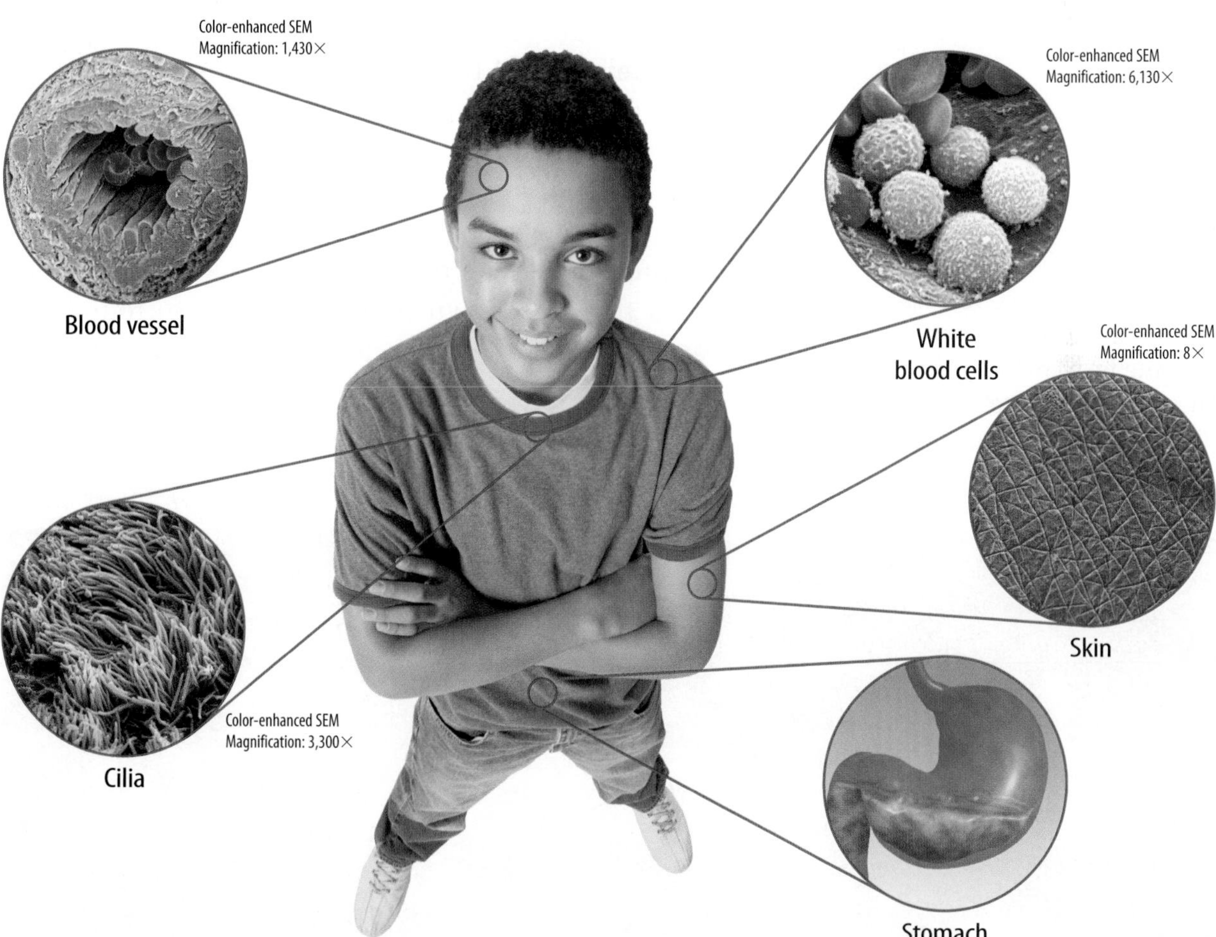

Figure 7 The skin, the respiratory system, the digestive system, and the circulatory system all support the immune system to provide the first line of protection against disease-causing pathogens. Blood vessels and white blood cells throughout the body help protect from pathogens.

Visual Check How does the respiratory system trap pathogens?

Circulatory System and Nervous System Your circulatory system also protects you from pathogens. Pathogens can be moved through the circulatory system to organs that fight infection. The nervous system and the circulatory system also work together and increase the body's temperature to fight pathogens more effectively. Certain foreign substances trigger the brain to increase body temperature. When this **occurs,** blood vessels narrow and a fever develops. Many pathogens cannot survive at this higher temperature. For those that do survive, the fever brings another line of defense. The fever also stimulates white blood cells, which are part of the second-line defenses against pathogens.

ACADEMIC VOCABULARY
occur
(verb) to come into existence

Second-Line Defenses

Sometimes pathogens get past the defenses of the skin, the respiratory system, the digestive system, and the circulatory system. When they do, the next line of defense goes into action. Like the first-line defense, second-line defenses are nonspecific, fighting against any type of pathogen.

White Blood Cells Recall that the spongy tissue in the center of your bones is called bone marrow. This is where white blood cells form. These cells attack pathogens. White blood cells flow through the circulatory system. However, they do most of their work attacking pathogens in the fluid outside blood vessels. They fight infection several different ways. Some white blood cells, such as the one shown in **Figure 8,** can surround and destroy bacteria directly. Others release chemicals that make it easier to kill the pathogens. Another type of white blood cell produces proteins that destroy viruses and other foreign substances that get past the first-line defenses.

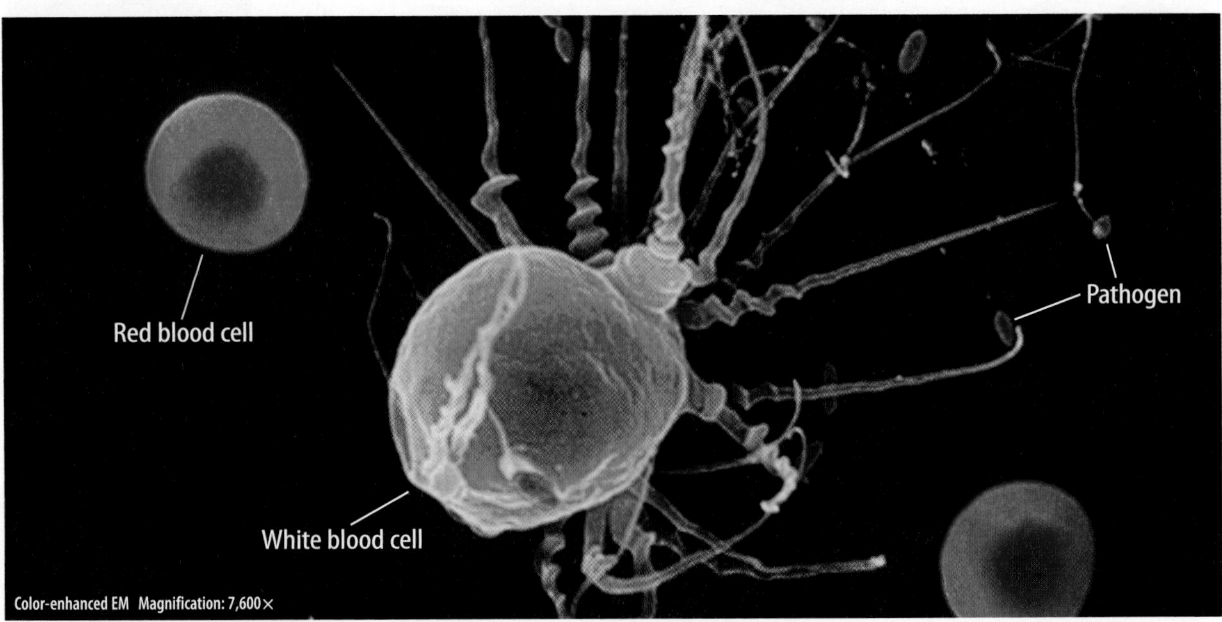

Figure 8 White blood cells fight pathogens that get past the first-line defenses. This white blood cell can digest pathogens and damaged cells.

The Inflammatory Response

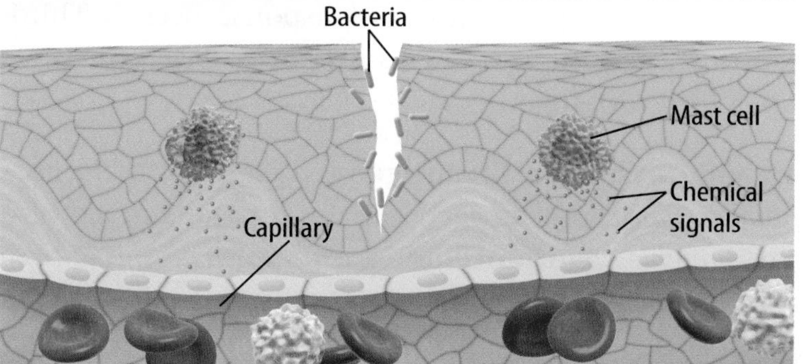

1 When skin is torn or cut, the damaged tissue triggers the inflammatory response. Special cells called mast cells release chemical signals into the surrounding tissue.

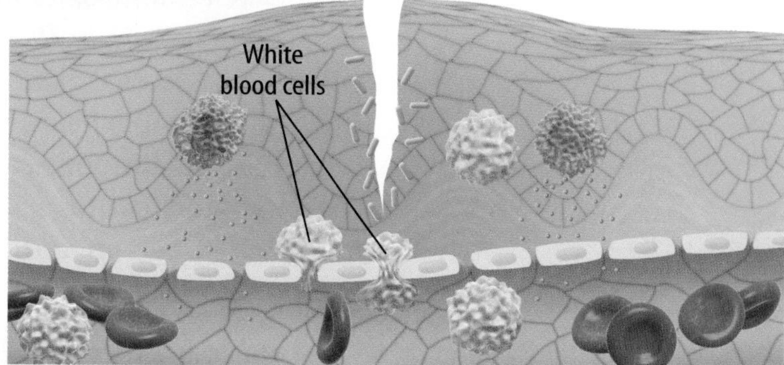

2 Some chemical signals attract more white blood cells to the area. Other chemical signals cause the capillary to dilate. White blood cells and plasma leak into the tissue, causing swelling.

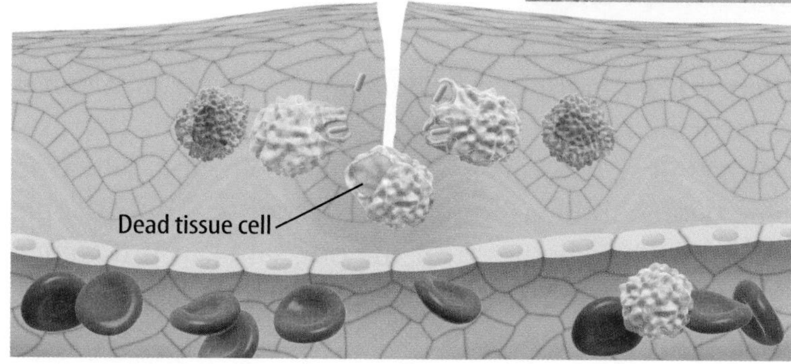

3 The white blood cells surround and take in the bacteria and any dead cells. As the tissue heals, the white blood cells and the plasma flow back into the capillary, and swelling decreases. The area returns to normal.

Figure 9 Inflammation is another nonspecific response to pathogens.

Inflammatory Response When you have an injury, your body produces an inflammatory response, causing inflammation. **Inflammation** *is a process that causes the area to become red and swollen.* If the injury is to the surface of the skin, you can observe the inflammatory response, as shown in **Figure 9**. First, damaged cells release a protein that signals the capillaries to dilate, or widen. Blood flow to the area increases and the injury site becomes red and warmer than the surrounding area. Second, plasma and white blood cells leak into the area, causing swelling. Third, the white blood cells break down damaged cells and destroy any bacteria that might have entered the wound. The inflammatory response cleans the area of the injury and keeps the infection from spreading. The inflammation enables the damaged tissue to heal.

Reading Check Explain the inflammatory response.

Word Origin

inflammation
from Latin *inflammare*, means "to set on fire"

Third-Line Defenses

If first- and second-line defenses do not destroy all invading pathogens, another type of immune response occurs. Third-line defenses are specific to foreign substances. Often the three lines of defense work together.

Antigens and Antibodies *An* **antigen** *is a substance that causes an immune response.* An antigen can be on the surface of a pathogen. *Proteins called* **antibodies** *can attach to the antigen and make it useless.* Certain white blood cells, called B cells and T cells, form antibodies. **B cells** *form and mature in the bone marrow and secrete antibodies into the blood.* **T cells** *form in the bone marrow and mature in the thymus gland. They produce a protein antibody that becomes part of a cell membrane.* Antibodies match with specific antigens, as shown in **Figure 10.** Once your body has developed antibodies to an antigen, it can respond rapidly when the same pathogen invades your body again. This information is stored in antibodies on white blood cells called memory B cells.

Key Concept Check How do the parts of the immune system work together?

An **allergy** *is an overly sensitive immune response to common antigens.* Most people do not produce antibodies to the proteins in dog saliva. However, the antigens in dog saliva do cause some people to have an immune response. These people have an allergy. Their bodies treat the dog saliva as if it were a pathogen. Inflammation and increased mucus production are common immune responses for people with allergies.

Figure 10 Antibodies are produced as a result of a specific immune response to particular antigens on pathogens.

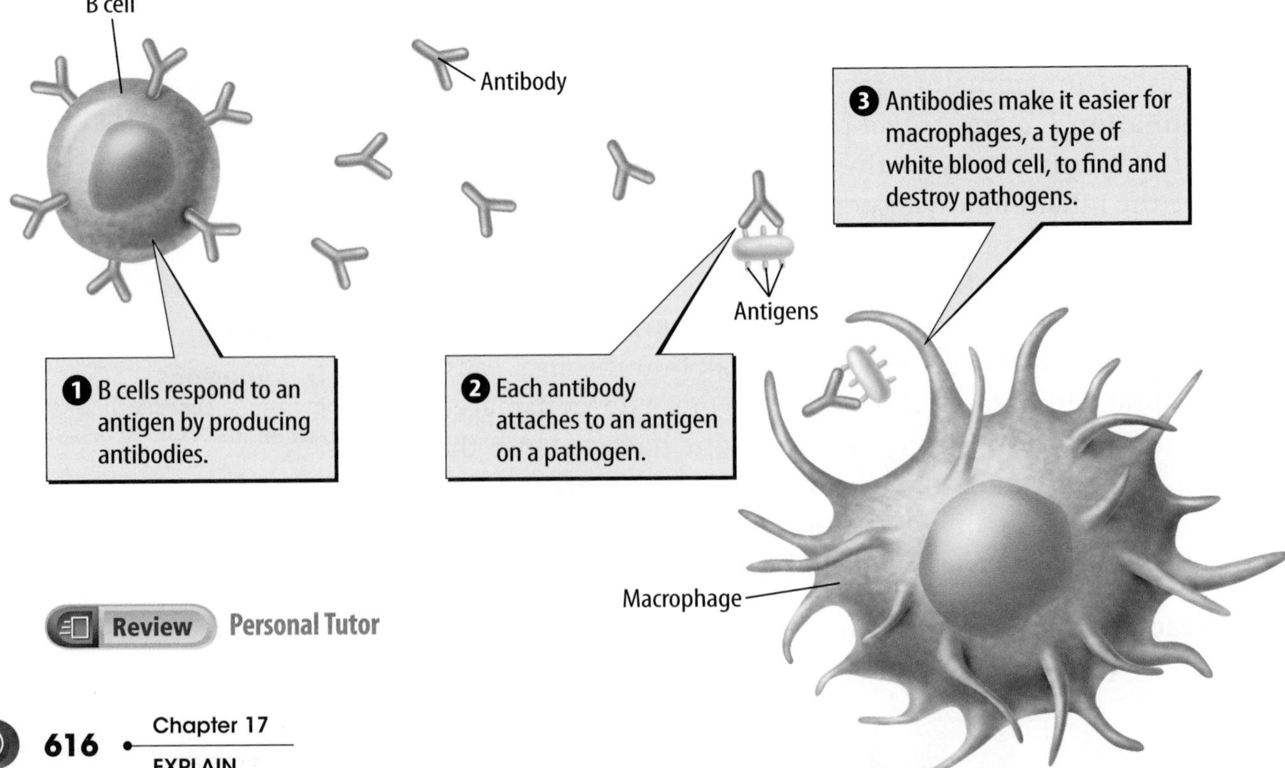

Review Personal Tutor

Immunity

The resistance to specific pathogens is **immunity.** There are two types of immunity—active immunity and passive immunity.

Active Immunity *Your body produces antibodies in response to an antigen in* **active immunity.** Your body recognizes the antigen, and the matching antibodies respond quickly. You can develop active immunity through illness or infection. After an illness or an infection is over, antibodies remain in your body. Because of this, you usually get certain diseases, such as chicken pox, only once. However, you can catch a cold many times because many different cold viruses cause similar symptoms.

You can also develop antibodies if you are exposed to an antigen through a vaccination. *A* **vaccination** *is weakened or dead pathogens placed in the body, usually by injection or by mouth.* A vaccination causes the body to develop specific antibodies that can rapidly fight a pathogen's antigens when exposed to them. **Table 1** lists the effects of vaccinations on the average annual number of cases of some diseases.

Passive Immunity You can also become resistant to specific antigens through passive immunity. **Passive immunity** *is the introduction of antibodies that were produced outside the body.* A fetus can get antibodies from its mother. Injections of some antibodies are available for adults. Passive immunity is temporary—the body does not continue to make these antibodies.

The Immune System and Homeostasis

You are exposed to many different pathogens every day. The immune system works to maintain your body's homeostasis. Body systems, including the circulatory system and respiratory system, work together and protect against invaders.

Key Concept Check How does the immune system interact with other body systems?

Math Skills

Use Percentages

Table 1 shows that the number of tetanus cases fell from 1,300 cases before the vaccine was developed to 34 cases after the vaccine was developed. What percent change does this represent?

Subtract the starting value from the final value.

$$34 - 1{,}300 = -1{,}266$$

Note that the value is negative because the number of cases decreased.

Divide the difference by the starting value.

$$\frac{-1{,}266}{1{,}300} = -0.974$$

Multiply the answer by 100 and add a percent sign.

$$-0.974 \times 100 = -97.4\%$$

Practice

Based on the data in Table 1, what was the percent change in the cases of polio after the vaccine was discovered?

- Math Practice
- Personal Tutor

Table 1 Effects of Vaccinations — Concepts in Motion Interactive Table

Disease	Cases Before Vaccination (annual average in the United States)	Year Vaccination Was Developed	Cases After Vaccination (annual average in the United States)
Tetanus	1,300	1927	34
Polio	18,000	1955/1962	8
Measles	425,000	1963	90
Mumps	200,000	1967	610
Rubella	48,000	1970	345

Lesson 2 Review

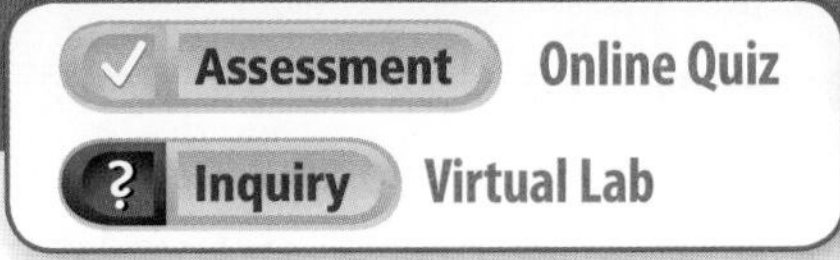

Visual Summary

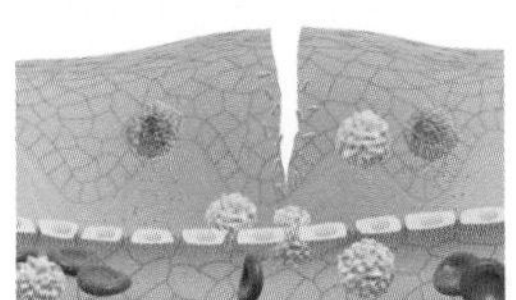

Inflammation may cause an injury to become warmer than the surrounding area due to increased blood flow to the area.

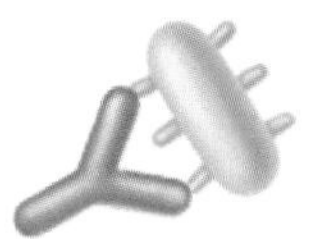

Antibodies produced by the white blood cells match with specific antigens, like a lock and key.

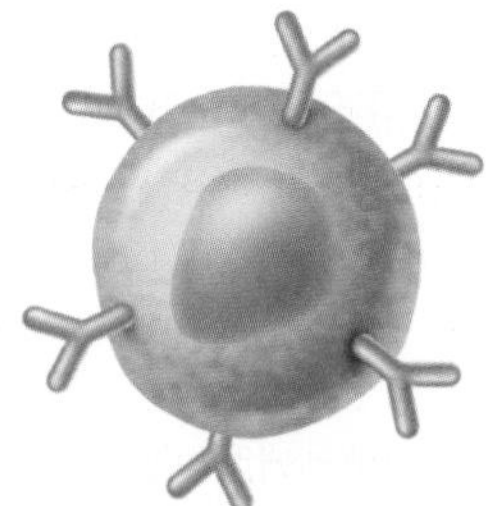

Immunity can be developed through different processes. If you have immunity to a particular pathogen, you will experience little or no effect from exposure to that pathogen.

FOLDABLES®

Use your lesson Foldable to review the lesson. Save your Foldable for the project at the end of the chapter.

What do you think NOW?

You first read the statements below at the beginning of the chapter.

3. The immune system helps keep the body healthy.

4. All immune responses are specific to the invading germs.

Did you change your mind about whether you agree or disagree with the statements? Rewrite any false statements to make them true.

Use Vocabulary

1. **Distinguish** between active immunity and passive immunity.
2. The ________ response includes swelling and heat near injured tissue.
3. **Define** the term *vaccination* in your own words.

Understand Key Concepts 

4. **List** three body systems that work with the immune system to form first-line defenses.
5. Which is a first-line defense?
   A. antibody
   B. hormone
   C. inflammation
   D. skin
6. **Explain** why antibodies are considered specific responses to pathogens.

Interpret Graphics

7. **Summarize** Copy and fill in the graphic organizer below to summarize the steps in the inflammatory response.

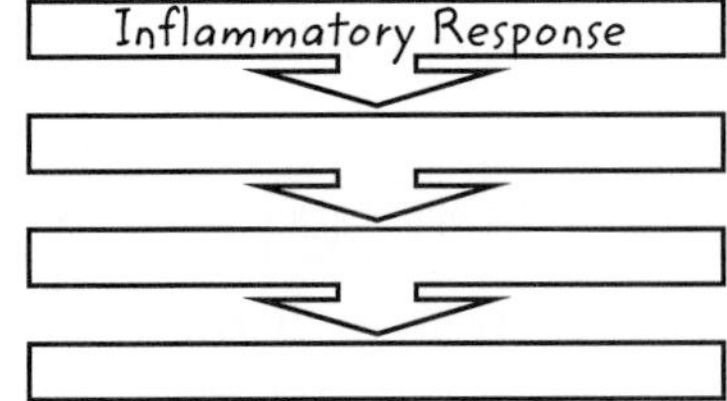

Critical Thinking

8. **Predict** what might happen to a person who had very few of the type of cell shown at right.

Math Skills 

Review
Math Practice

9. The number of diphtheria cases fell from 175,000 cases before the vaccine to one case after the vaccine was developed. What percent change does this represent?

The Victory over Chicken Pox

SCIENCE & SOCIETY

Vaccines are helping win the war against viruses.

Until recently, having chicken pox was just a part of growing up. The disease commonly occurs in children, usually before the age of 15.

Chicken pox is highly contagious, passing easily from person to person. The chicken pox virus, varicella-zoster, produces a rash of red spots and small blisters that can appear all over the body. The fluid-filled blisters crust over and become very itchy. Within a few days, the spots and blisters disappear. It takes about one week for the disease to run its course.

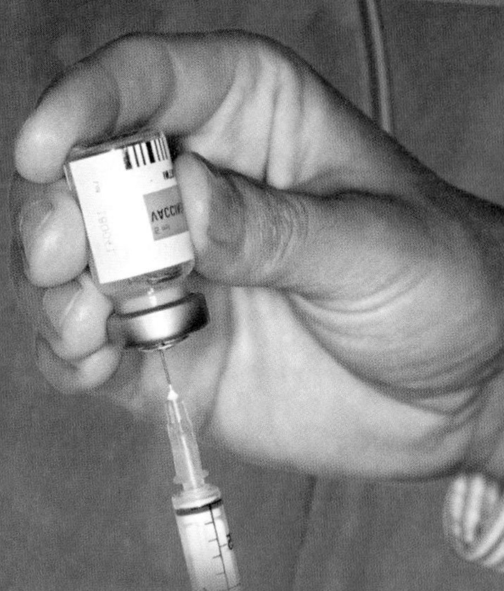

▲ **The chicken pox vaccine is 80 to 90 percent effective in preventing the disease.**

◄ **The chicken pox rash usually appears on the face, chest, and back.**

Chicken pox is contagious beginning 1–2 days before the rash appears until all the blisters form scabs. During this period, the virus can be spread through direct contact with the sores or through the air. Those who have not had chicken pox can become infected easily.

In 1995, a chicken pox vaccine was introduced in the United States, and the number of cases of chicken pox reported each year has declined sharply. It is estimated that almost 80 percent of young children develop immunity after one dose of vaccine. Over 90 percent of older children and adults develop immunity after the second dose.

◄ **Varicella-zoster virus**

It's Your Turn

REPORT Identify three diseases that were common 50 years ago but are not common today due to vaccines. Write a report about the diseases, when vaccines were developed against them, and how the incidences of the diseases changed.

Lesson 3

Staying Healthy

Reading Guide

Key Concepts 
ESSENTIAL QUESTIONS

- How can healthful habits and healthful choices affect diseases?
- How do sanitation practices affect human health?
- How can chemicals affect the human body?

Vocabulary

antibiotic p. 624

chemotherapy p. 624

Multilingual eGlossary

Inquiry Why Wash?

Why do surgeons wash their hands before an operation even though they wear gloves? Can hand washing keep you healthy?

Launch Lab

10 minutes

Where might bacteria be?

The people who work in the cafeteria in your school probably wear aprons, hairnets, and gloves. They might also wash dishes and wipe the counters with a bleach solution to kill bacteria. Why do they take such precautions?

1. Read and complete a lab safety form.
2. Go to your assigned station where surfaces, **utensils,** and **other objects** representing cookware have been coated with a "bacterial" solution of invisible **fluorescent detergent.**
3. Using a **cloth** and a **cleaning solution containing bleach,** clean the objects at your station as well as you can in 5 minutes.

   ⚠ Be careful with the bleach solution; it will stain your clothes.
4. When you are finished, your teacher will examine your station with a **black light,** which will show any remaining "bacteria."

Think About This

1. How well did you clean your station? Did you miss any bacteria?
2. How could you change your cleaning methods to clean all surfaces and objects of bacteria?
3. **Key Concept** Why do you think it is important to keep a kitchen clean, especially in a school or a restaurant?

Healthful Habits

Imagine you are sitting in class, and the person next to you sneezes. Fortunately, she covers her nose and mouth with her hand. After sneezing, she picks up a pencil. Just then, you realize you need to borrow a pencil. She hands you her pencil. Will you get her cold? What could you do to make that less likely?

Pathogens passed from person to person make infectious diseases such as colds and flu very common. Personal hygiene can limit the spread of these pathogens. For example, good hygiene includes using a tissue or handkerchief when you sneeze and then washing your hands. This lessens the chance you will spread your germs to others. Good hygiene can protect you from getting an infectious disease, too.

Pathogens are less likely to get past your first-line defenses if you wash your hands before you eat and avoid putting objects, such as pencils, in your mouth. Why do you think surgeons scrub their hands, as shown in the lesson opener photo, even though they wear gloves during surgical procedures?

FOLDABLES®

Make a vertical half-book from a sheet of paper. Use it to record information about habits and choices that can help you stay healthy.

Figure 11 You can make healthful choices when planning for a day at the beach.

 **Visual Check** Identify the healthful choices.

Healthful Choices

In addition to good personal hygiene, other everyday choices, like those shown in **Figure 11,** can help keep you healthy. Choices that affect environmental conditions can also protect you from many infectious and noninfectious diseases.

Diet Think about the foods you ate this week. If you eat a healthful diet, your immune system can react more efficiently against pathogens. A healthful diet can even protect you against noninfectious diseases such as osteoporosis and heart disease. A healthful diet, a healthful weight, and regular exercise have been linked with overall disease prevention.

Reading Check How can a healthful diet protect against disease?

Sun Protection Skin cancer is a noninfectious disease. The ultraviolet (UV) rays from the Sun damage skin **cells** and can cause them to reproduce uncontrollably. Sunscreen blocks the UV rays and limits the damage from sunlight. Wearing a hat, long sleeves, pants, and sunglasses also helps protect you against UV damage.

SCIENCE USE V. COMMON USE
cell
Science Use the basic unit of life
Common Use a room in a monastery or prison

Alcohol and Tobacco Lung cancer is one of the most deadly cancers. Most cases of lung cancer are related to smoking or working in environments with poor air quality. Many other cancers are related to excessive drinking of alcohol and to smoking. Healthful choices include not smoking or chewing tobacco and limiting or avoiding alcoholic beverages.

 **Key Concept Check** How can healthful habits and healthful choices affect diseases?

Health and Sanitation

Improved cleanliness in schools, hospitals, and public areas has increased overall health in our communities. In the mid-1800s, hospitals were dirty, overcrowded places. Patients were rarely bathed, and linens were rarely washed. Pathogens caused infections in most patients. One nurse, Florence Nightingale, is credited with improving cleanliness in hospitals. She understood there was a connection between cleanliness and health.

Figure 12 Sanitation has improved health by reducing exposure to pathogens.

Food Preparation

Improved sanitation in food preparation has also led to better health. Employees must wash their hands regularly, as indicated in **Figure 12,** and keep equipment clean. Inspections are performed regularly to catch problems early and protect consumers from most pathogens.

Waste Management

In the mid-1300s, there were no plumbing or sewer systems. People in European cities often dumped their personal waste and garbage in the streets. Today modern landfills and sewer systems keep our streets and households much cleaner. This cleanliness slows the spread of infectious diseases.

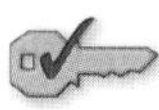

Key Concept Check How do sanitation practices affect human health?

Inquiry MiniLab

20 minutes

How clean are your hands?

Washing your hands properly can keep you from getting sick or spreading illnesses. How clean are your hands after you wash them?

1. Read and complete a lab safety form.
2. Have your partner thoroughly cover both of your hands, including your wrists and under your fingernails, with **washable paint.** Let your hands dry for 2 minutes.
3. Hold your hands over a **sink.** Have your partner put a drop of **liquid soap** on one of your hands and turn on the **warm water.** Have your partner cover your eyes with a **bandana.**
4. Wash your hands for 20 seconds. Have your partner remove the bandana from your eyes. Have him or her draw your hands on a **sheet of paper** and record any areas that still have paint.

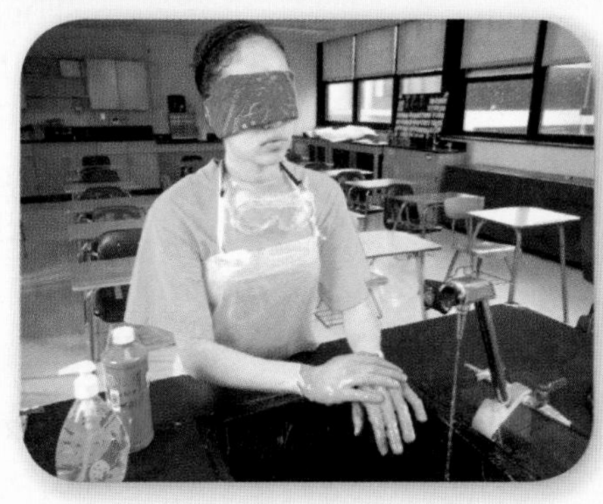

5. Wash any remaining paint from your hands.
6. Switch roles and repeat steps 2–5.

Analyze and Conclude

1. **Describe** how well you washed your hands.
2. **Key Concept** Explain why hand washing is important to keep from getting sick or spreading disease.

Figure 13 Many chemicals are beneficial to our health.

Health and Chemicals

Chemicals, like those shown in **Figure 13,** can be beneficial for people. Some chemicals, such as the ones in sunscreen that block UV rays, protect us from noninfectious diseases. Chemicals used to make vitamin supplements improve nutrition, which helps the immune system fight disease.

WORD ORIGIN

antibiotic
anti–, means "against"; and Greek *biotikos*, means "fit for life"

Other chemicals are used in medicines. **Antibiotics** *are medicines that stop the growth and reproduction of bacteria.* Chemicals are also used to destroy cancer cells. *These medicines, used in a type of treatment called* **chemotherapy,** *kill the cells that are reproducing uncontrollably.*

These and many other chemicals, such as paints and pesticides, might make lives easier. But if they are not disposed of properly, they can harm our health. Some people choose to use chemicals that are harmful to health. For instance, more than 50 of the chemicals in cigarettes have been linked to cancer.

Health and the Environment

Some chemicals that are harmful to our health, such as lead, are in our environment. Before 1978, lead was used in many paints. If the dried paint flaked, it released lead into the air. Inhaling lead-contaminated air can cause noninfectious kidney and nervous system diseases.

Some objects containing harmful chemicals are safe until the object is broken. For example, when ceiling and floor tiles containing asbestos are broken, asbestos fibers are released. People who are often exposed to such chemicals might develop cancer.

Key Concept Check How can chemicals affect the human body?

Lesson 3 Review

Assessment Online Quiz

Visual Summary

Developing healthful habits and making healthful choices is one of the best ways you can stay healthy.

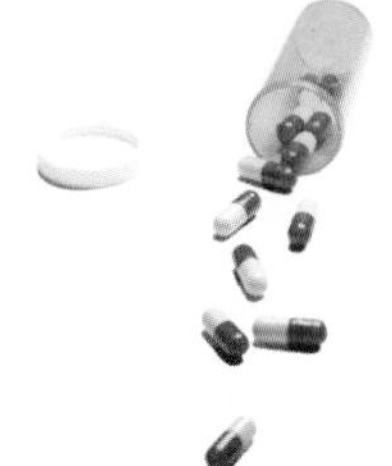

Chemicals are used in medicines such as antibiotics. These may be used for common bacterial infections, such as strep throat or ear infections.

Life choices, such as whether or not you eat a healthful diet, can influence the development and severity of diseases.

FOLDABLES®

Use your lesson Foldable to review the lesson. Save your Foldable for the project at the end of the chapter.

What do you think NOW?

You first read the statements below at the beginning of the chapter.

5. Exercise and sleep can help keep you healthy.

6. Chemicals make you sick and should not be used.

Did you change your mind about whether you agree or disagree with the statements? Rewrite any false statements to make them true.

Use Vocabulary

1. A treatment that uses chemicals to kill cancer cells is called ________.

2. Medicines that kill bacteria are called ________.

Understand Key Concepts 

3. **List** two healthful life choices.

4. Which chemical can be harmful to your health?

   A. antibiotic
   B. asbestos
   C. sunscreen
   D. vitamin

Interpret Graphics

5. **Examine** the photo on the right and identify the healthful practices you see.

6. **Create** a table like the one below. In the first column, list four chemicals or materials containing chemicals. In the second column, indicate whether the chemical is beneficial or harmful. In the third column, describe how the chemical benefits or harms people.

Chemical	Beneficial or Harmful?	How?

Critical Thinking

7. **Justify** Imagine you are Florence Nightingale. Justify your plan to increase cleanliness in hospitals.

8. **Evaluate** your personal hygiene in your daily routine. What could you do to limit your exposure to pathogens?

Inquiry Lab Day 1: 40 minutes; Days 3 and 7: 20 minutes each

Materials

rotting apple

fresh apples (4)

potato peeler

cotton swabs

rubbing alcohol

self-sealing plastic bags (5)

camera

Also needed: labels, paper towels

Safety

Can one bad apple spoil the bunch?

Just like your skin protects the tissue underneath from infection, an apple's skin protects the inside of the apple from infection by harmful organisms.

Question

How is an apple's skin like your skin?

Procedure

1. Read and complete a lab safety form.
2. Use aseptic techniques to prepare your work area. Wash your hands, thoroughly disinfect the entire work area, and do not recontaminate. Put on gloves.
3. Label the plastic bags 1 through 5. Take a picture of a fresh apple. Place the apple in bag 1.
4. Insert a cotton swab into the decayed tissue from the rotting apple. Spread the tissue on a fresh apple and let it dry. Take a picture of the apple. Place the apple in bag 2.
5. Use the end of a potato peeler to carefully make two holes about 2 cm in diameter in the peels of the remaining three apples. Take a picture of one apple with holes. Place the apple in bag 3.
6. Insert a cotton swab into the decayed tissue of the rotten apple and then into one hole in one of the remaining apples. Repeat for the remaining holes in the last two apples. Take a picture of one apple. Place the apple in bag 4.

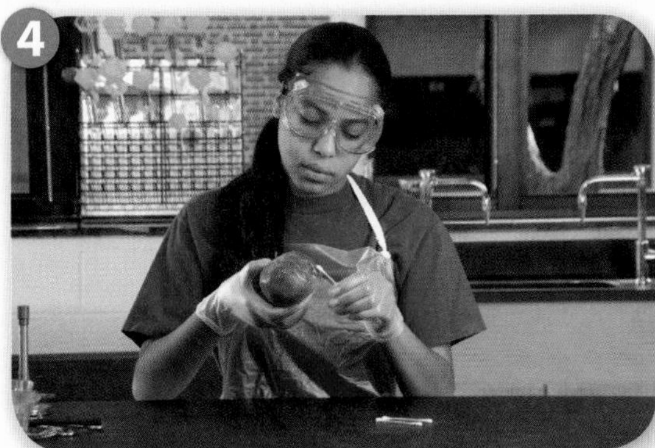
4

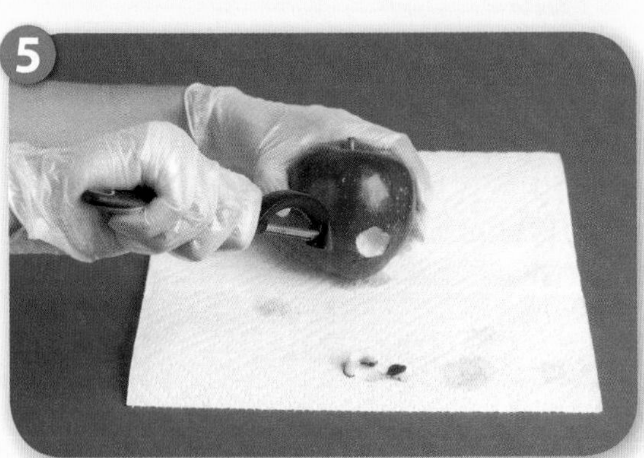
5

7. Use a cotton swab to disinfect the holes in the last apple with rubbing alcohol. Take a picture of the disinfected apple. Place the apple in bag 5.
8. Seal all bags and put them in a dark place.
9. After 3 days, note the condition of each apple. Do not open the plastic bags. In your Science Journal, record your observations in a data table like the one shown at right. Take a picture of each apple.
10. After 7 days, repeat step 10.

Apple Observations		
Condition	Day 3	Day 7
Fresh, no decay		
Fresh with decay		
Holes, no decay		
Holes with decay		
Holes with decay and rubbing alcohol		

Analyze and Conclude

11. **Describe** How does each apple look at the start of this activity, after 3 days, and after 7 days?
12. **Explain** What are the differences in the conditions of the apples?
13. **The Big Idea** Relate what you have learned about rotting in apples to how the immune system protects humans from infection and helps maintain homeostasis.

Communicate Your Results

Display your data on a poster. Include the photos of each of the apples at the beginning of the activity, after 3 days, and after 7 days.

Inquiry Extension

There are other factors that contribute to apples rotting besides infection by microorganisms. How can you tell if the rotting in this case was caused by microorganisms or other factors? Propose a study using microscope observations that might help establish the role of microorganisms in rotting fruit.

Lab Tips

- ☑ Remember that hand washing is an important part of aseptic techniques.
- ☑ Use good aseptic techniques to prepare your work area prior to this lab.

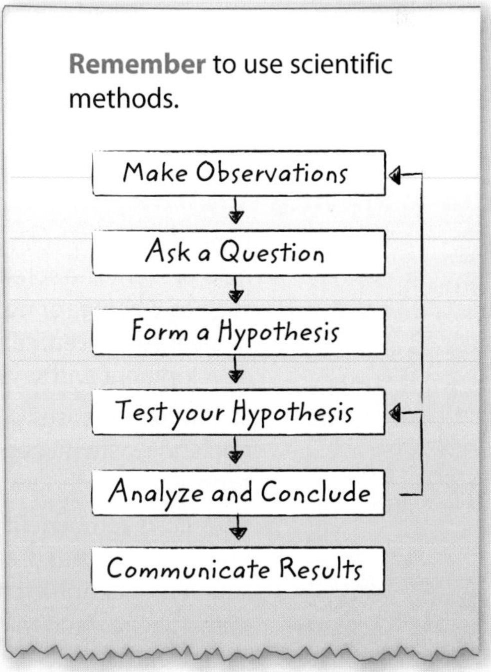

Chapter 17 Study Guide

The immune system protects the body against infections and diseases.

Key Concepts Summary

Lesson 1: Diseases

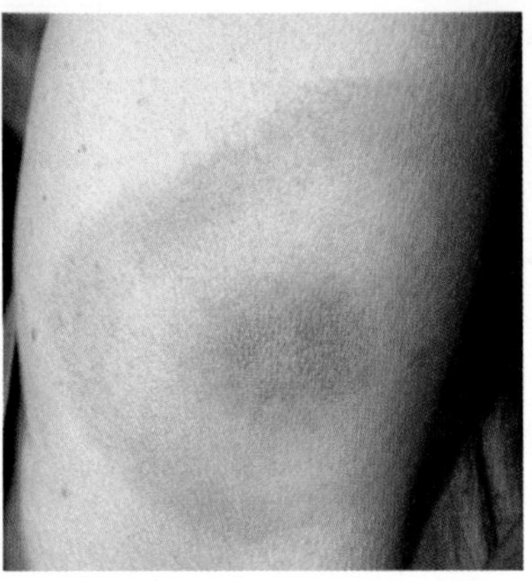

- Diseases can result from infection by **pathogens**, heredity, or the environment.
- **Infectious diseases** are caused by pathogens and are spread from an infected organism or the environment to another organism. **Noninfectious diseases** are not caused by pathogens and are not spread from one organism to another.

Vocabulary

pathogen p. 601
pasteurization p. 603
infectious disease p. 605
vector p. 605
noninfectious disease p. 606
cancer p. 607

Lesson 2: The Immune System

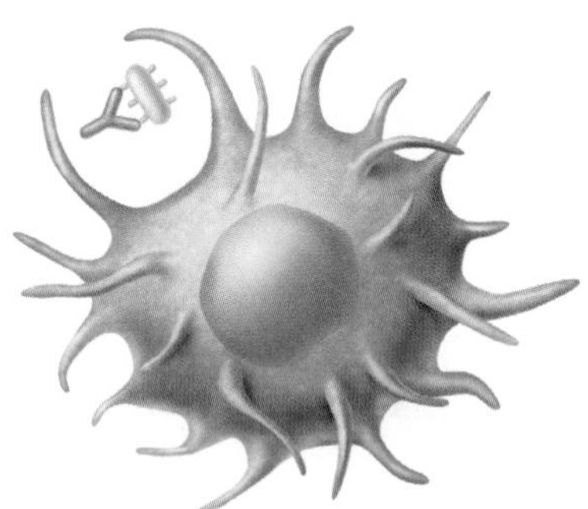

- The immune system protects against and defends the body from disease.
- Your body has first-line, second-line, and third-line defenses against pathogens.
- The immune system works with other body systems, including the circulatory system, the respiratory system, and the digestive system, to protect against invaders.

Vocabulary

inflammation p. 615
antigen p. 616
antibody p. 616
B cell p. 616
T cell p. 616
allergy p. 616
immunity p. 617
active immunity p. 617
vaccination p. 617
passive immunity p. 617

Lesson 3: Staying Healthy

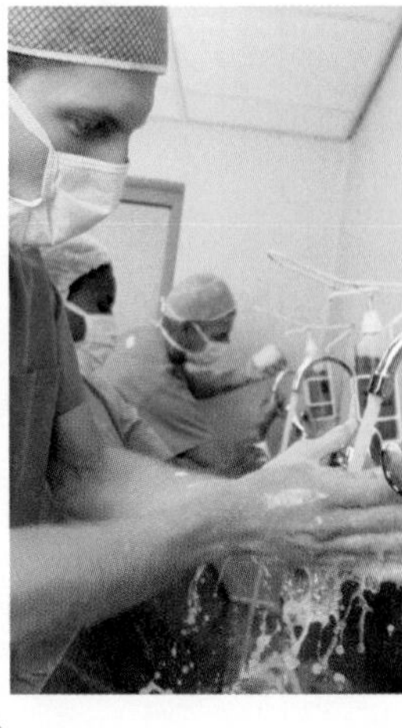

- Healthful habits, such as hand washing, can help prevent the spread of disease. Life choices, such as whether or not you eat a healthful diet or wear sunscreen, can influence the development and severity of diseases.
- Sanitation practices, such as safe food preparation and waste management, limit human exposure to pathogens and toxic substances.
- Chemicals can benefit human health when used as medicines, treatments for disease, and supplements. Some chemicals are harmful to human health and might cause diseases such as cancer.

Vocabulary

antibiotic p. 624
chemotherapy p. 624

Review
- Personal Tutor
- Vocabulary eGames
- Vocabulary eFlashcards

FOLDABLES® Chapter Project

Assemble your lesson Foldables as shown to make a Chapter Project. Use the project to review what you have learned in this chapter.

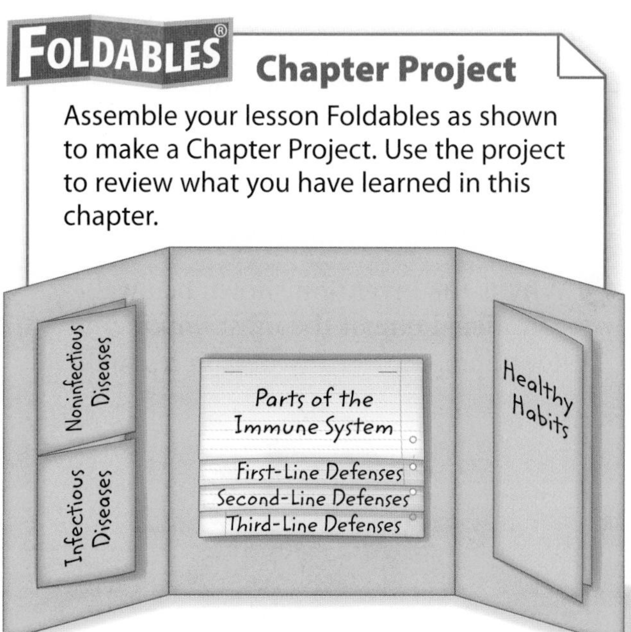

Use Vocabulary

1. A disease in which cells multiply uncontrollably is called ________.
2. Use the term *vector* in a sentence.
3. Define *pasteurization* in your own words.
4. Distinguish between antibodies and antigens.
5. In your own words, define *immunity.*
6. An overly sensitive immune response to common antigens is a(n) ________.
7. Use the term *antibiotics* in a sentence.
8. Differentiate between antibiotics and chemotherapy.

Link Vocabulary and Key Concepts

Interactive Concept Map

Copy this concept map, and then use vocabulary terms from the previous page to complete the concept map.

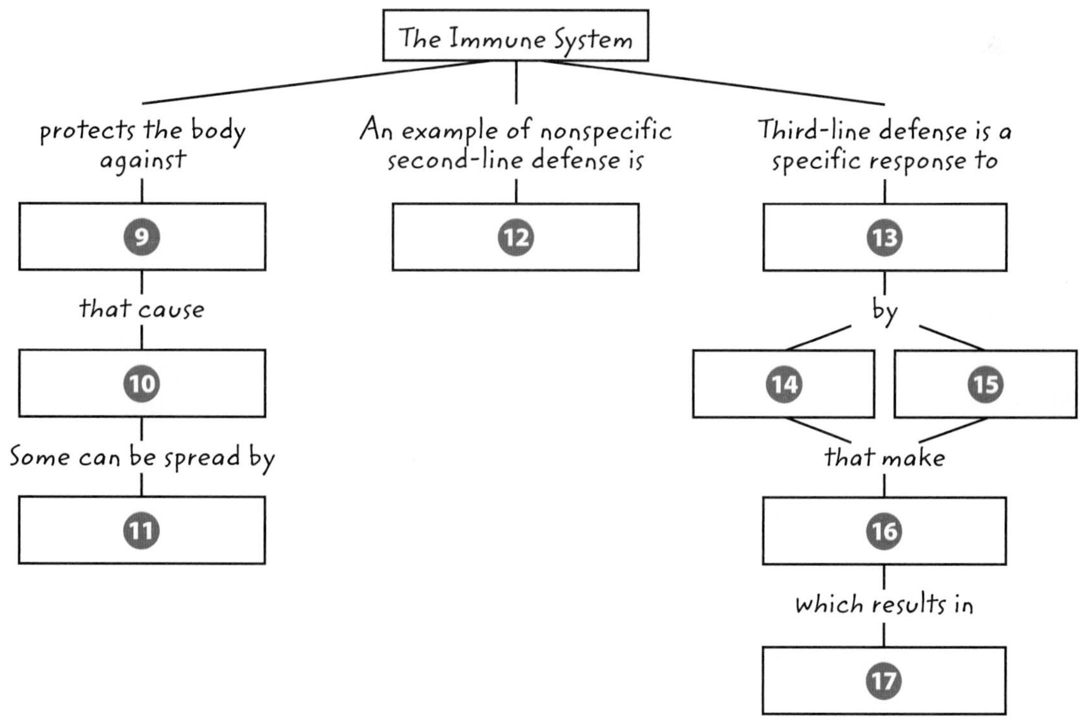

Chapter 17 Review

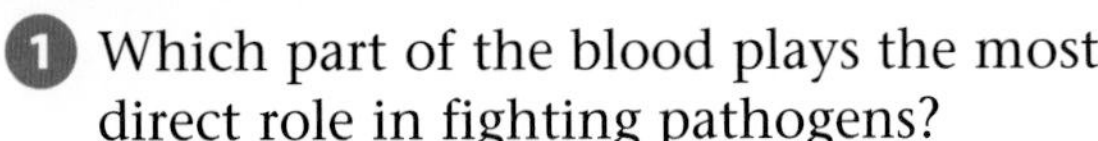

Understand Key Concepts

1. Which part of the blood plays the most direct role in fighting pathogens?
   A. plasma
   B. platelets
   C. red blood cells
   D. white blood cells

2. What is illustrated below?

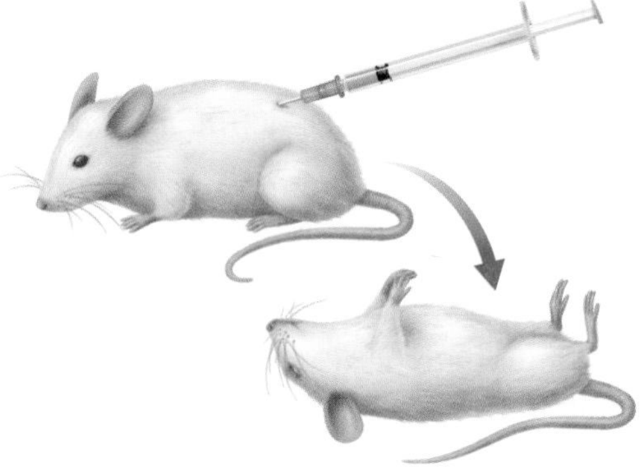

   A. Bacteria can be treated with heat.
   B. Bacteria do not infect other animals.
   C. Bacteria cause disease in healthy animals.
   D. One type of bacteria causes several different diseases.

3. Which has NOT led to improved health?
   A. cleaner hospitals
   B. lead paint
   C. better waste management
   D. cleaner food preparation tools

4. Which chemical might be helpful to the human body?
   A. asbestos
   B. antibiotic
   C. lead
   D. pesticide

5. Which is NOT part of the nonspecific, first-line defenses against pathogens?
   A. antibody
   B. cilia
   C. mucus
   D. skin

6. Inflammation is a common response to pathogens. Which would NOT be a common part of the inflammatory response at the site of an injury?
   A. bruising
   B. reddening
   C. swelling
   D. warmth

7. Why is the invention shown below considered one of the most important developments in early disease research?

   A. It showed people how to pasteurize liquids.
   B. It showed people that vaccines were effective.
   C. It showed people that microorganisms existed outside of wounds.
   D. It showed people that some diseases had a genetic component.

8. Which is NOT a healthful lifestyle choice?
   A. exercising regularly
   B. smoking cigarettes
   C. using sunscreen
   D. washing hands

9. Which is true about John Snow's research?
   A. He developed the first vaccine.
   B. He invented the first microscope.
   C. He made surgery safer for patients.
   D. He mapped a cholera outbreak.

10. Which is caused by a virus?
   A. chicken pox
   B. malaria
   C. pneumonia
   D. ringworm

Assessment

Online Test Practice

Critical Thinking

11 **Compose** a letter to explain the causes and symptoms of cystic fibrosis to a friend of the family who is concerned about his or her child developing cystic fibrosis.

12 **Role-Play** Choose the most important researcher from among John Snow, Joseph Lister, Robert Koch, and Edward Jenner. Defend your choice.

13 **Evaluate** the following statement, using the data in the table below to support your conclusion: Every person has not been vaccinated for common diseases, such as tetanus, polio, measles, mumps, and rubella.

Effects of Vaccinations

Disease	Rate Before Vaccination	Rate After Vaccination Developed
Tetanus	1,300	34
Polio	18,000	8
Measles	425,000	90
Mumps	200,000	610
Rubella	48,000	345

14 **Categorize** these parts of the immune system as first-line, second-line, or third-line defenses: cilia, white blood cells, antibodies, skin, mucus, inflammation.

15 **Plan and implement** a survey to determine when and how often the students in your class wash their hands.

16 **Design and create** a poster to remind the students in your school to make healthful life choices.

Writing in Science

17 **Write** a paragraph analyzing the differences in the causes of most childhood non-infectious diseases compared to other noninfectious diseases.

REVIEW THE BIG IDEA

18 Explain how the immune system helps the body maintain homeostasis.

19 The photo below shows a cancer cell being attacked by T cells. How does your immune system react when it detects an invader?

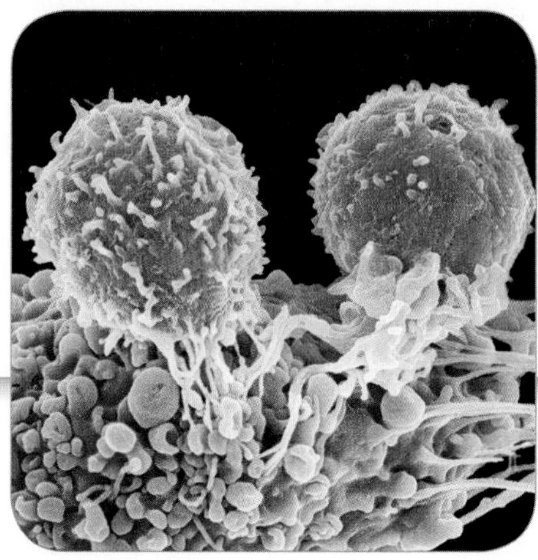

Math Skills

Math Practice

Use Percentages

20 The average annual number of rubella cases before the vaccine was developed was 48,000. There were 345 cases after the vaccine was developed. What percent change does this represent?

21 The average annual number of cases before vaccines were developed was 425,000 for measles and 200,000 for mumps. After the vaccines were developed, there were 90 cases of measles and 610 cases of mumps. Which vaccine was most effective in reducing cases of the disease?

Standardized Test Practice

Record your answers on the answer sheet provided by your teacher or on a sheet of paper.

Multiple Choice

1 Which would a doctor *exclude* as the cause of her patient's noninfectious disease?

A environmental conditions

B inherited traits

C lifestyle choices

D transmitted pathogens

Use the diagram below to answer question 2.

Immune Response

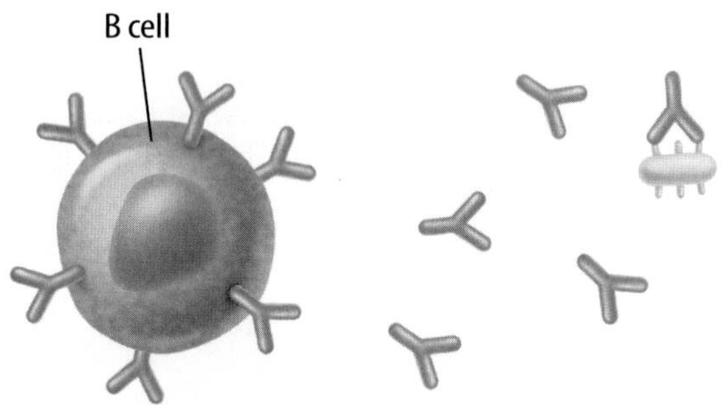

2 What are the Y-shaped objects in the diagram?

A antibodies

B antigens

C bacteria

D pathogens

3 Which first-line defense systems use acids to kill pathogens?

A circulatory and respiratory

B digestive and integumentary

C nervous and circulatory

D respiratory and digestive

4 Which is directly linked to increased risk of skin cancer in humans?

A acid rain

B asbestos insulation

C sunlight exposure

D water pollution

Use the diagram below to answer question 5.

Stage 1 - The Inflammatory Response

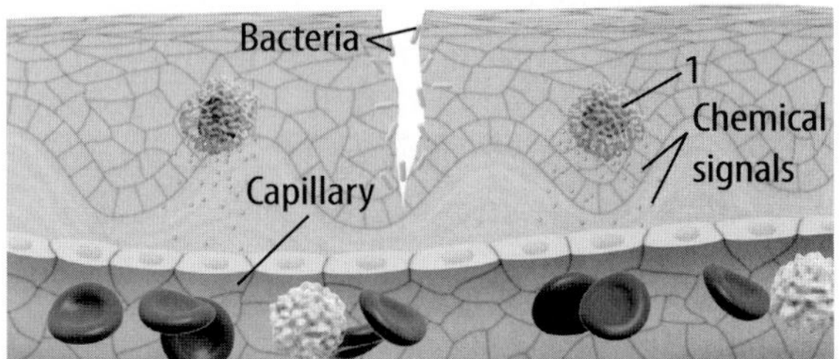

5 Which does *1* in the diagram represent?

A host cell

B mast cell

C sheath cell

D stem cell

6 In stage 2 of the inflammatory response, white blood cells and plasma leak into the area. What happens as a result?

A bacteria destruction

B capillary dilation

C protein release

D tissue swelling

7 How does sanitation improve health?

A aids white blood cell production

B increases vitamin absorption

C reduces pathogen exposure

D stimulates blood circulation

8 How do vitamin supplements contribute to health?

A They block UV rays.

B They improve nutrition.

C They kill bacteria.

D They kill cancer cells.

Use the table below to answer question 9.

Bacteria are present in ALL organisms with the disease but NOT in healthy organisms.
Bacteria must reproduce in the lab.
Sample of bacteria must cause disease in healthy animals.
Lab-grown pathogen is identical to original.

9 Which scientist developed the rules in the table above?

A Koch

B Lister

C Pasteur

D Snow

10 What kills bacteria in the pasteurization process?

A antiseptic

B heat

C isolation

D pressure

11 What is a disease-carrying organism that does NOT develop the disease?

A antigen

B B cell

C T cell

D vector

Constructed Response

Use the diagram below to answer questions 12 and 13.

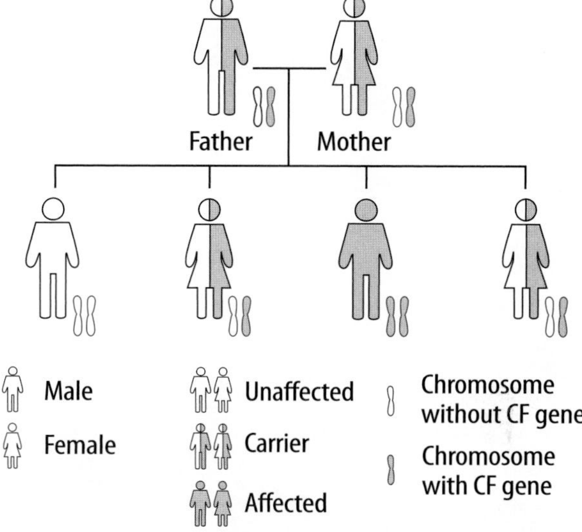

12 Based on the diagram above, why is cystic fibrosis a noninfectious disease? Use the terms *allele* and *recessive trait* to describe how someone contracts this disease.

13 Why might the parents in the diagram above be unaware that their children could contract cystic fibrosis? Suppose the parents had only one child. Will that child have cystic fibrosis? Explain.

14 List three sanitation practices. How do they promote health?

15 Explain the difference between active and passive immunity. How do vaccinations contribute to human immunity?

NEED EXTRA HELP?															
If You Missed Question...	1	2	3	4	5	6	7	8	9	10	11	12	13	14	15
Go to Lesson...	1	2	2	3	2	2	3	3	1	1	1	1	1	3	2

Chapter 18

Control and Coordination

THE BIG IDEA How do the nervous and endocrine systems help maintain the body's homeostasis?

Inquiry How does she do it?

This softball player must be able to sense activity and react quickly. The batter quickly swings the bat as the ball speeds toward her.

- How will this softball player react to try to hit the ball thrown toward her?
- What is happening in her brain and body to prepare for the action?
- How are these activities essential for survival in daily life?

Get Ready to Read

What do you think?

Before you read, decide if you agree or disagree with each of these statements. As you read this chapter, see if you change your mind about any of the statements.

1. The nervous system contains two parts—the central nervous system and the peripheral nervous system.
2. The autonomic nervous system controls voluntary functions.
3. A human has five senses that detect his or her environment.
4. The senses of smell and hearing work together.
5. Positive feedback systems in humans help maintain homeostasis.
6. Endocrine glands secrete hormones.

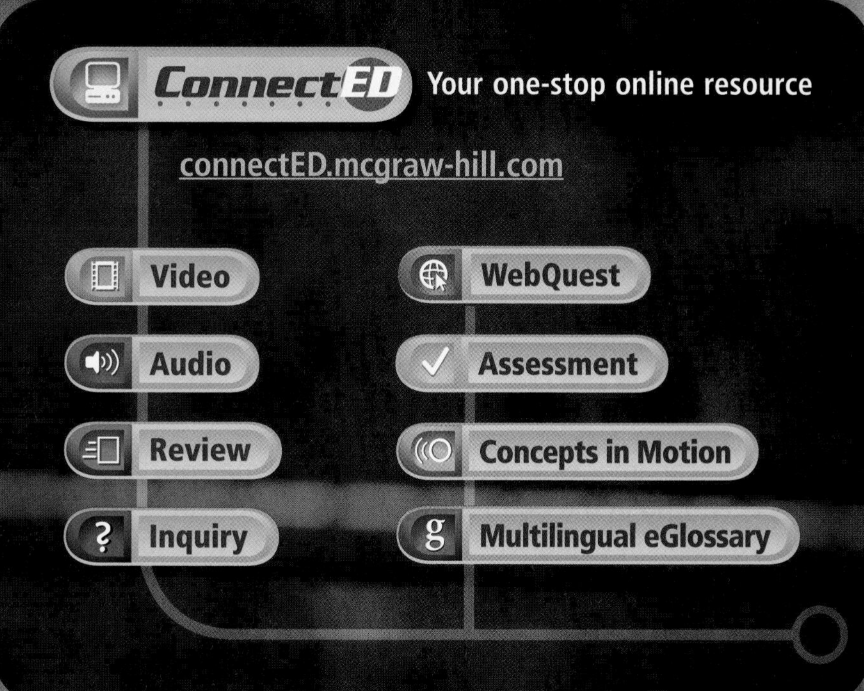

Lesson 1

Reading Guide

Key Concepts 
ESSENTIAL QUESTIONS

- What does the nervous system do?
- How do the parts of the nervous system work together?
- How does the nervous system interact with other body systems?

Vocabulary

nervous system p. 637
stimulus p. 638
neuron p. 639
synapse p. 639
central nervous system p. 640
cerebrum p. 640
cerebellum p. 640
brain stem p. 641
spinal cord p. 641
peripheral nervous system p. 641
reflex p. 642

The Nervous System

Inquiry What is memory?

These brain scan images all belong to the same person. Activity occurring in the brain is colored red, yellow, and orange. The person being scanned was asked to listen to and memorize words as the scans were taking place. What might the scans tell you about how the brain is working as the person hears different words?

Inquiry Launch Lab

5 minutes

Can you make your eyes blink or dilate?

Are you telling your body to breathe right now? Many functions the human body performs are not under your control. These functions, such as your heart beating, are usually important to your survival. Can you always control your muscles?

1. Sit facing a partner. Take turns trying to blink your eyes. Record your observations in your Science Journal.
2. Face your partner again. Try to make your pupils dilate, or get bigger. Record your observations.
3. Cover your eyes with your hand for 30 seconds. Make sure to completely block the light from your eyes.
4. Have your partner look at your eyes as you take your hand away. Record the observation in your Science Journal.
5. Repeat steps 3 and 4 using a **mirror** to observe your own eyes when you open them. Record what you observe.

Think About This

1. Why do you think it might be important that you do not need to think about blinking or breathing, even though you can make yourself do those actions?
2. **Key Concept** What do you think the purpose of dilating the eye might be? Why might it be helpful to not have to think about dilating your eyes?

Functions of the Nervous System

Have you ever had goose bumps form on your arms when you were cold? These bumps form because muscle cells in your skin respond to the cold temperature. As the muscle cells contract, or shorten, bumps form, and the hairs on your arms rise up. The hairs trap air, which helps to insulate the skin. This helps you feel warmer. How did the muscle cells know to contract? When you first felt the cold, a message was sent to your brain. After the message was processed, the brain sent a message to your skin's muscle cells, and goose bumps formed.

The part of an organism that gathers, processes, and responds to information is called the **nervous system.** Your nervous system receives information from your five senses—vision, hearing, smell, taste, and touch. You will read more about the five senses in Lesson 2.

The nervous system functions very quickly. It can receive information, process it, and respond in less than one second. In fact, signals received by the nervous system can travel as fast as some airplanes. This is around 400 km/h.

FOLDABLES®

Make a horizontal three-tab book and label it as shown. Use the book to organize your notes about the functions of the nervous system.

Gathering Information | Responding to Stimuli | Maintaining Homeostasis
The Nervous System

Figure 1 The goalie gathers information about the puck and responds by moving to block it from entering the goal. His body maintains homeostasis when his heart rate and breathing return to normal.

Visual Check What is the stimulus that the goalie is reacting to?

WORD ORIGIN

stimulus
from Latin *stimulare,* means "goad, urge"

Gathering Information

Have you ever seen a goalie react quickly to block a hockey puck from entering the goal? The sight of an object, such as the approaching puck in **Figure 1,** is a stimulus (STIHM yuh lus) (plural, stimuli). *A* **stimulus** *is a change in an organism's environment that causes a response.* The goalie's nervous system gathers and interprets the sight of the puck approaching and causes his body to react by raising his arm to block the shot.

Responding to Stimuli

How would you react if you saw the puck approaching the goal in **Figure 1?** Some people might move quickly to block the shot, while others might turn away to avoid being hit. These reactions are ways that the nervous system enables people to respond to a stimulus from the environment. Since the nervous system receives many stimuli at the same time, the type of response depends on how the information is processed.

Maintaining Homeostasis

Think again about the event shown in **Figure 1.** While responding to the stimuli of the approaching puck, the goalie's nervous system causes his heart and breathing rates to increase. This helps make his reaction time faster. People continually react to changes in their environments. Their nervous systems help maintain homeostasis, or the regulation of their internal environments. For example, the goalie's nervous system must signal his heart and breathing to slow down to restore homeostasis once he has blocked the shot.

Key Concept Check What are some of the tasks performed by the nervous system?

Neurons

The basic functioning units of the nervous system are called nerve cells, or **neurons** (NOO rahnz). Neurons help different parts of your body communicate with each other. Without looking down, how do you know whether you are walking on sand or pavement? Neurons in your feet connect to other neurons that send information to your brain about the surface. As shown in **Figure 2,** neurons have three parts—dendrites (DEN drites), a cell body, and an axon (AK sahn). A dendrite receives information from another neuron or from another cell in your body. A cell body processes information. An axon sends information out to another neuron or cell in your body.

Types of Neurons

There are three types of neurons that work together. They send and receive information throughout your body. Sensory neurons send information about your environment to your brain or spinal cord. Motor neurons send information from your brain or spinal cord to tissues and organs in your body. Interneurons connect sensory and motor neurons, much like a bridge connects two different areas of land.

Synapses

The gap between two neurons is called a **synapse** (SIH naps), as shown in **Figure 2.** Most neurons communicate across synapses by releasing chemicals. The chemicals carry information from the axon of one neuron to a dendrite of another neuron. This is similar to the way a baton is passed between runners in a relay race. Most synapses are between an axon of one neuron and a dendrite of another neuron. Information is usually transmitted in only one direction.

Figure 2 Information travels through the nervous system when chemical signals are released by the axon of one neuron and received by a dendrite on another neuron.

Neurons and Synapses Concepts in Motion Animation

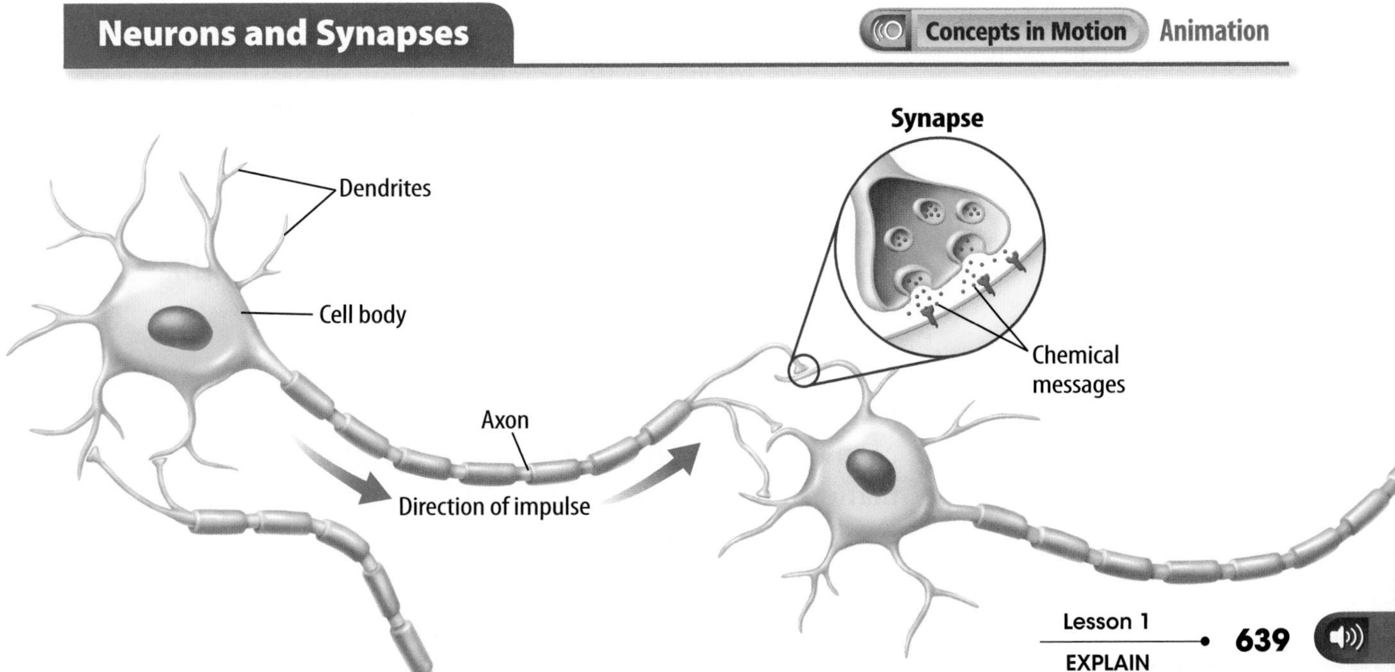

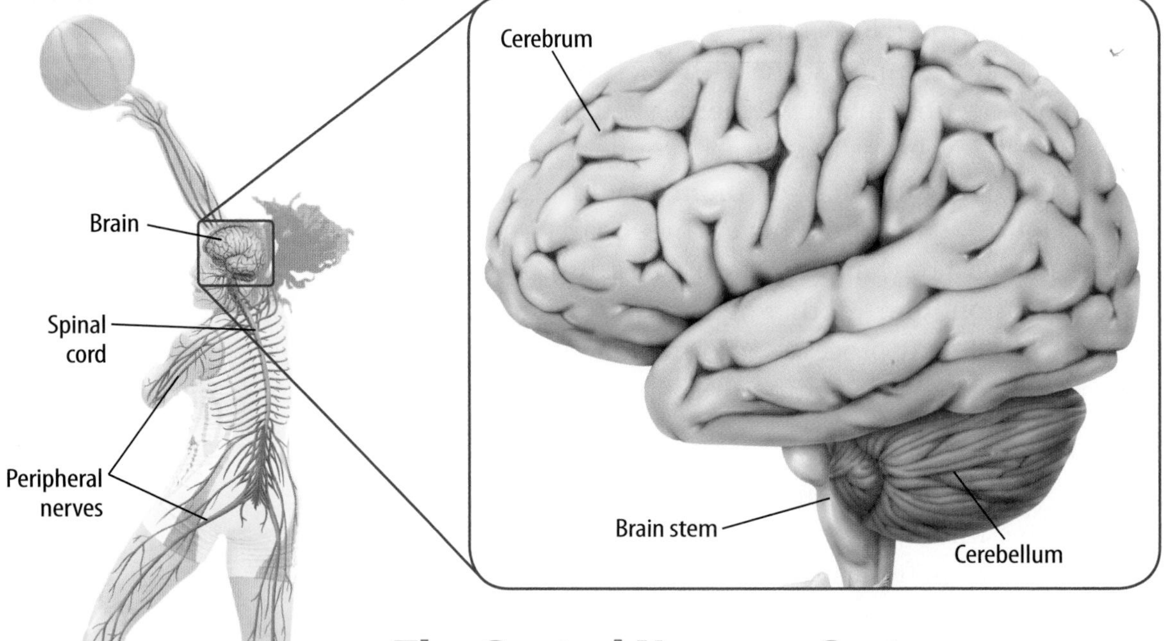

Figure 3 The nervous system consists of the central nervous system (CNS) and the peripheral nervous system (PNS). The brain is part of the CNS and has three main parts with specialized functions.

The Central Nervous System

As shown in **Figure 3,** your nervous system has two parts—the central nervous system and the peripheral (puh RIH frul) nervous system. You will read about the peripheral nervous system later in this lesson. *The* **central nervous system** (CNS) *is made up of the brain and the spinal cord.* The CNS receives, processes, stores, and transfers information.

The Brain

Much like a general is the commander of an army, the brain is the control center of your body. Your brain receives information, processes it, and sends out a response. The brain also stores some information as memories.

The Cerebrum *The part of the brain that controls memory, language, and thought is the* **cerebrum** (suh REE brum). The cerebrum also processes touch and visual information. It is the largest and most complex part of the brain. As shown in **Figure 3,** the surface of the cerebrum has many folds. These folds enable a large number of neurons to fit into a small space. If you could unfold the cerebrum, you would find that it has the surface area of a large pillowcase.

The Cerebellum *The part of the brain that coordinates voluntary muscle movement and regulates balance and posture is the* **cerebellum** (ser uh BEH lum). The cerebellum also stores information about movements that happen frequently, such as tying a shoe or pedaling a bicycle. This enables you to do repetitive things faster and with more accuracy.

The Brain Stem Some functions, such as digestion and the beating of your heart, are involuntary—they happen without your controlling them. *The area of the brain that controls involuntary functions is the* **brain stem,** shown in **Figure 4.** It also controls sneezing, coughing, and swallowing. It connects the brain to the spinal cord.

The Spinal Cord

The **spinal cord,** shown in **Figure 4,** *is a tubelike structure of neurons.* The neurons extend to other areas of the body. This enables information to be sent out and received by the brain. The spinal cord is like an information highway. Just like cars travel on a highway from one city to another, neurons in the spinal cord send information back and forth between the brain and other body parts. Bones called vertebrae protect the spinal cord.

Reading Check What are the two parts of the central nervous system?

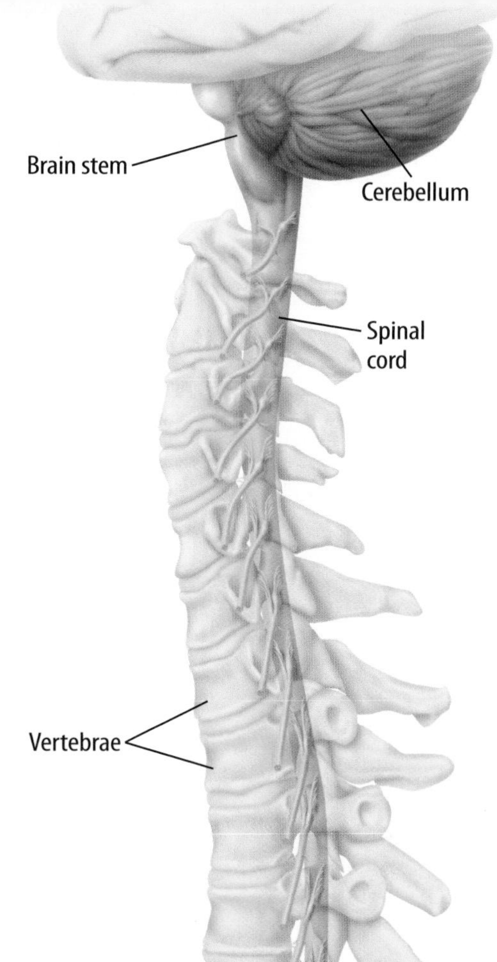

Figure 4 The spinal cord relays information between the brain and the PNS.

The Peripheral Nervous System

Recall that the nervous system is made of both the CNS and the peripheral nervous system. *The* **peripheral nervous system** (PNS), shown in **Figure 3,** *has sensory neurons and motor neurons that transmit information between the CNS and the rest of the body.*

Like the CNS, the PNS also has two parts—the somatic system and the autonomic system. The somatic system controls skeletal muscles. Neurons of the somatic system communicate between the CNS and skeletal muscles and cause voluntary movements, such as picking up a book. The autonomic system controls smooth muscles and cardiac muscles. It regulates involuntary actions, such as dilating blood vessels and the beating of your heart.

Key Concept Check How do the PNS and the CNS interact?

Math Skills

Review
Math Practice
Personal Tutor

Use Proportions

A nerve impulse from your hand travels at about 119 m/s. How long does it take the signal from your hand to reach your spinal cord if the distance is 0.40 m? You can use proportions to solve the problem.

Set up a proportion. $\frac{119\text{ m}}{1\text{ s}} = \frac{0.40\text{ m}}{y\text{ s}}$

Cross multiply. $119\text{ m} \times y\text{ s} = 0.40\text{ m} \times 1\text{ s}$

Solve for the y by dividing both sides by 119 m. $y = \frac{0.40\text{ m·s}}{119\text{ m}}$

$y = 0.003\text{ s}$

Practice

One giraffe neuron has an axon 4.6 m long that extends from its toe to the base of its neck. How long will it take a nerve impulse to travel this distance at a speed of 75 m/s?

Nervous System Health

A healthy nervous system is necessary for maintaining homeostasis. The nervous system can be damaged by infections and diseases. The most common way the nervous system is damaged is by **physical** injuries.

Academic Vocabulary
physical
(adjective) relating to material things

Physical Injuries

Falling, being in an automobile accident, and being hurt while participating in sports are some of the ways that you can injure and harm the nervous system. The injured nerves can no longer send and receive signals. This stops communication between the CNS and the PNS. When this happens, paralysis can occur. Paralysis is the loss of muscle function and sometimes a loss of feeling.

Preventing Injuries

Imagine that you are walking barefoot and you step on a sharp object. Without thinking, you quickly lift your foot. You do not think about moving your foot, it just happens. *An automatic movement in response to a stimulus is called a* **reflex.** Reflexes are fast because, in most cases, the information goes only to the spinal cord, not to the brain, as shown in **Figure 5.** This fast response protects us from injuries because it takes less time to move away from harm. A reflex often occurs before the brain knows that the body was in danger. However, once nerve signals reach the brain after the response, you feel pain.

Reading Check Why are reflexes fast?

Figure 5 A reflex enables you to respond to stimuli quickly.

The Path of a Reflex

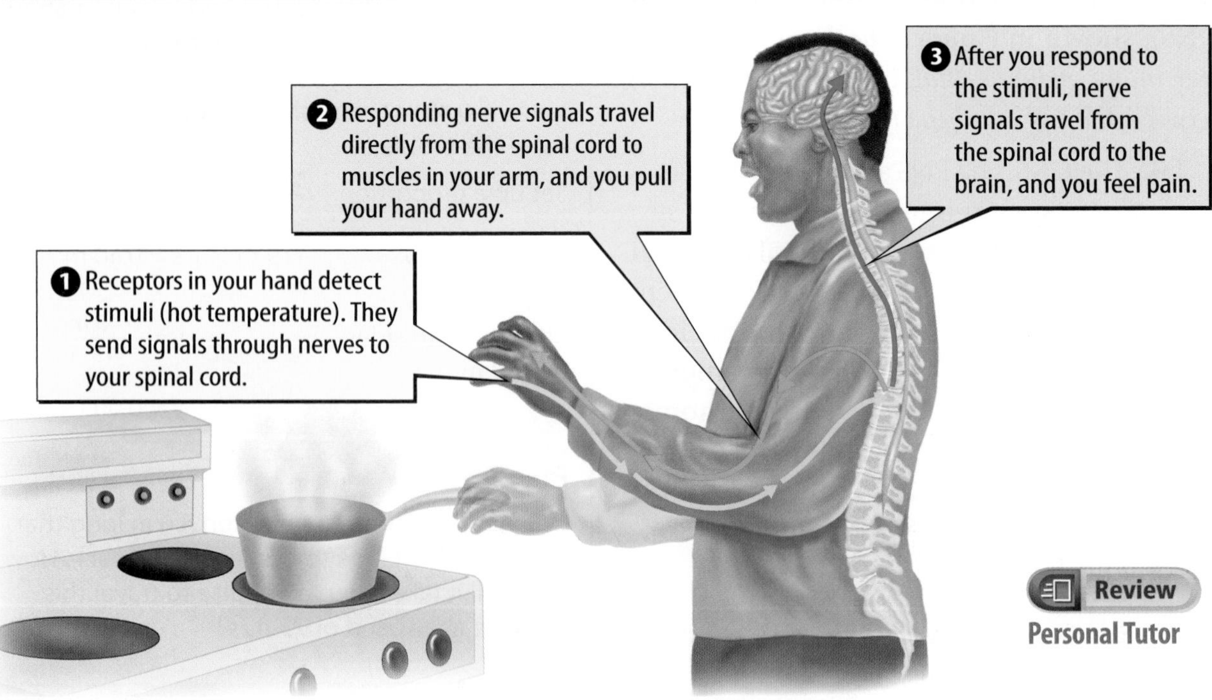

Review
Personal Tutor

Drugs

In addition to physical injuries, the nervous system can also be affected by substances you take into your body. Drugs are chemicals that affect the body's functions. Many drugs affect the nervous system by either speeding up or slowing down the communication between neurons.

Some pain medicines slow down this communication so much that they stop pain stimuli from reaching the brain. A drug that slows down the communication between neurons is called a depressant. Some people avoid drinking beverages that contain caffeine in the evening because it keeps them awake. Caffeine speeds up the communication between neurons. A drug that speeds up neuron communication is called a stimulant.

The Nervous System and Homeostasis

Why do you shiver when you get cold? It's because your nervous system senses the cold temperature and signals your muscles to contract rapidly in order to warm your body. Your body maintains homeostasis by receiving information from your environment and responding to it. The nervous system is vital to sensing changes in your environment. The nervous system signals other systems, such as the digestive, endocrine, and circulatory systems, to make adjustments when needed.

Key Concept Check Give an example of how the nervous system works with another body system to maintain homeostasis.

Inquiry MiniLab

20 minutes

How quickly can you improve your reaction time?

During a baseball game, a batter has less than one second to see the pitch, decide to try to hit the ball, and aim his swing. Through practice, he can improve his reaction time, which improves his performance. Can you improve your reaction time in just five trials?

1. Read and complete a lab safety form.
2. In your Science Journal, design a data sheet for five trials.
3. Hold a **metric ruler** vertically from the top. Have your partner position his or her open index finger and thumb just below the bottom edge of the ruler without touching it.
4. With the ruler slightly above the catcher's fingers, release the ruler without warning. Your partner should try to catch the ruler as quickly as possible by closing his or her thumb and index finger together. Record the distance the ruler fell.
5. Repeat steps 3 and 4 for four more trials.
6. Switch places with your partner, and repeat steps 3–5. Then create a graph to show the results for your five trials.
7. Record your results on a class data sheet.

Analyze and Conclude

1. **Determine** the mean, median, mode, and range for your results and the class results.
2. **Predict** what factors might affect reaction time for this activity.
3. **Key Concept** How is the ability of your nervous system to react quickly helpful to your survival?

Lesson 1 Review

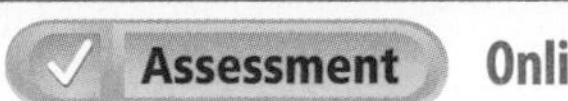

Visual Summary

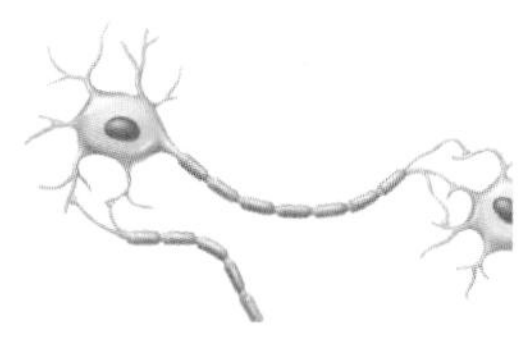

The nervous system gathers and interprets stimuli using a system of neurons that connect throughout the body.

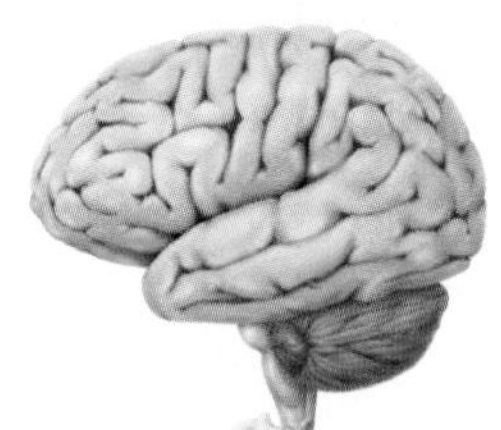

The central nervous system receives, processes, stores, and transfers information.

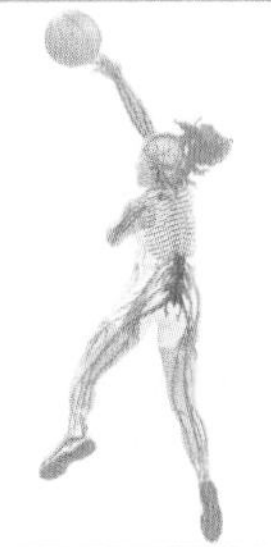

The peripheral nervous system is made up of the neurons that transmit information between the CNS and the rest of the body.

FOLDABLES®

Use your lesson Foldable to review the lesson. Save your Foldable for the project at the end of the chapter.

What do you think NOW?

You first read the statements below at the beginning of the chapter.

1. The nervous system contains two parts—the central nervous system and the peripheral nervous system.
2. The autonomic nervous system controls voluntary functions.

Did you change your mind about whether you agree or disagree with the statements? Rewrite any false statements to make them true.

Use Vocabulary

1. **List** the parts of the central nervous system.
2. **Distinguish** between a neuron and a synapse.
3. **Use the terms** *autonomic* and *somatic* in a sentence.

Understand Key Concepts 

4. Which causes an organism to react?
   A. CNS
   B. PNS
   C. stimulus
   D. synapse
5. **Compare** the functions of the three parts of the brain.
6. **Explain** how a spinal cord injury may prevent movement.

Interpret Graphics

7. **Summarize** Copy and fill in the graphic organizer below to summarize how the nervous system receives, processes, and responds to a stimulus.

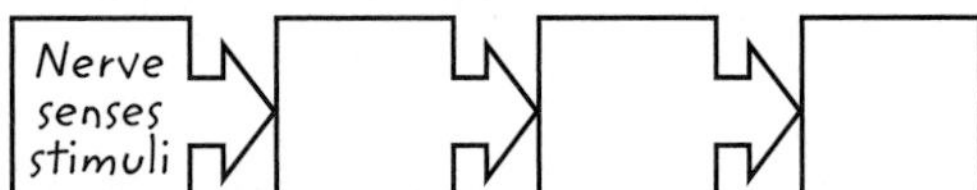

Critical Thinking

8. **Describe** Look at the synapse shown below. Describe the path of a signal from one neuron to another.

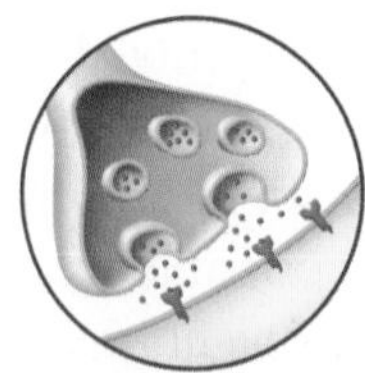

Math Skills

Math Practice

9. The longest axon in a whale can be 18.2 m long. If it takes 0.26 s for a signal to travel the length of this axon, how fast is the nerve impulse traveling in m/s?

How does the nervous system respond to a stimulus?

Neuroscientists can model how information moves between neurons in the body. The chemicals that send messages between neurons are too small to see, so a neuroscientist might make a model that shows how the messages are transmitted.

Learn It

A **model** is a representation of how something functions. It might be an object, a drawing, a calculation, a diagram, or something else that explains and describes how things work. Scientists often model biological activities in order to better understand them.

Try It

1. Read and complete a lab safety form.
2. Form groups of five. Stand in a line with the four other students so that you can reach the hand of the person next to you. Do not move or talk. The student at the front of the line represents the neurons in the fish's head and eyes. The student at the end of the line represents the neurons in the fish's tail. Only the fish's tail—the last student in line—can make the fish swim.
3. Imagine that your fish sees a tasty shrimp in the water. In order to eat the shrimp, the head must pass a message to the tail telling it to swim toward the shrimp. The student that represents neurons in the fish's head should send the message to the tail by squeezing the hand of the next student in line.
4. Continue passing the message between neurons (students). When the student representing neurons in the tail gets the message, this student should move forward. Observe how long it takes for the message to reach the tail and make the fish swim toward the shrimp.

Apply It

5. Take one student out of the middle of the line, leaving a gap. Now imagine that the fish sees a shark that wants to eat it. What happens when the head sends a message to the tail to swim away from the shark? What happens to the fish?
6. Take two more students out of the middle of the line, making the line only two cells long—the fish's head and tail. Stand so that the head and tail are next to one another and hold hands. Now how long does it take to send a message to swim away from the shark? Why does the message move more quickly?
7. **Key Concept** What type of neuron does the student at the fish's head represent? What type of neuron does the student at the fish's tail represent? When students squeeze hands to pass messages down the line of neurons, what function are they modeling?

Lesson 2

Reading Guide

Key Concepts
ESSENTIAL QUESTIONS

- How do you learn about your environment?
- What is the role of the senses in maintaining homeostasis?

Vocabulary

sensory system p. 647
receptor p. 647
retina p. 648
eardrum p. 650

Multilingual eGlossary

Video

- BrainPOP®
- What's Science Got to do With It?

The Senses

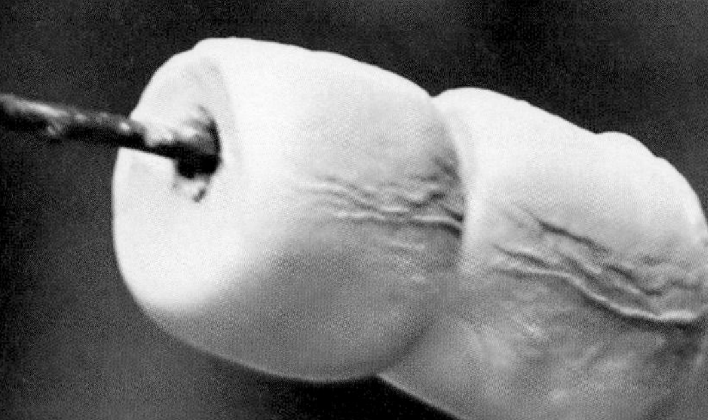

Inquiry Using All Five Senses?

Have you ever toasted marshmallows over a campfire? What did you see? What smells did you experience? What sounds did you hear? Which sensations did you feel? What tastes do you remember? How do these things together make up a campfire experience?

Launch Lab

10 minutes

Does your nose help you taste food? 

Sometimes you might hold your nose when you don't want to taste something you're eating. Does this really work? Your senses all function because of nervous impulses that send information. Are your senses of taste and smell connected?

1. Read and complete a lab safety form.
2. Hold your nose closed while your teacher walks around and places a **food cube** on your **plate**.
3. While still holding your nose, place the food cube in your mouth and chew it.
4. In your Science Journal, write down what the food tasted like to you and what you think the food is.
5. Repeat steps 2–4 with a different food cube. Record your observations.
6. Let go of your nose and chew another sample of both food cubes. Record your observations.

Think About This

1. What did the two food cubes taste like when you held your nose?
2. What did the two food cubes taste like when you were not holding your nose?
3. **Key Concept** How do both senses together affect your ability to learn about your environment?

You and Your Environment

Recall from Lesson 1 that your nervous system enables your body to receive information about your environment, process the information, and react to it. Your nervous system is constantly responding to many different types of stimuli. However, your body has to receive a stimulus before it can respond to one. How does this happen?

The **sensory system** *is the part of your nervous system that detects or senses the environment.* A human uses five senses—vision, hearing, smell, taste, and touch—to detect his or her environment. What senses might you use around a campfire, like the one in the photo on the opposite page? You might see the flames of the campfire, hear the crackle of the wood burning, smell the smoke, feel the warmth of the fire, and taste the cooked marshmallows. *All parts of the sensory system have special structures called* **receptors** *that detect stimuli.* Each of the five senses uses different receptors.

Word Origin
sensory
from Latin *sentire*, means "to perceive, feel"

Key Concept Check Describe one way your senses help you to learn about the environment.

Vision Concepts in Motion Animation

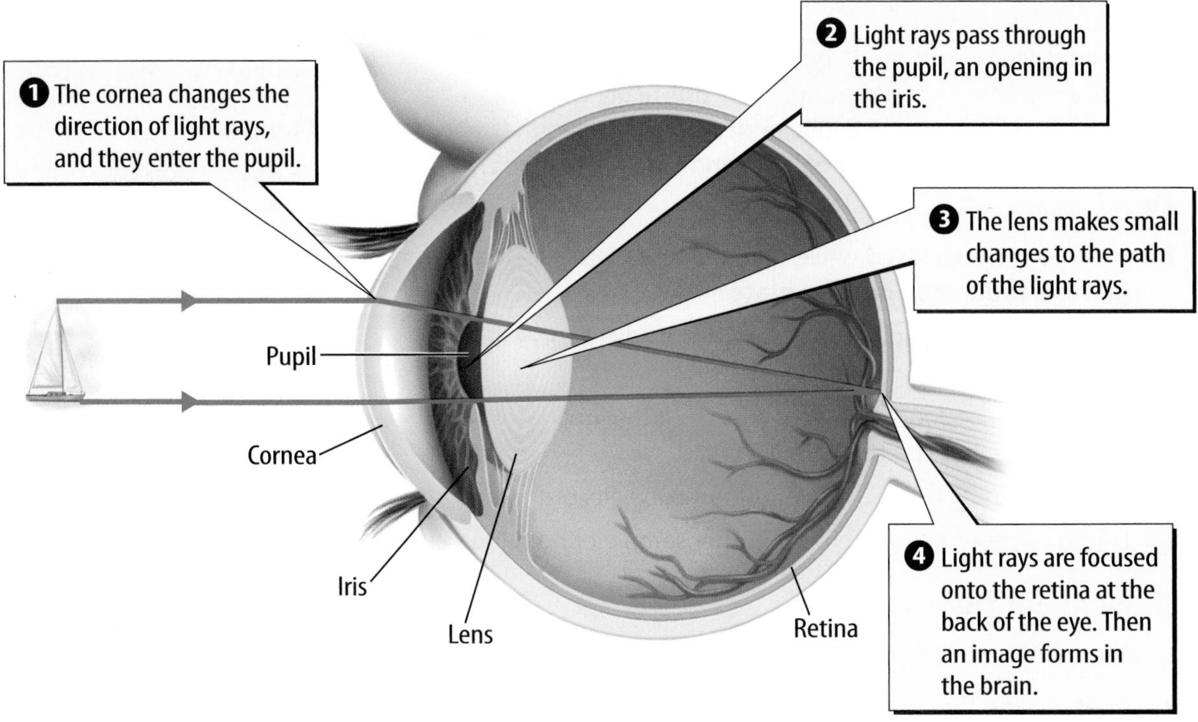

Figure 6 Your eyes contain photoreceptors that receive and interpret light signals from the environment.

FOLDABLES®

Use three sheets of notebook paper to make a layered book. Label it as shown. Use it to record information about the five senses.

Vision
Hearing
Taste
Smell
Touch
The Five Senses

Vision

Have you ever wondered how your eyes work? Your sense of vision lets you see things that are close, such as the words on this page, and objects that are far away, such as a star in the night sky. The visual system uses photoreceptors in the eye to detect light and create vision.

The Parts of the Eye

As shown in **Figure 6,** light enters the eye through the cornea (KOR nee uh), a thin membrane that protects the eye and changes the direction of light rays. The colored part of your eye is the iris (I rus). After light passes through the cornea, it goes through an opening formed by the iris called the pupil. The iris controls the amount of light that enters the eye by changing the size of the pupil. In bright light, the iris constricts, making the pupil smaller and letting in less light. In dim light, the iris relaxes, making the pupil larger and letting in more light. Light then travels through a clear structure called the lens. As shown in **Figure 6,** the lens works with the cornea and focuses light. *The* **retina** (RET nuh) *is an area at the back of the eye that has two types of cells—rod cells and cone cells—with photoreceptors.* Rod cells detect shapes and low levels of light. They are important for night vision. Cone cells detect color and function best in bright light.

How You See

In order to see, light that enters your eyes has to be detected by the **rods** and cones in the retina. The rods and cones detect information about the colors and shapes of objects from the light that enters the eyes. The retina then sends that information as electric signals through the optic nerve to the brain. The brain uses the information and creates a picture of what you are seeing.

Science Use v. Common Use

rod

Science Use one of the photoreceptors in the eye that distinguish the shapes of objects

Common Use any long, cylinder-shaped object

Focusing Light

The lens and the cornea work together and change the direction of the light that enters the eye. As shown in **Figure 6,** both the lens and the cornea are curved. These curved shapes change the direction of light and focus it onto the retina. Why do some people need glasses to see well? If corneas or lenses are not curved exactly right, the eyes will have trouble focusing images, as shown in **Figure 7.** If a person's eyes are longer than normal, the person is nearsighted and has trouble seeing images that are far away. If a person's eyes are shorter than normal, the person is farsighted and has trouble seeing images that are close up. Glasses or contacts are used to correct vision problems by correctly focusing the light on the retina.

Reading Check Which parts of vision are rods and cones responsible for?

Figure 7 The lens and the cornea focus light. Vision problems occur when they are not curved correctly.

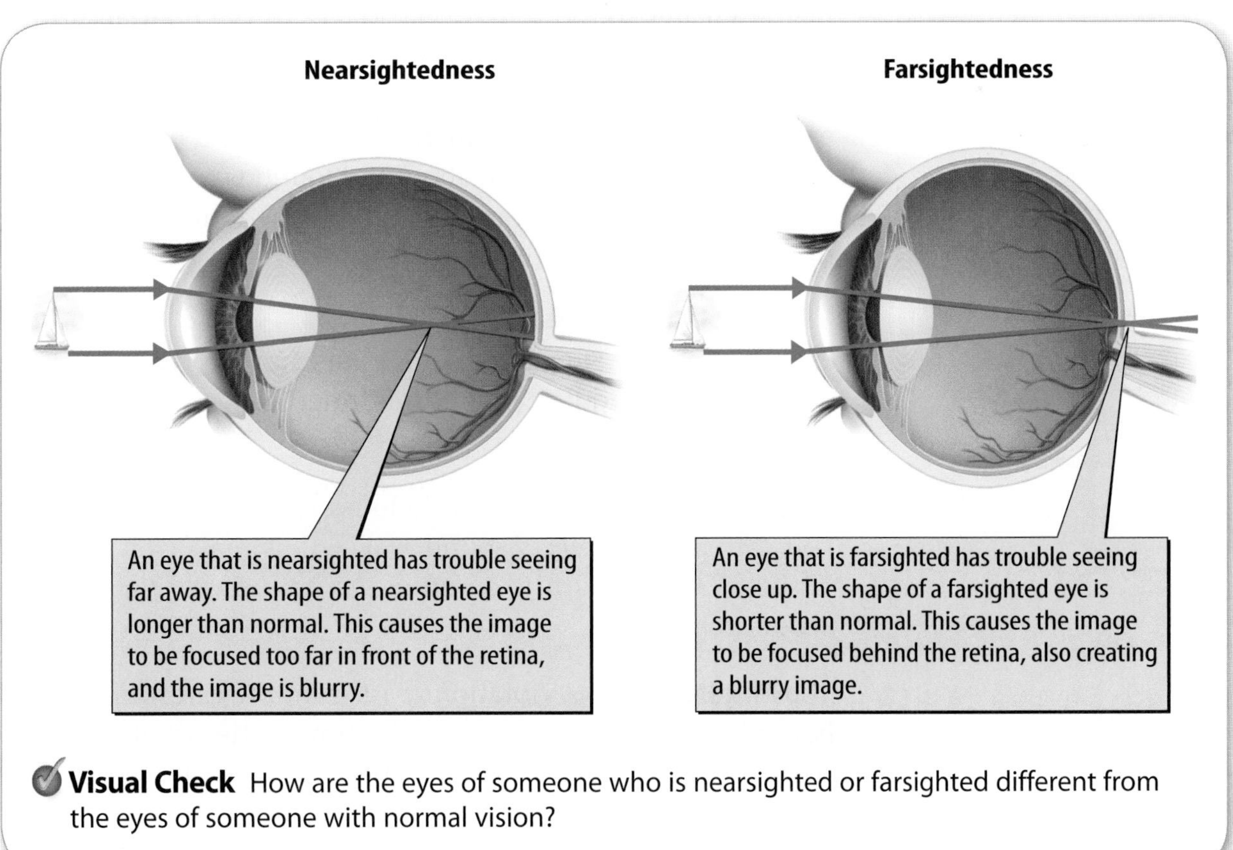

Visual Check How are the eyes of someone who is nearsighted or farsighted different from the eyes of someone with normal vision?

Concepts in Motion Animation

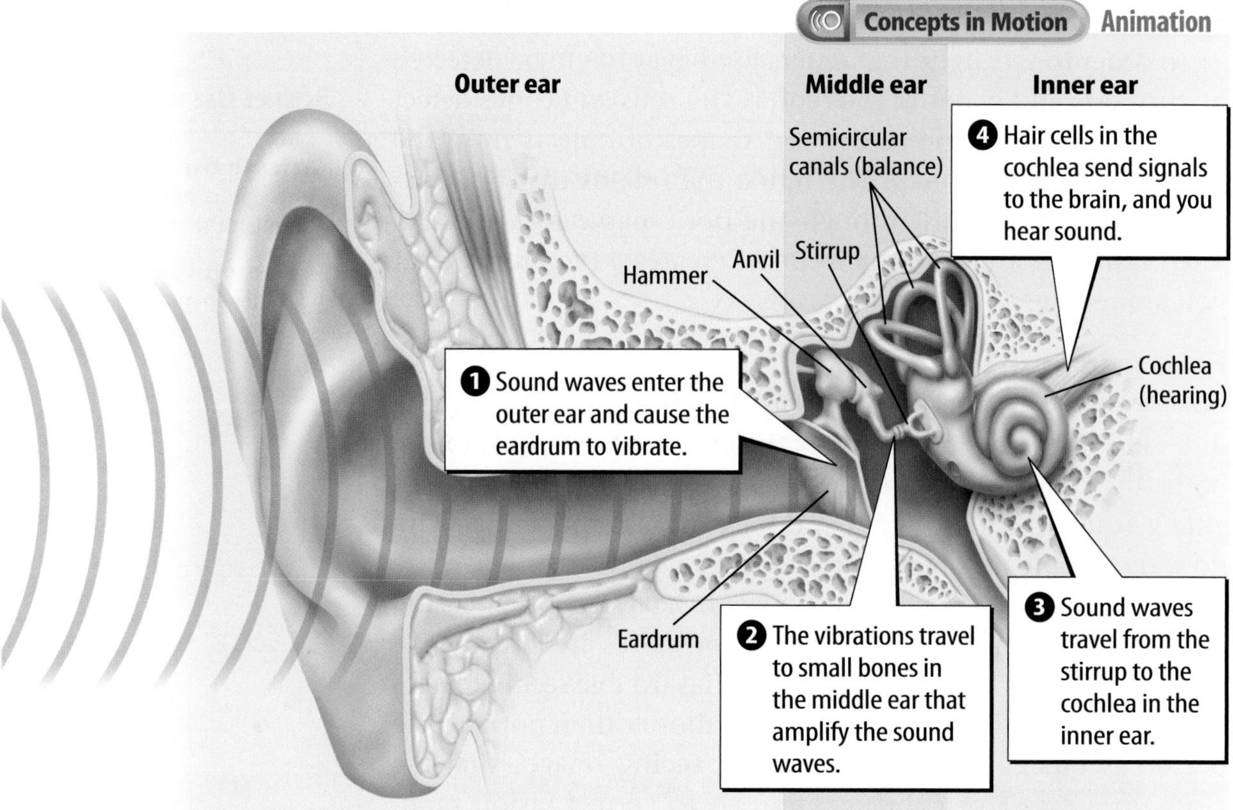

Figure 8 Scientists have identified three specialized regions in the human ear.

Hearing

How do you hear the sounds around you? The vibration of matter creates sound **waves** that travel through air and other substances. Sound waves that enter the ear are detected by auditory (AW duh tor ee) receptors. As waves travel within the ear, they are amplified, or increased, and move hair cells. The hair cells send information about the sound waves to the brain. The brain processes information about the loudness and tone of the sound, and you hear.

REVIEW VOCABULARY

wave
a disturbance in a material that transfers energy without transferring matter

The Parts of the Ear

As shown in **Figure 8,** human ears have three areas—the outer ear, the middle ear, and the inner ear—each with a special function.

The Outer Ear The outer ear includes the parts of the ear that you can see. It collects sound waves that make the eardrum vibrate. *The* **eardrum** *is a thin membrane between the outer ear and the inner ear.*

The Middle Ear The vibrations pass from the eardrum to small bones in the middle ear—the hammer, the anvil, and the stirrup. The movement of these bones amplifies the sound waves.

The Inner Ear The part of the ear that detects sound is the inner ear. The inner ear converts sound waves into messages that are sent to the brain. The structure shown in **Figure 8** that looks like a snail shell is called the cochlea (KOHK lee ah). The cochlea is filled with fluid. When sound waves reach the cochlea, they make the fluid vibrate. The liquid in the cochlea moves much like the way that hot chocolate moves in a cup when you blow on it. The moving fluid bends hair cells, which send messages to the brain for processing.

 **Reading Check** What are the three areas of the ear?

The Ear and Balance

In addition to detecting sound waves, the inner ear has another function. Parts of the inner ear, called the semicircular canals, help maintain balance. Like the cochlea, the semicircular canals contain fluid and hair cells. Whenever you move your head, the fluid moves, which moves the hair cells, as shown in **Figure 9.** Information about the movement of the hair cells is sent to the brain. The brain then signals muscles to move your head and body in order to maintain your balance.

Figure 9 Fluid in your inner ear helps maintain your balance by sensing changes in the position of your head.

The Ear and Balance

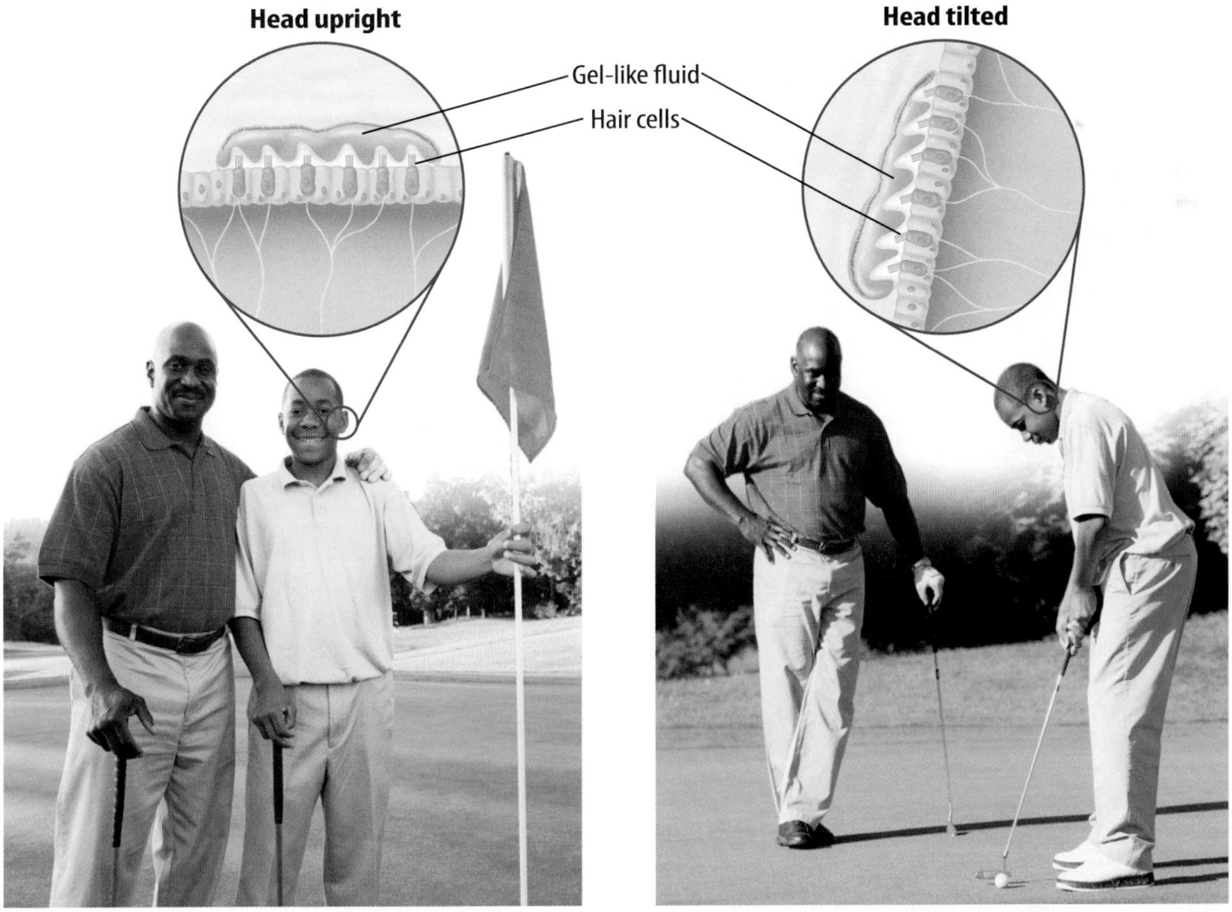

Smell

How are you able to distinguish between the citrus aroma of an orange and the sulfur odor of rotten eggs? Humans have hundreds of different receptors for detecting odors. Some dogs have over 1,000 different odor receptors and can be used to track things by smell. Odors are molecules that are detected by chemical receptors, called chemoreceptors (kee moh rih SEP turz), in your nose. These receptors send messages to the brain. The brain then processes the information about the odor. A smell might make you feel hungry, or it could trigger a strong memory or feeling.

Taste

The sense of taste also relies on chemoreceptors. Chemoreceptors on your tongue detect chemicals in foods and drinks. The receptors then send messages to the brain for processing. Chemoreceptors on the tongue are called taste buds. Taste buds, such as the ones shown in **Figure 10,** can detect five different tastes: bitter, salty, sour, sweet, and a taste called umami (oo MAH mee). Umami is the taste of MSG (monosodium glutamate), a substance often used in processed foods. **Figure 10** illustrates how the chemoreceptors in your nose and mouth work together to help you taste foods.

Taste and Smell

Figure 10 Taste and smell stimuli are both detected by chemoreceptors. Chemoreceptors detect chemicals in the substances you eat and drink and in the odors you breathe.

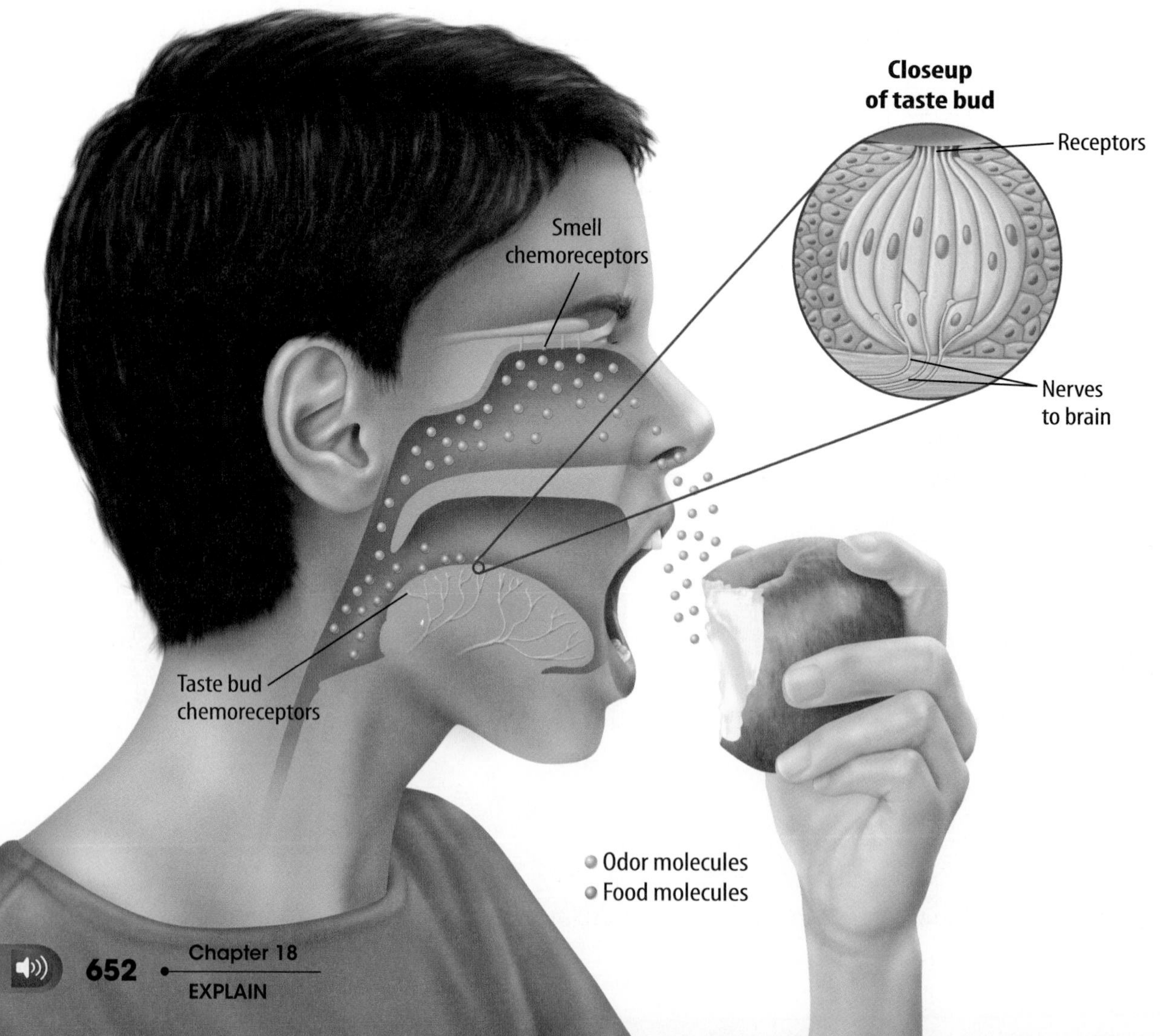

Touch

Like all the other senses, the sense of touch also uses special receptors that detect the environment. Touch receptors in your skin can detect temperature, pain, and pressure. For example, skin receptors can detect the difference between a light tap and a poke. These receptors are all over your body. Some areas, such as the palms of your hands and the soles of your feet, have lots of receptors. The middle of your back has fewer receptors. Just like the other four senses, touch receptors send messages to the brain for processing.

The Senses and Homeostasis

Gathering information about the environment is important for your survival. The five senses collect information about the environment and send it to your nervous system. Your brain is then able to respond and activate your body to maintain homeostasis. Whether it is sensing temperature changes, finding food and water, avoiding harmful environments, or detecting other important stimuli, your senses are the vital first step.

Key Concept Check Why are senses important in maintaining homeostasis?

Inquiry MiniLab

20 minutes

What area of your skin is most sensitive?

Your skin has millions of nerve endings. Although this might not be a surprise, you might not realize that different areas of the body are more sensitive to touch and pain than other areas. Which regions of the skin are most sensitive to touch? How does the sense of touch help maintain homeostasis?

1. Read and complete a lab safety form.
2. In your Science Journal, design a data sheet to include experimental results for both you and your partner. You will record the sensitivity of the following eight areas of the body: fingertip, back of hand, elbow, toes, top of foot, knee, back of neck, and nose.
3. Blindfold your partner with a **clean tube sock or scarf**. Lightly touch one of the areas of your partner's skin with a **thin, washable marker.** Leave a small, colored mark where you touched your partner's skin.
4. Using a **different colored marker,** have your partner try to touch the same point that was previously marked by your pen.
5. Test all eight areas of the body.
6. Use a **metric tape measure** to determine the distance from the point you marked to the point your partner touched.
7. Record your results and your partner's results in your data table. Graph both sets of results from least distance to greatest distance between the actual point of touch and the guessed point of touch. Label your graph.

Analyze and Conclude

1. **Calculate and analyze** what part of your body has the least distance between the actual point and the guessed point. Which body part has the greatest distance?
2. **Compare** the results for both you and your partner. Predict the results for all class members.
3. **Key Concept** Analyze how the sensitivity of different areas of your skin help you maintain homeostasis.

Lesson 2 Review

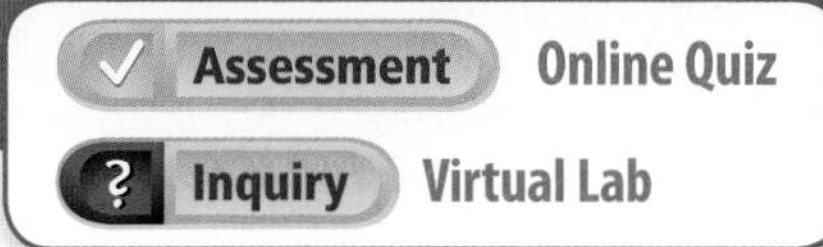

Visual Summary

The sensory system is part of the nervous system. It uses receptors in five senses and collects information about the environment.

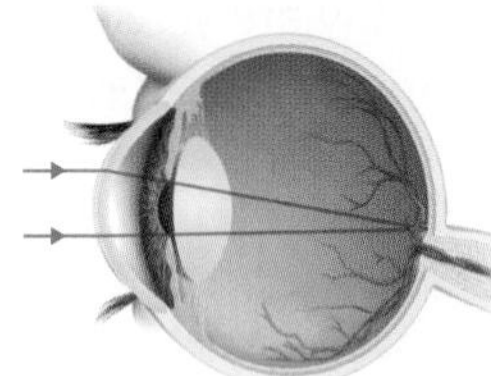

Your eyes detect light as it passes through the cornea, the pupil, and the lens, and then to the retina.

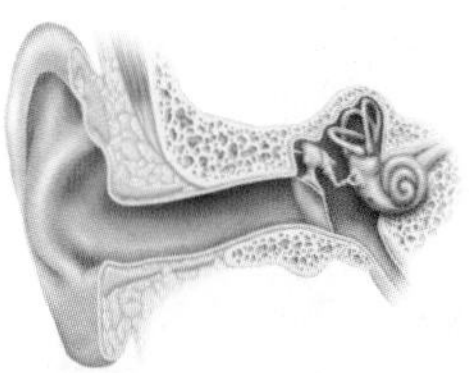

The eardrum is a structure between the outer and middle ear. It vibrates in the presence of sound waves.

FOLDABLES®

Use your lesson Foldable to review the lesson. Save your Foldable for the project at the end of the chapter.

What do you think NOW?

You first read the statements below at the beginning of the chapter.

3. A human has five senses that detect his or her environment.

4. The senses of smell and hearing work together.

Did you change your mind about whether you agree or disagree with the statements? Rewrite any false statements to make them true.

Use Vocabulary

1. **Name** the part of the ear that collects sound waves from the environment.
2. **Define** the term *receptor* using your own words.
3. **Use the word** *retina* in a sentence.

Understand Key Concepts 

4. Which is NOT a sensory system?
   A. digestion
   B. hearing
   C. smell
   D. vision
5. **Explain** the relationship between the five senses and homeostasis.
6. **Contrast** the different types of stimuli that can be detected by touch receptors.

Interpret Graphics

7. **Evaluate** the relationship between the senses of smell and taste using the figure at right as a guide.

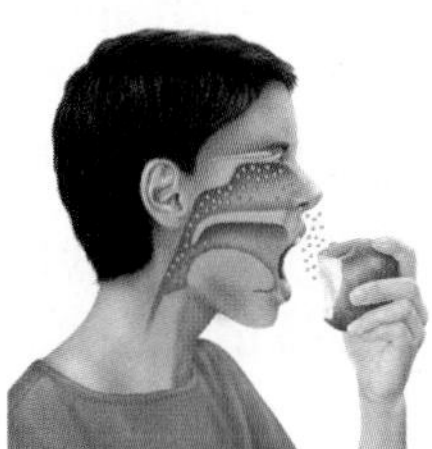

8. **Organize** Copy and fill in the table below to organize information about the five senses.

Sense	Stimulus

Critical Thinking

9. **Examine** why it might be difficult to taste food when you have a cold.
10. **Infer** how an ear infection might lead to problems with balance.

HOW IT WORKS

Night Vision Goggles

Would you like to be able to see in the dark?

You might already know that vision begins with light entering your eyes. What happens when it is dark? If there is no light available to enter your eyes, how can any device make it possible to see?

Even when it is dark, there is light all around you. Although you can't see it, almost everything gives off, or emits, infrared light. Objects also reflect some infrared light, in the same way that they reflect visible light. Night vision goggles work by collecting that infrared light and converting it to visible light.

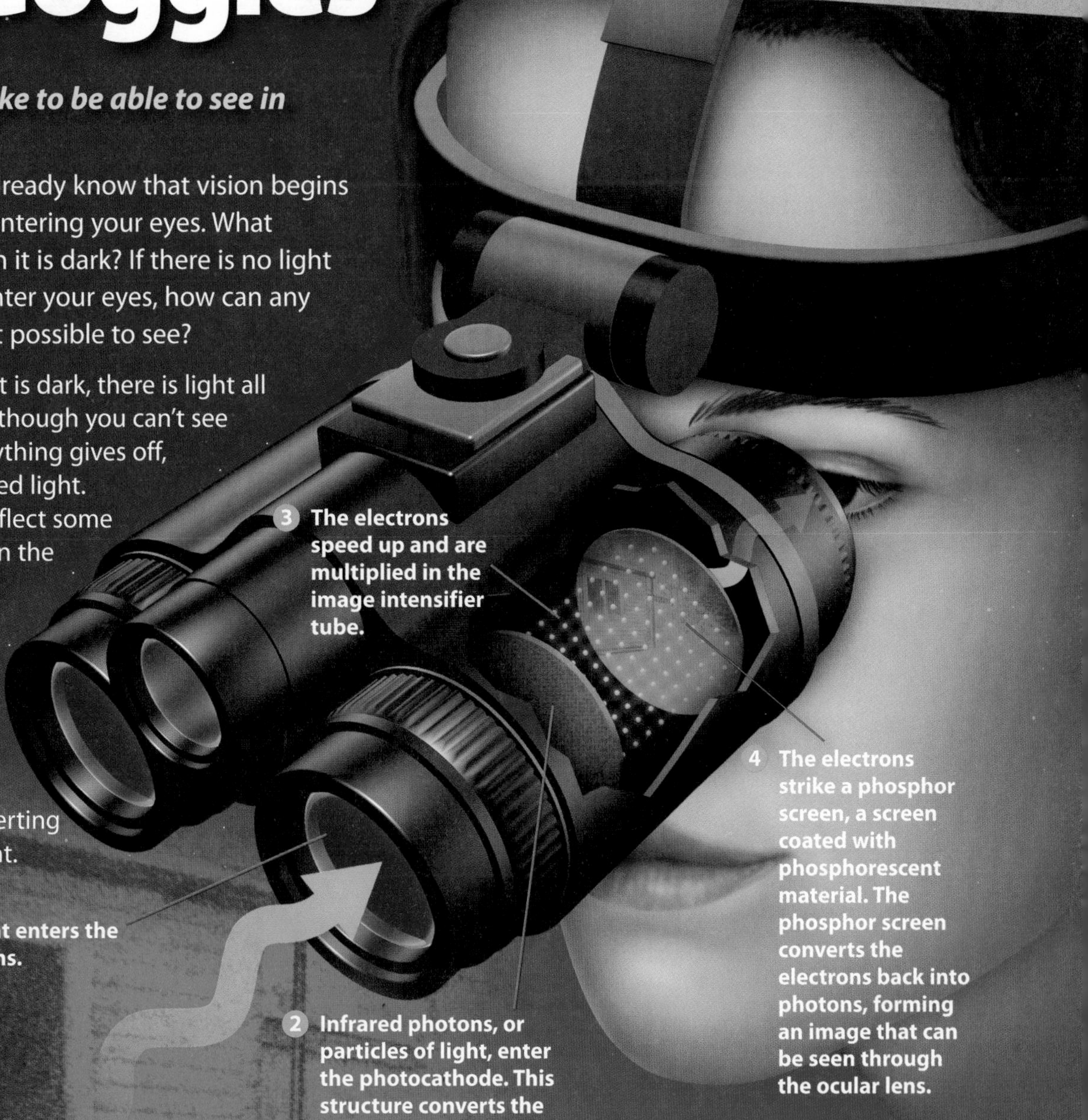

1 Infrared light enters the objective lens.

2 Infrared photons, or particles of light, enter the photocathode. This structure converts the pattern of infrared photons into a pattern of electrons.

3 The electrons speed up and are multiplied in the image intensifier tube.

4 The electrons strike a phosphor screen, a screen coated with phosphorescent material. The phosphor screen converts the electrons back into photons, forming an image that can be seen through the ocular lens.

It's Your Turn

REPORT How can owls see more clearly in dim light than humans can? Research rods and cones in the eye and what they have to do with vision. Use what you discover to make a "How It Works" diagram about rods and cones in the eye.

Lesson 3

Reading Guide

Key Concepts 

ESSENTIAL QUESTIONS

- What does the endocrine system do?
- How does the endocrine system interact with other body systems?

Vocabulary

endocrine system p. 657

hormone p. 657

negative feedback p. 660

positive feedback p. 660

g Multilingual eGlossary

The Endocrine System

Inquiry Blue Butterfly?

This image might look like a blue and green butterfly, but it is actually an image of a person's butterfly-shaped thyroid gland. The image was taken after a dose of radioactive material was given to the patient. The radioactive material collected in thyroid tissue.

Launch Lab

10 minutes

What makes your heart race?

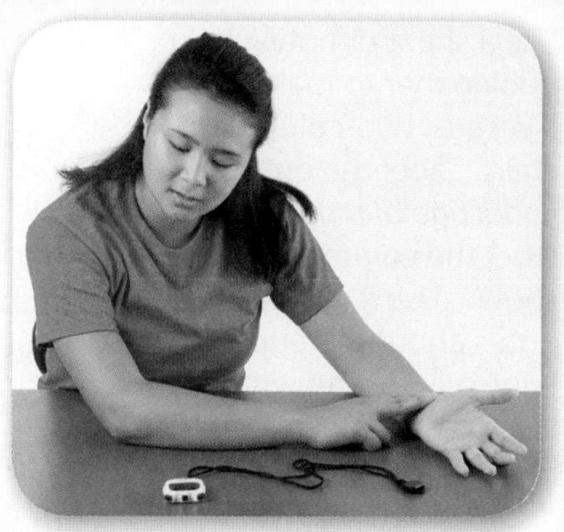

Have you ever felt your heart start to race? Maybe something startled you or you were nervous or frightened. Usually this sensation lasts only for a few minutes. What causes your heart to pound so fast at times?

1. Read and complete a lab safety form.
2. Find your pulse by holding your first and second fingers of one hand on the inside of the wrist of your other hand. Count heartbeats for 15 seconds. Record this number in your Science Journal. Multiply this number by four to find your average heartbeats per minute.
3. Your teacher will attempt to startle you. Immediately afterward, count your heartbeats again for 15 seconds and multiply by 4.
4. Wait five minutes. Calculate your heartbeats per minute again.

Think About This

1. What changes did you note in your heart rate after your teacher startled you?
2. What happened to your heartbeat when you checked five minutes later?
3. **Key Concept** Why do you think your heart reacted to the noise even though you knew it was coming? Why do you think this change occurred?

Functions of the Endocrine System

Like the nervous system, your body sends messages using another system called the endocrine (EN duh krun) system. *The* **endocrine system** *consists of groups of organs and tissues that release chemical messages into the bloodstream.*

Endocrine tissues that secrete chemical molecules are called endocrine glands. The thyroid gland on the previous page is one of the glands in the endocrine system. The chemical messages sent by the endocrine system are called hormones. *A* **hormone** *is a chemical that is produced by an endocrine gland in one part of an organism and is carried in the bloodstream to another part of the organism.*

The messages sent by the endocrine system are transmitted less rapidly than messages sent by the nervous system. The chemical messages usually are sent to more cells and last longer than the messages sent by the nervous system. For example, a message from the nervous system might cause a body movement. A message from the endocrine system might cause your body to grow taller over a period of time.

Key Concept Check What does the endocrine system do?

WORD ORIGIN

hormone
from Greek *hormon,* means "that which sets in motion"

FOLDABLES®

Use a sheet of notebook paper to make a vertical two-column chart. Label the outside *Endocrine Glands* and label the columns as shown. Use it to take notes about the endocrine glands and what their hormones do.

Endocrine Glands and Their Hormones

Hormones move from endocrine glands to other parts of the body in the bloodstream. Hormones affect the functions of organs and tissues by carrying messages to specific cells called target cells. Hormones recognize their target cells because the target cells have certain receptor proteins on or inside them. The hormone recognizes the receptor proteins and attaches to them. A hormone is like a key that wanders around and tries locks until it finds one that it fits and unlocks it. Once a hormone finds its target cell, it binds to a receptor protein and delivers its chemical message. The target cell responds by taking a specific action.

Figure 11 illustrates the path that a hormone takes from its production in an endocrine gland until it finds and delivers its message to the target cell. There are many different hormones with various purposes that are produced and distributed throughout your body. The endocrine glands and the hormones they produce are shown in **Figure 12.**

The Path of a Hormone

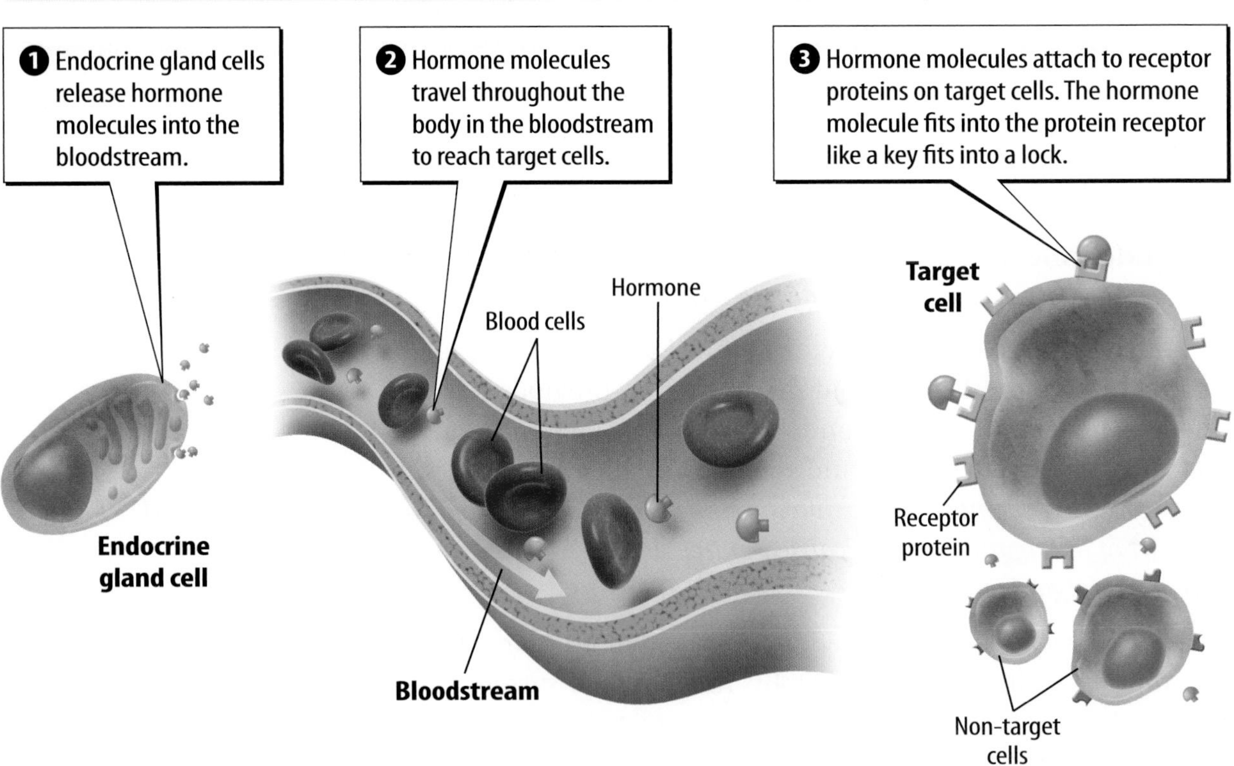

Figure 11 Hormones move through the bloodstream until they encounter the specific cells they are targeted to affect.

Visual Check Why doesn't the green hormone molecule attach to non-target cells?

Endocrine Glands and Hormones

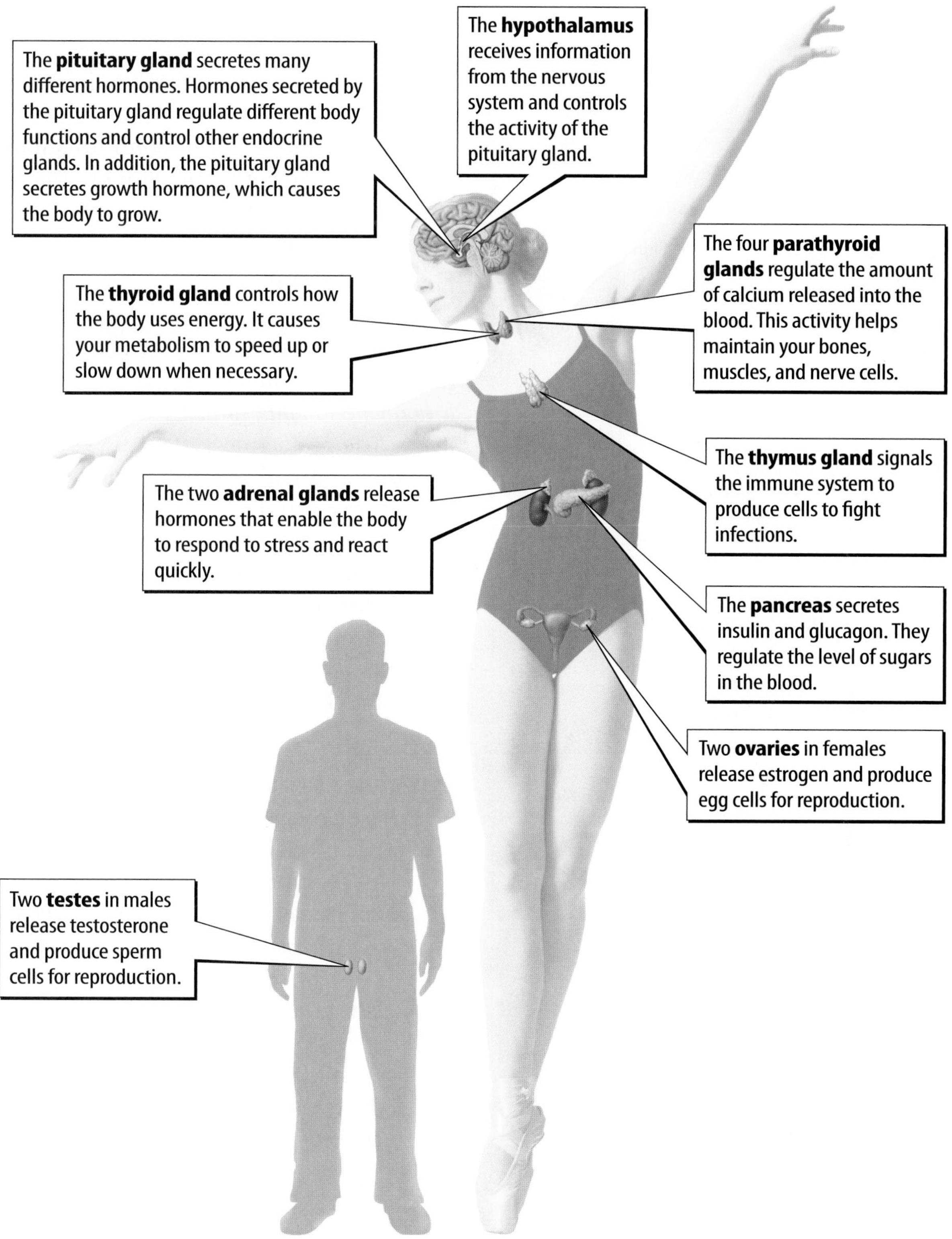

Figure 12 The endocrine system is made up of different glands that secrete hormones to maintain homeostasis in the body.

Visual Check Which endocrine gland affects metabolism?

The Endocrine System and Homeostasis

You might recall from Lesson 1 that your nervous system responds to changes in your environment and maintains homeostasis. Glands also help maintain homeostasis by releasing hormones in response to stimuli. Hormones change the function of other tissues and organs in the body that regulate internal conditions.

Negative Feedback Systems

Organisms can maintain homeostasis using negative feedback systems. **Negative feedback** *is a control system where the effect of a hormone inhibits further release of the hormone.* The endocrine system uses negative feedback and controls the amount of hormone that a gland releases. This is similar to the way a thermostat in a house monitors temperature. The thermostat signals the furnace to turn on when the temperature in the house drops below a preset temperature. After the house warms to the preset temperature, the thermostat signals the furnace to shut off.

Positive Feedback Systems

Whereas negative feedback helps maintain homeostasis, positive feedback does not. **Positive feedback** *is a control system in which the effect of a hormone causes more of the hormone to be released.* Because positive feedback does not help maintain homeostasis, your body uses fewer positive feedback systems. Childbirth is one example of a positive feedback system. Labor begins when a hormone called oxytocin is released. Oxytocin causes contractions. The contractions cause more oxytocin to be released. **Figure 13** shows how both negative feedback systems and positive feedback systems work within the endocrine system.

Key Concept Check How do other body systems work with the endocrine system to maintain homeostasis?

Feedback Systems

Negative Feedback System

Body is in homeostasis.

Signal to stop releasing hormone

Body uses hormone. More hormone is needed.

Signal to release hormone

Positive Feedback System

Hormone causes body to react and need more hormone.

Signal to release hormone

Hormone causes body to react and need more hormone.

Signal to release hormone

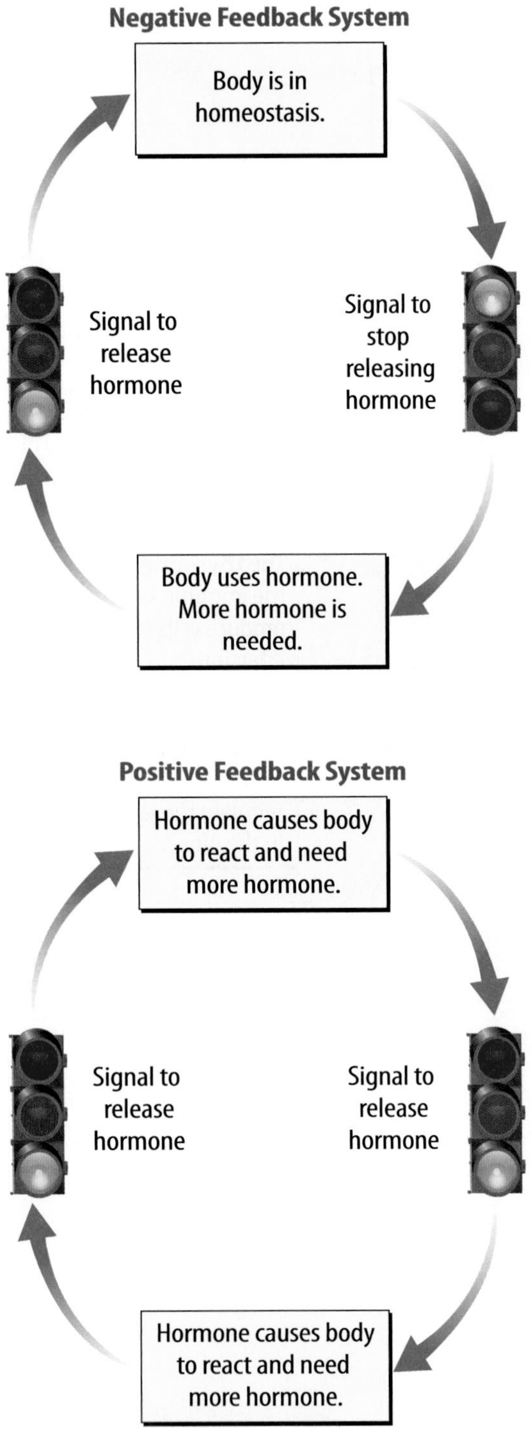

Figure 13 Negative feedback systems are used to maintain homeostasis while positive feedback systems increase the production of hormones.

Visual Check How do positive and negative feedback systems work?

Lesson 3 Review

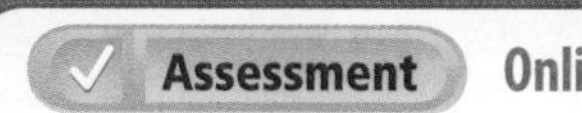

Visual Summary

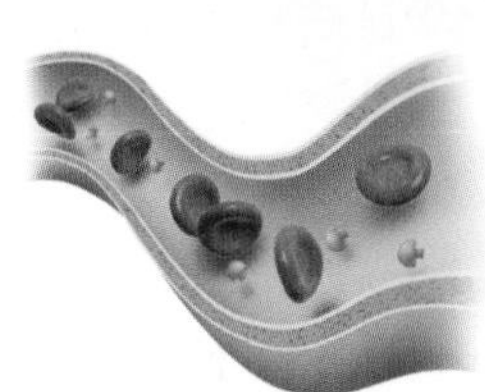

The endocrine system is made up of glands that secrete chemical hormones. They send messages throughout the body and maintain homeostasis.

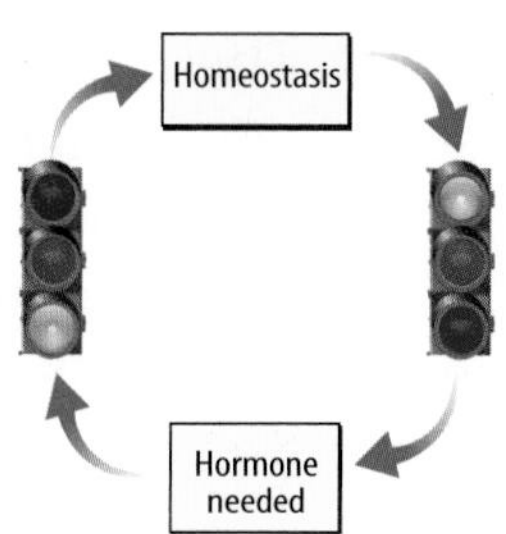

Negative feedback systems maintain homeostasis by inhibiting the release of a hormone as part of the effect of that hormone.

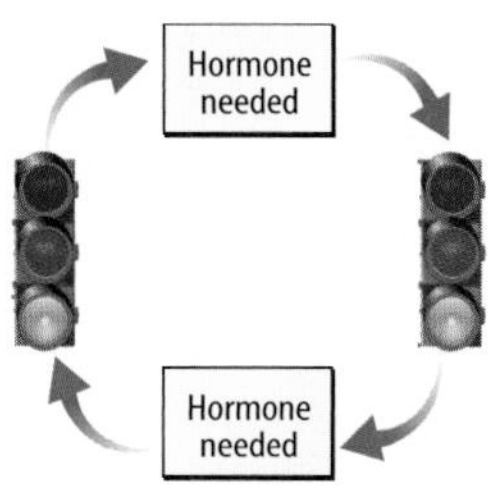

Positive feedback systems increase the effects caused by a hormone by signaling more of that hormone to be released.

FOLDABLES®

Use your lesson Foldable to review the lesson. Save your Foldable for the project at the end of the chapter.

What do you think NOW?

You first read the statements below at the beginning of the chapter.

5. Positive feedback systems in humans help maintain homeostasis.

6. Endocrine glands secrete hormones.

Did you change your mind about whether you agree or disagree with the statements? Rewrite any false statements to make them true.

Use Vocabulary

1. **Define** the phrase *endocrine system.*
2. **Distinguish** between negative feedback and positive feedback.
3. **Describe** the function of a hormone using your own words.

Understand Key Concepts 

4. Hormones travel to other parts of the body using the ________ system.
   A. circulatory **C.** nervous
   B. digestive **D.** skeletal
5. **Explain** the relationship between the nervous system and the endocrine system.
6. **Relate** negative feedback systems to maintaining homeostasis.

Interpret Graphics

7. **Describe** the function of the endocrine gland shown below.

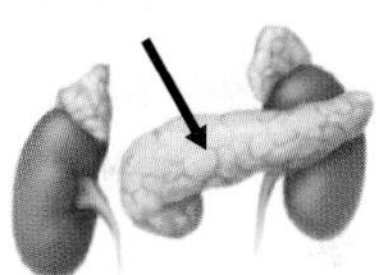

8. **Explain** Copy the graphic organizer below. Fill in information about the role of oxytocin in the positive feedback system of childbirth.

Critical Thinking

9. **Predict** what might happen if negative feedback did not function correctly in one of the endocrine glands.
10. **Explain** why doctors often check the activity of the thyroid glands of a person who is feeling tired and sluggish frequently.

Modeling a Negative Feedback Cycle

Materials

creative construction materials

Safety

The activity of cells and chemicals in your body is important. However, cells and molecules are so small that it is helpful to use models to simulate their activities. Cells use the process of negative feedback to maintain homeostasis. This happens when hormones are released and deliver a chemical message to specific target cells. The message directs the cells to take a specific action to cause an effect in the body. It also signals the other cells to stop producing the hormone. You will use materials to model the cells, hormones, and receptor sites that make up the pathway in a negative feedback cycle. Your model should show how some hormones use receptor sites to identify target cells and avoid the wrong cells.

Question

How do specific hormones find and enter the proper cells to cause an effect?

Procedure

1. Read and complete a lab safety form.
2. Review the diagram of the negative feedback cycle and the list of hormones.
3. Observe the working model your teacher demonstrates for the class. How do the hormone and the cell interact to ensure that the hormone is delivered to the proper cell to cause the necessary effect?

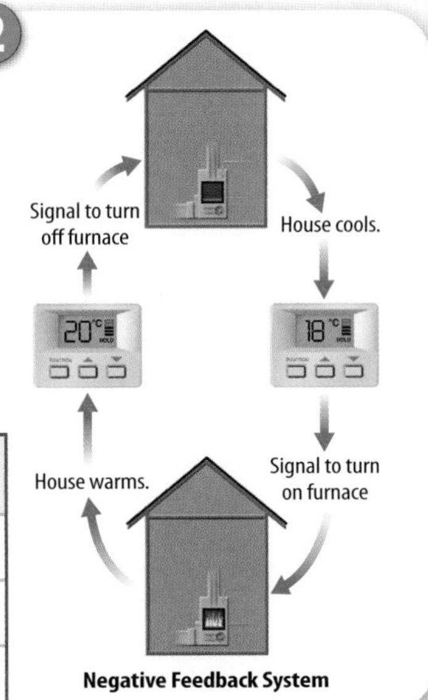

Negative Feedback System

Endocrine gland	Hormone
Adrenal glands	Adrenaline
Hypothalamus	Oxytocin
Ovaries	Estrogen
Pancreas	Insulin
Parathyroid glands	Parathyroid hormone
Pituitary gland	Growth hormone
Testes	Testosterone
Thyroid gland	Thyroxine

4. Form small groups. Your teacher will hand out an assortment of materials to each group.
5. Choose one of the hormone examples and create a working model that demonstrates the pathway that the hormone and the target cells undergo as part of negative feedback.
6. Label the parts of your model with the following tags: *hormone, hormone receptor, target cell*.

Analyze and Conclude

7. **Think Critically** How does your model demonstrate the delivery of hormones to the proper cells in order to cause an effect?
8. **The Big Idea** How does the activity you modeled interact with other systems in the body to maintain homeostasis?

Communicate Your Results

Draw a diagram of your model on the chalkboard in the form of a negative feedback cycle. Then demonstrate your model, showing how it follows the process of the negative feedback pathway.

Inquiry Extension

Imagine that the different models built in your class all make up different hormones and cells in the endocrine system. What will happen if you attempt to use the hormone model from another group to activate your cell? Take turns trying the hormone models from other groups with your cell model. Describe the results and what they mean.

Lab Tips

- ☑ Remember that your model is meant to represent the functions of the negative feedback system. It is not meant to resemble the appearance of the things it represents.
- ☑ The putty can be used to attach some of the parts together.

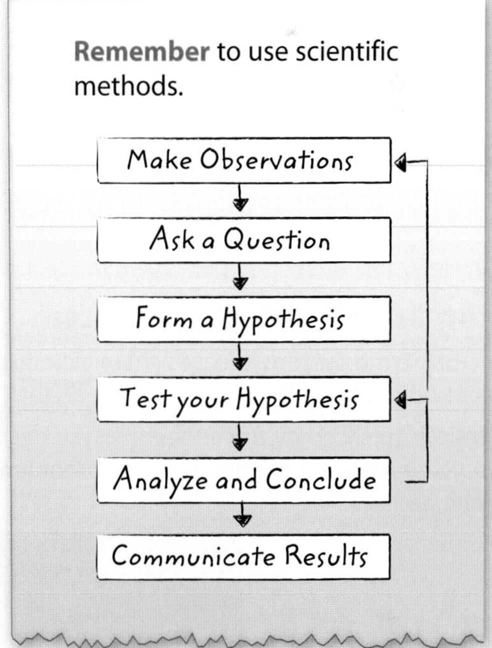

Chapter 18 Study Guide

The nervous and endocrine systems work to sense the environment and coordinate body functions.

Key Concepts Summary

Lesson 1: The Nervous System

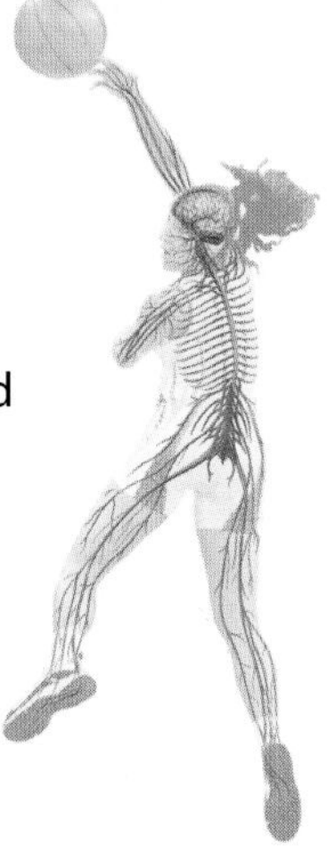

- The **nervous system** gathers information from the environment, processes it, and signals the body to respond.
- The nervous system is made up of the **central nervous system** and **peripheral nervous system.** The peripheral nervous system gathers and transmits information to and from the central nervous system, which processes the information.
- The nervous system helps other body systems maintain homeostasis by responding to the environment.

Vocabulary

nervous system p. 637
stimulus p. 638
neuron p. 639
synapse p. 639
central nervous system p. 640
cerebrum p. 640
cerebellum p. 640
brain stem p. 641
spinal cord p. 641
peripheral nervous system p. 641
reflex p. 642

Lesson 2: The Senses

- The **sensory system** provides information about the environment.
- The senses help maintain homeostasis by gathering information about changes in the environment so the body can respond.

Vocabulary

sensory system p. 647
receptor p. 647
retina p. 648
eardrum p. 650

Lesson 3: The Endocrine System

- The **endocrine system** releases chemical messages that control or affect body functions.
- Chemicals released by the endocrine system cause other body systems to react to changes in the environment to maintain homeostasis.

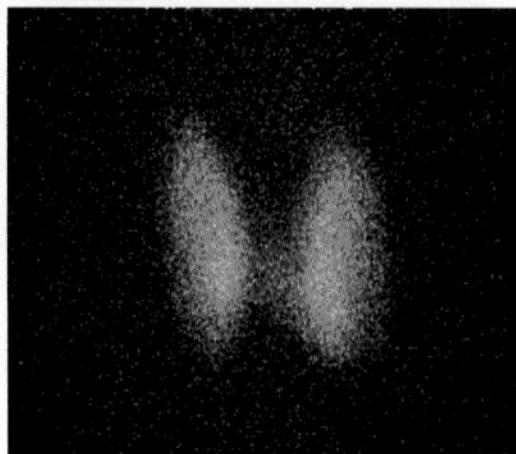

Vocabulary

endocrine system p. 657
hormone p. 657
negative feedback p. 660
positive feedback p. 660

Study Guide

- Personal Tutor
- Vocabulary eGames
- Vocabulary eFlashcards

FOLDABLES® Chapter Project

Assemble your lesson Foldables as shown to make a Chapter Project. Use the project to review what you have learned in this chapter.

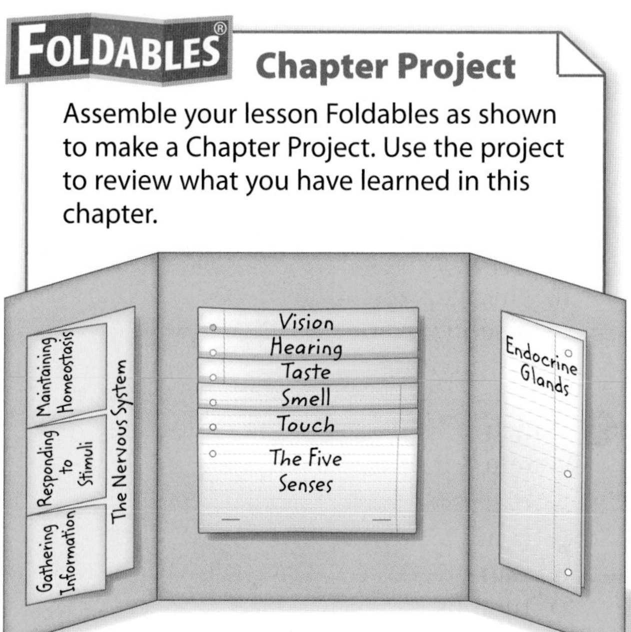

Use Vocabulary

1. The ________ nervous system is made up of the somatic and autonomic systems.
2. Voluntary muscle movement and coordination is managed by the ________ part of the brain.
3. The ________ in the eye contains photoreceptors.
4. The fluid-filled structure in the inner ear that helps maintain balance is the ________.
5. Chemicals produced by endocrine glands are called ________.
6. The effect of a hormone on a ________ feedback system is to cause it to release more of the hormone.

Link Vocabulary and Key Concepts

Copy this concept map, and then use vocabulary terms from the previous page to complete the concept map.

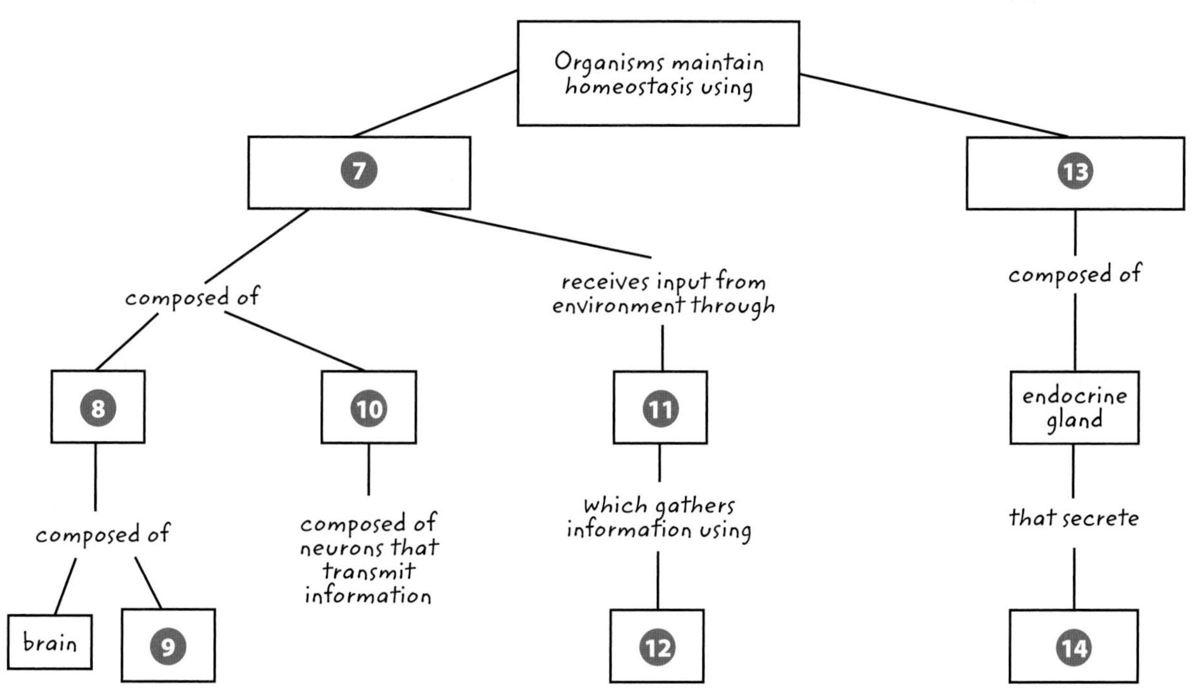

Chapter 18 Review

Understand Key Concepts 

1. Which part of the brain controls language, memory, and thought?
   A. brain stem
   B. cerebellum
   C. cerebrum
   D. spinal cord

2. Which is NOT a part of the CNS?
   A. brain stem
   B. cerebellum
   C. somatic system
   D. spinal cord

3. What is the gap between the two neurons shown below?

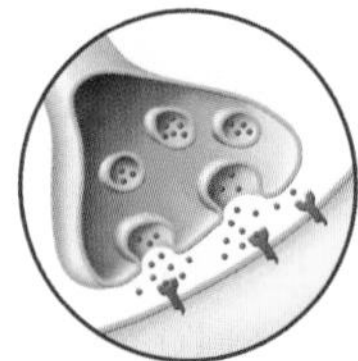

   A. motor
   B. reflex
   C. stimulus
   D. synapse

4. Which is NOT a part of the eye?
   A. cochlea
   B. cornea
   C. pupil
   D. retina

5. Which senses work together and determine how a person perceives flavor?
   A. smell and taste
   B. taste and hearing
   C. taste and touch
   D. vision and taste

6. Which part of the eye focuses light?
   A. iris
   B. lens
   C. pupil
   D. retina

7. Which system transports hormones to other parts of the body?
   A. circulatory system
   B. digestive system
   C. nervous system
   D. reproductive system

8. Which hormone do ovaries release?
   A. estrogen
   B. glucagon
   C. insulin
   D. testosterone

9. Which is NOT part of the endocrine system?
   A. pancreas
   B. ovary
   C. skin
   D. thymus

10. Which system is illustrated by the diagram below?

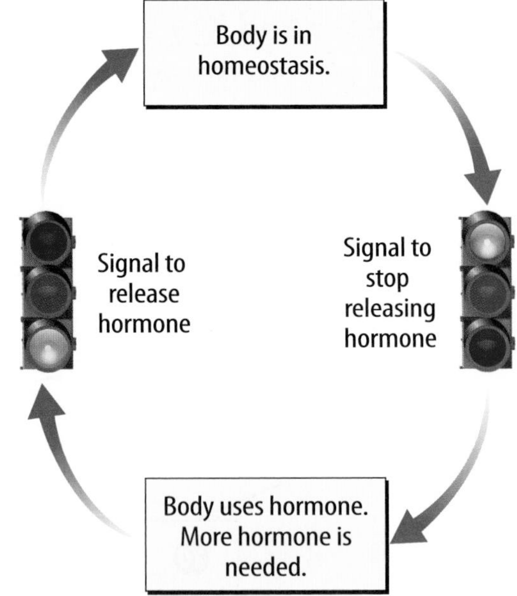

   A. autonomic system
   B. reflex system
   C. negative feedback system
   D. positive feedback system

Assessment
Online Test Practice

Critical Thinking

11 **Describe** how reflexes prevent injuries.

12 **Compare and contrast** the somatic nervous system and the autonomic nervous system.

13 **Investigate** the effect of a spinal cord injury on the function of the nervous system.

14 **Relate** how receptors in the nose take part in sensing food.

15 **Determine** Is the vision problem shown in the figure below nearsightedness or farsightedness? Explain your answer.

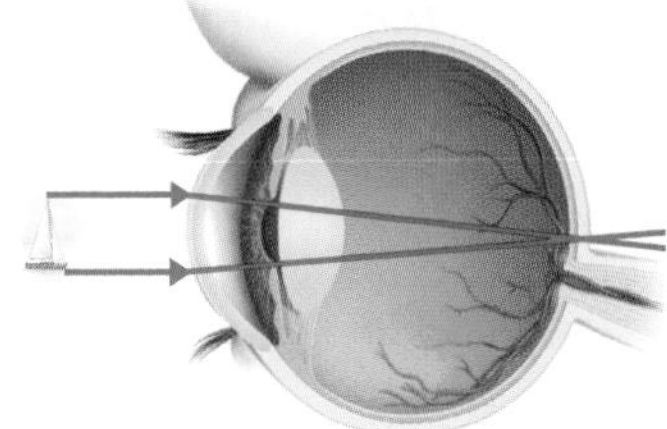

16 **Infer** why people sometimes feel sick when traveling on a boat.

17 **Evaluate** how negative feedback systems control the amount of hormone released by a gland.

18 **Infer** why positive feedback systems are not used to maintain homeostasis.

19 **Assess** why reflexes are controlled by the nervous system and not the endocrine system.

20 **Evaluate** how the endocrine system works with the circulatory system to maintain the body's homeostasis.

Writing in Science

21 **Write** a paragraph that describes a time when your nervous system helped you to survive. Describe what likely occurred in your body, including the use of your senses and the endocrine system during the event.

REVIEW THE BIG IDEA

22 **Describe** the relationship between the nervous system and the endocrine system in the body. Include some examples of how they maintain homeostasis.

23 **Explain** what is most likely happening in the nervous system and the endocrine system of the batter shown below. Describe how these activities help her accomplish her goal of hitting the ball.

Math Skills

Review
Math Practice

Use Proportions

Use the information below to answer questions 24 and 25.

A person's nerve axon from a toe to the spine is 1.1 m long. The nerve impulse that signals touch travels about 76.2 m/s.

24 How long does it take the nerve impulse for touch to travel from the toe to the spine?

25 Pain impulses travel more slowly—about 0.61 m/s. How long would it take the pain impulse to travel from the toe to the spine?

26 A reflex nerve impulse travels 100 m/s from a fingertip touching a hot object to the spine. The distance is 0.5 m. If a signal from the spine telling the finger to move travels back at the same speed, how long will it take the person to react?

Standardized Test Practice

Record your answers on the answer sheet provided by your teacher or on a sheet of paper.

Multiple Choice

Use the diagram below to answer question 1.

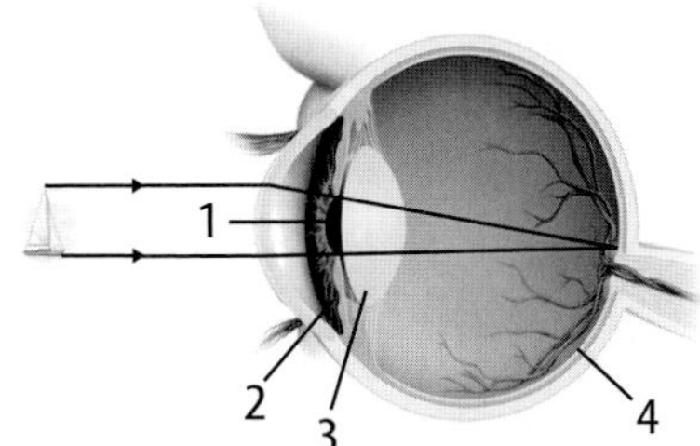

1 Which numbered structure focuses the light on the retina?

A 1

B 2

C 3

D 4

2 A doctor sees a patient who has a loss of balance from an illness. The doctor thinks the receptors for balance might be injured. Where are they located?

A inner ear

B middle ear

C nasal cavity

D spinal cord

3 Which statement about the nervous system is false?

A It gathers information.

B It maintains homeostasis.

C It responds to stimuli.

D It transports hormones.

4 Which relies on chemoreceptors?

A hearing

B sight

C taste

D touch

Use the graph below to answer question 5.

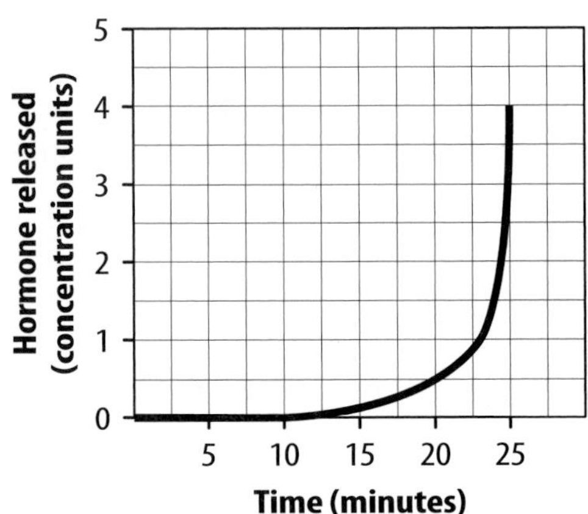

5 Which is likely to produce a curve similar to the one above?

A a reflex

B a synapse

C negative feedback

D positive feedback

6 Which system releases chemical messages to the bloodstream?

A circulatory

B digestive

C endocrine

D skeletal

7 Why does the nervous system tell the body to shiver?

A to imprint a memory on the brain

B to maintain homeostasis

C to slow neuron communication

D to stop PNS–CNS communication

Use the diagram below to answer question 8.

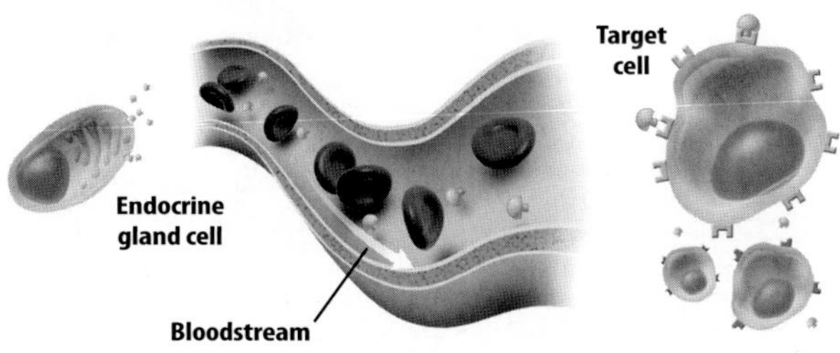

8 What is the path of a hormone as illustrated above?

A bloodstream → endocrine cell → target cell

B endocrine cell → bloodstream → target cell

C endocrine cell → target cell → bloodstream

D target cell → bloodstream → endocrine cell

9 What is the primary function of the senses?

A choosing pleasant scents, sounds, tastes, and textures

B gathering information about the environment

C making each organism slightly different from others

D providing ideas for organisms to communicate

Constructed Response

Use the table below to answer questions 10 and 11.

Part	Description
Neuron	
Synapse	
CNS	
PNS	

10 Briefly describe the parts of the nervous system listed in the table. Explain what the abbreviations *CNS* and *PNS* represent.

11 Using an example, such as a pain reflex, explain how the parts of the nervous system provided in the table above work together.

Use the table below to answer question 12.

Sense	Organ and Function

12 List the five human senses. Then identify each sensory organ and describe its function.

NEED EXTRA HELP?												
If You Missed Question...	1	2	3	4	5	6	7	8	9	10	11	12
Go to Lesson...	2	2	1	2	3	3	1	3	2	1	1	2

Chapter 19

Reproduction and Development

THE BIG IDEA

What are the stages of human reproduction and development?

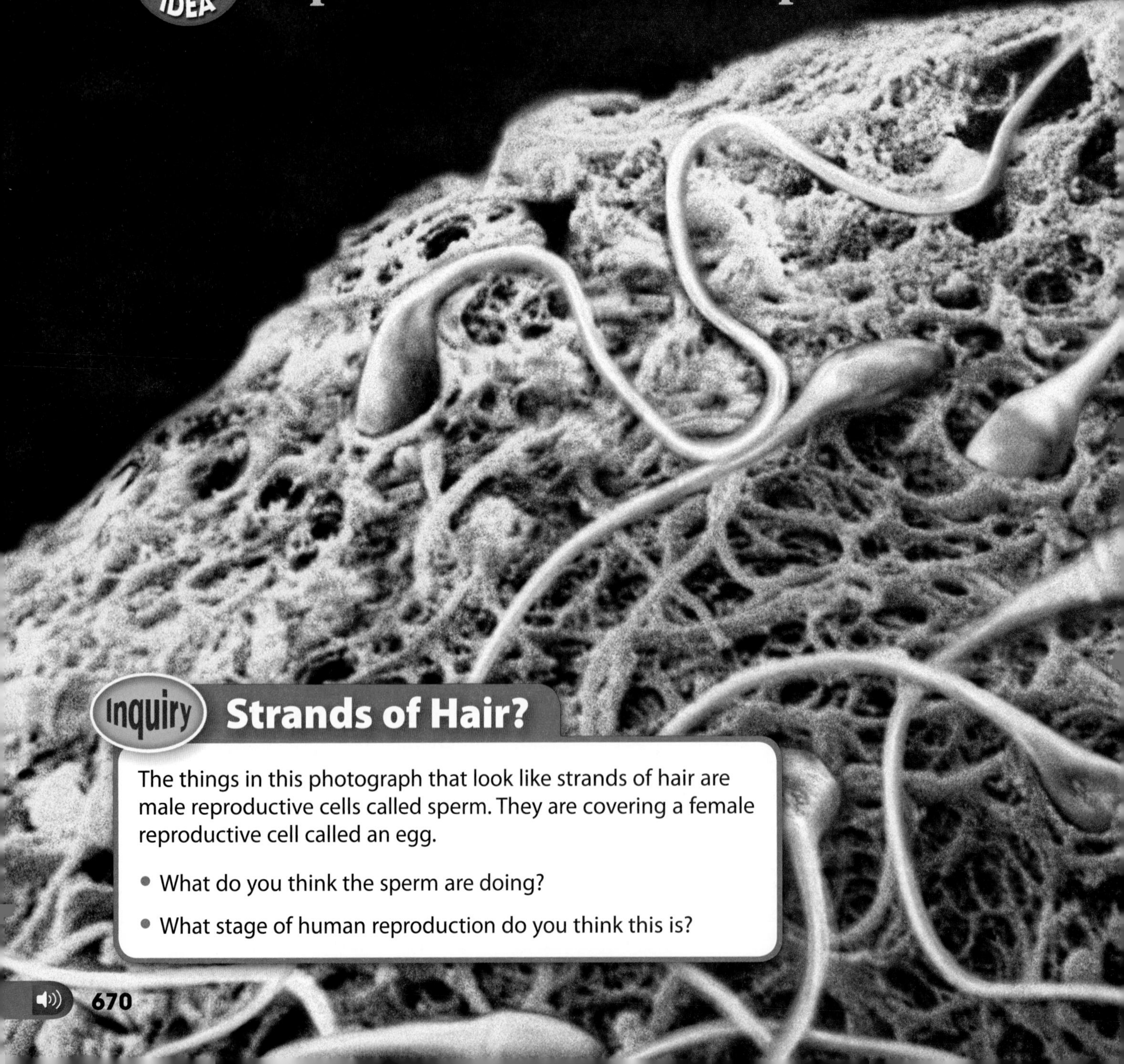

Inquiry Strands of Hair?

The things in this photograph that look like strands of hair are male reproductive cells called sperm. They are covering a female reproductive cell called an egg.

- What do you think the sperm are doing?
- What stage of human reproduction do you think this is?

Get Ready to Read

What do you think?

Before you read, decide if you agree or disagree with each of these statements. As you read this chapter, see if you change your mind about any of the statements.

1. Reproduction ensures that a species survives.
2. The male reproductive system has internal and external parts.
3. The menstrual cycle occurs in males and females.
4. Eggs are fertilized in the ovary.
5. Lead is a nutrient that helps a fetus develop.
6. Puberty occurs during adolescence.

Lesson 1

Reading Guide

Key Concepts
ESSENTIAL QUESTIONS

- What does the reproductive system do?
- How do the parts of the male reproductive system work together?
- How do the parts of the female reproductive system work together?
- How does the reproductive system interact with other body systems?

Vocabulary

sperm p. 673
egg p. 673
testis p. 674
semen p. 674
penis p. 674
ovary p. 676
vagina p. 676
menstrual cycle p. 678
ovulation p. 678
fertilization p. 679

Multilingual eGlossary

The Reproductive System

Inquiry A Red Ball?

The round object is a human egg being released from the ovary. Are eggs released often? What is shown here is part of the reproductive process in the human female. In this lesson, you will learn about the process.

Launch Lab

15 minutes

How do male and female gametes compare?

At some point in their lives, except in rare instances, humans will all make sex cells, or gametes. Males produce sperm, and females produce eggs. When a sperm cell and an egg cell unite, fertilization occurs.

1. Read and complete a lab safety form.
2. Obtain a **microscope** and **prepared slides of sperm cells and egg cells.** Carefully handle the microscope according to your teacher's instructions.
3. Observe the slides under the magnification power specified by your teacher.
4. Sketch a sperm cell and an egg cell in your Science Journal.

Think About This

1. Compare and contrast the appearance of an egg cell and the appearance of a sperm cell.
2.  **Key Concept** Why do you think there is a difference in the appearance of the male and female reproductive cells?

Functions of the Reproductive System

You have read about many of the organ systems that enable you to grow and respond to changes in the environment. Organ systems such as the nervous system, the skeletal system, the circulatory system, and the digestive system are all important for an individual's survival. But what organ system ensures that the human species survives?

A reproductive system is a group of tissues and organs. It enables the male and female reproductive cells to join and form new offspring. Like other animals, human males produce **sperm,** *the male reproductive cells,* and human females produce **eggs,** *the female reproductive cells.*

During reproduction, a sperm joins with an egg, as shown in **Figure 1.** This usually happens inside a female's reproductive system. Once joined, part of the female's reproductive system nourishes the developing human. In order to understand how humans develop, we must first learn the parts of the male and female reproductive systems.

 **Key Concept Check** What does the reproductive system do?

Figure 1 When a sperm cell combines with an egg cell, development of a new human being begins.

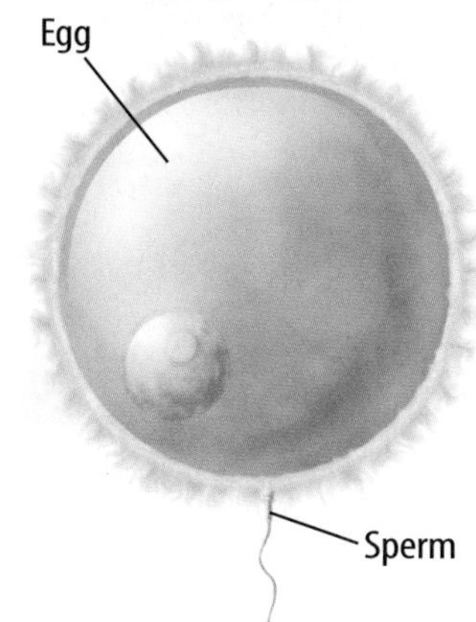

The Male Reproductive System

The main function of the male reproductive system is to produce and transport sperm to the female reproductive system. As shown in **Figure 2,** there are many parts of the male reproductive system, each with a unique function. *The* **testis** (TES tihs; plural, testes) *is the male reproductive organ that produces sperm.* A male's two testes are inside an external saclike **structure** called the scrotum (SKROH tum). Sperm development can occur only at a temperature that is lower than normal body temperature. Because the scrotum is outside the male's body, it is at a temperature slightly lower than normal body temperature.

Science Use v. Common Use
structure
Science Use cells and tissues arranged in a definite pattern
Common Use a building

Once sperm develop, they move to a tube called the sperm duct and are stored. During storage, sperm mature and develop the ability to swim. As mature sperm move through the remainder of the male reproductive system, they mix with fluids produced by several glands. *This mixture of sperm and fluids is called* **semen** (SEE mun). Semen also contains nutrients that provide sperm with energy. Semen leaves the body through the penis. *The* **penis** *is a tubelike structure that delivers sperm to the female reproductive system.* The path of sperm is shown in **Figure 3.**

Reading Check What does semen contain?

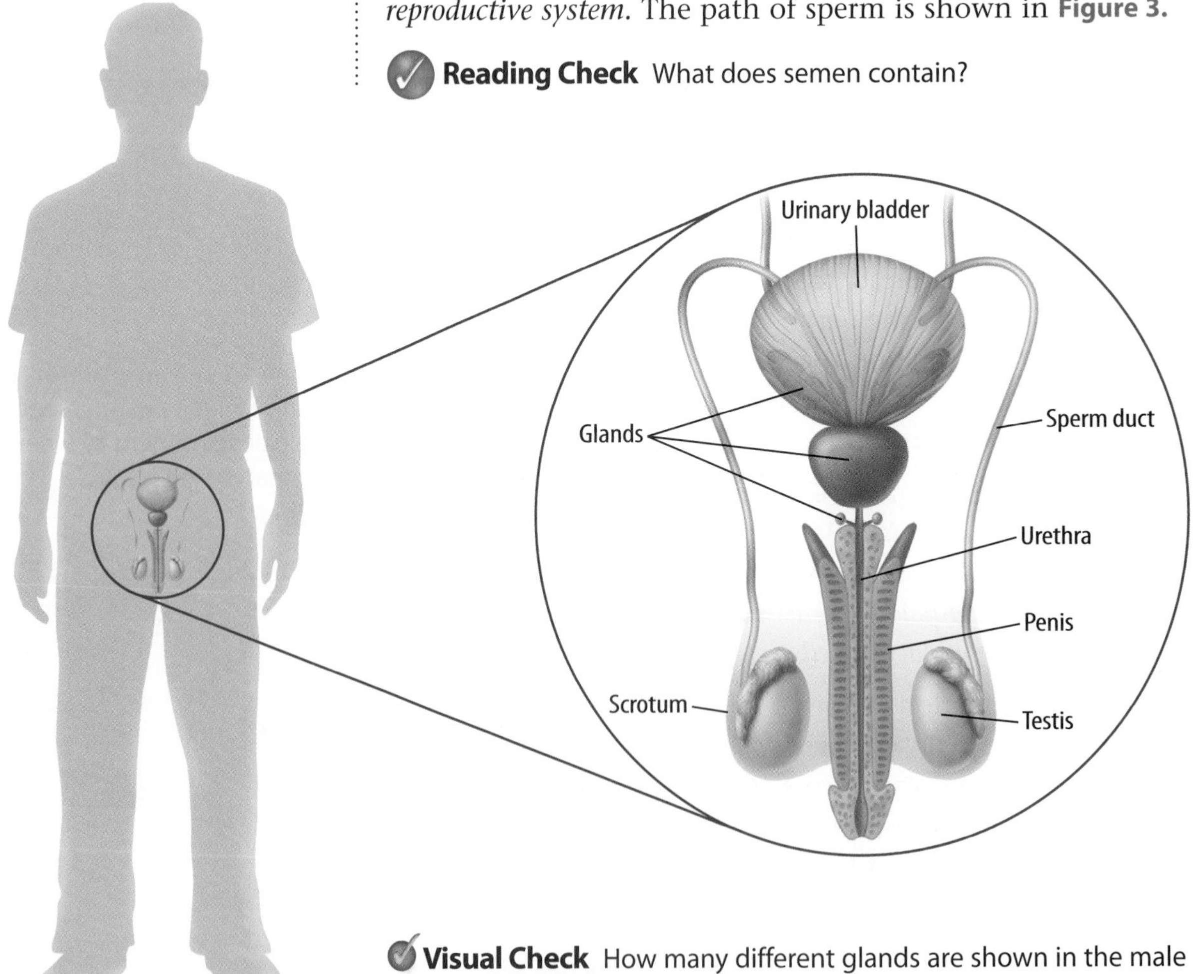

Figure 2 The male reproductive system has many parts, each with a unique function.

Visual Check How many different glands are shown in the male reproductive system?

Sperm

Now that you have learned about the parts of the male reproductive system, let's take a closer look at sperm. When mature, sperm can join with an egg. As illustrated in **Figure 3,** a mature sperm cell has three main parts—a head, a midpiece, and a tail. The head contains DNA and substances that help the sperm join with an egg. The midpiece contains organelles called mitochondria. Recall that mitochondria are the cell organelles that process food molecules and release energy. This energy enables movement of a sperm's tail. The tail is a long, slender structure that whips back and forth and moves the sperm. Although semen contains millions of sperm, only one sperm joins with an egg.

Key Concept Check How do the parts of the male reproductive system work together?

Male Reproductive System

Figure 3 Mature sperm leave the body through the penis.

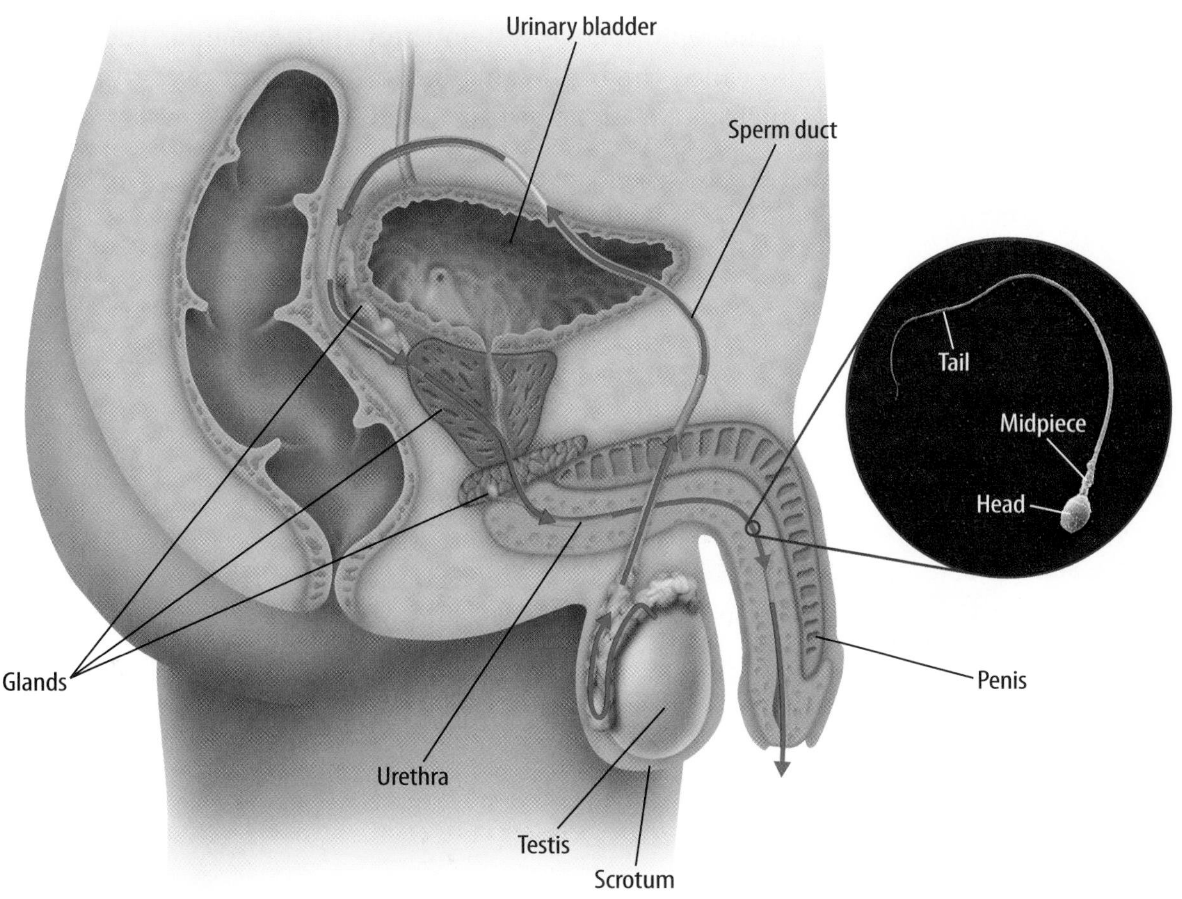

The Female Reproductive System

As shown in **Figure 4,** the female reproductive system also has many structures, each with a unique function. Recall from **Figure 2** that the male reproductive system has both internal and external parts. In contrast, all the parts of the female reproductive system are inside the body.

Recall that both the male and female reproductive systems produce reproductive cells. Males produce sperm, and females produce immature eggs called oocytes (OH uh sites). *An* **ovary** (OH vah ree) *is an organ where oocytes are stored and mature.* A mature oocyte is called an egg, or ovum (plural, ova). Just as males have two testes, females have two ovaries.

Word Origin
ovary
from Latin *ovum,* means "egg"

About once a month, an ovary releases an egg. After an egg is released, it enters the fallopian (fuh LOH pee un) tube. Short, hairlike structures called cilia move the egg through the fallopian tube toward the uterus. If the egg is fertilized by a sperm, the uterus provides a nourishing environment for a fertilized egg's development. You will read more about how humans develop in Lesson 2.

The part of the female reproductive system that connects the uterus to the outside of the body is the **vagina.** Sperm enter the female reproductive system through the vagina. The vagina is also called the birth canal because a baby moves through this structure during its birth.

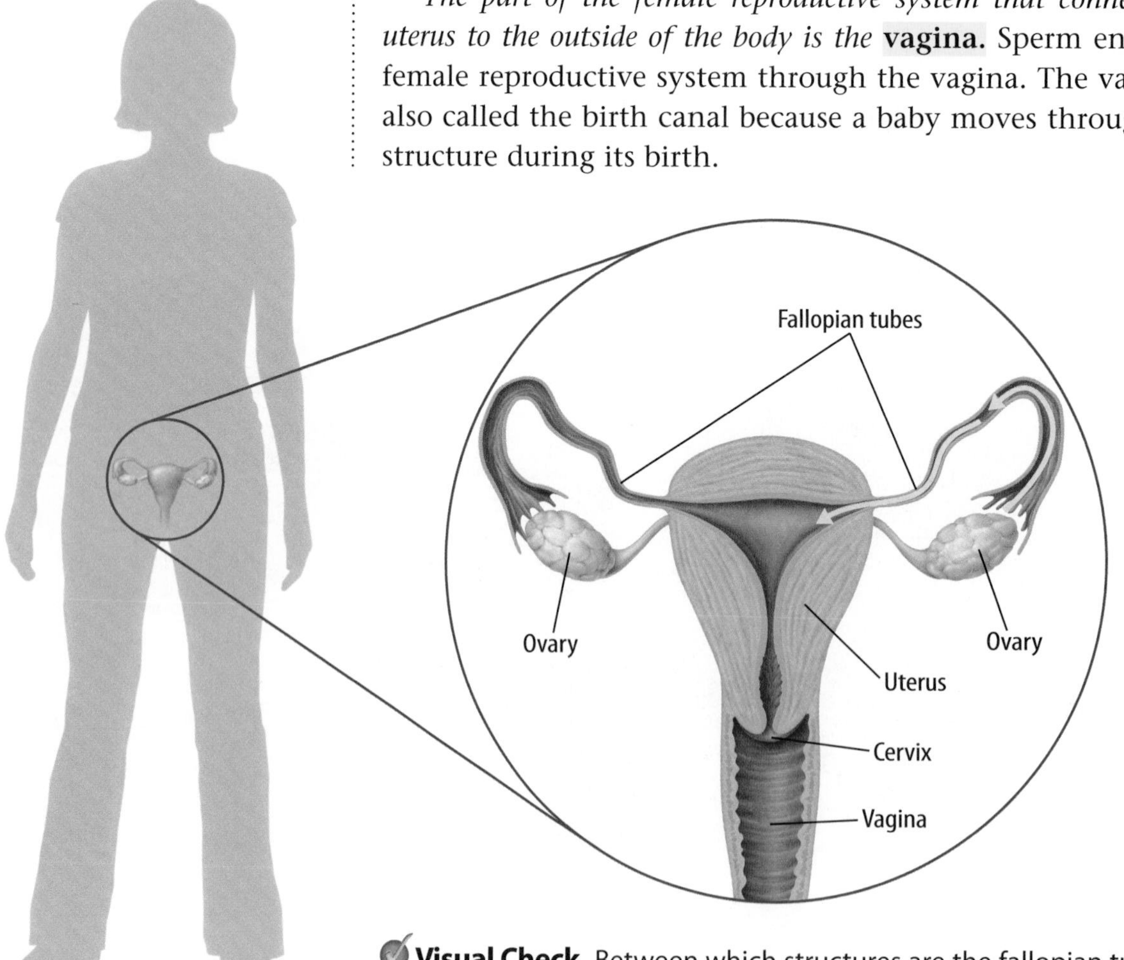

Figure 4 The female reproductive system has many parts, each with a unique function.

Visual Check Between which structures are the fallopian tubes located?

The Egg

Unlike sperm, which are long and slender when mature, eggs are large and round when mature. In fact, an egg is about 2,000 times larger than a sperm. Like sperm, eggs contain DNA. However, unlike sperm, an egg is filled with substances that provide it with nourishment. Another important difference between males and females is that a male releases millions of sperm in semen, but a female usually releases only one egg at a time. As shown in **Figure 5,** each oocyte in an ovary is surrounded by follicle cells. The follicle cells release hormones that help the oocytes develop into eggs. The changes that occur as oocytes develop and the release of an egg are also shown in **Figure 5.**

 **Reading Check** How do mature sperm and eggs differ?

Female Reproductive System

Figure 5 A follicle increases in size as it prepares to release an egg.

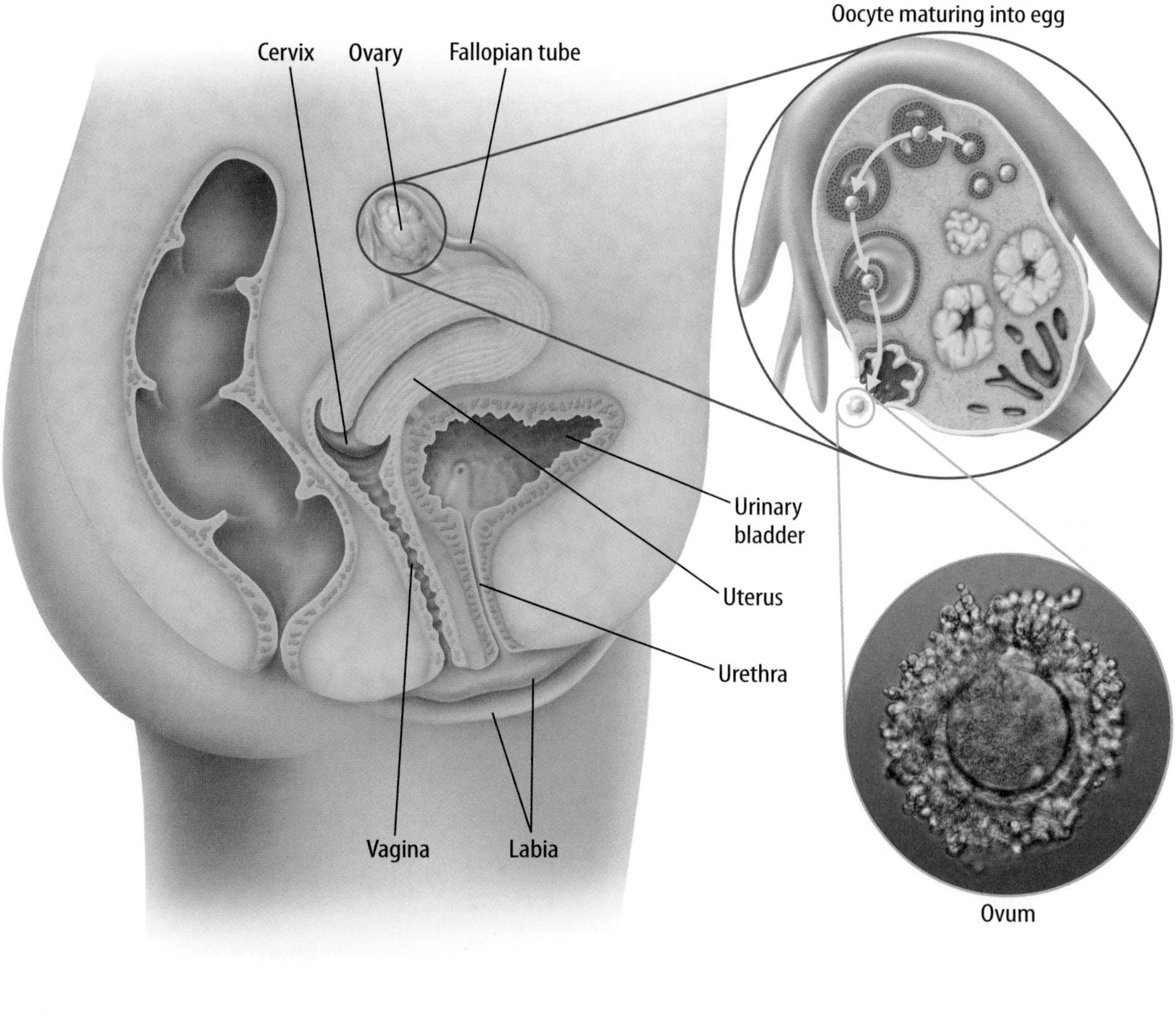

The Menstrual Cycle

A female usually releases one egg at a time. Because the function of the reproductive system is to produce a new human, an egg is released only when the uterus is prepared to nourish it. *The ovaries and the uterus go through reproductive-related changes called the* **menstrual** (MEN stroo ul) **cycle.** The menstrual cycle is caused by chemical signals called hormones. One menstrual cycle is about 28 days long and can be divided into three phases. It is called a cycle because the phases repeat in the same order and in about the same amount of time.

FOLDABLES®

Make a three-tab book from a sheet of notebook paper. Label it as shown. Use it to organize information about the phases of the menstrual cycle.

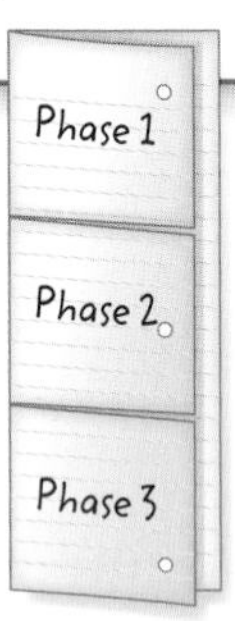

Phase 1 The menstrual cycle begins with a process called menstruation (men stroo WAY shun). During menstruation, tissue, fluid, and blood cells pass from the uterus through the vagina and are removed from the body. Menstruation usually lasts about five days.

Phase 2 In the next phase of the menstrual cycle, the tissue lining the uterus thickens, as shown in **Figure 6.** In the ovary, several oocytes begin maturing at the same time. After about a week, usually only one egg survives. *Near the end of this phase, hormones cause an egg to be released from the ovary in a process called* **ovulation.**

Figure 6 The tissue lining the uterus changes in thickness during the menstrual cycle.

Concepts in Motion Animation

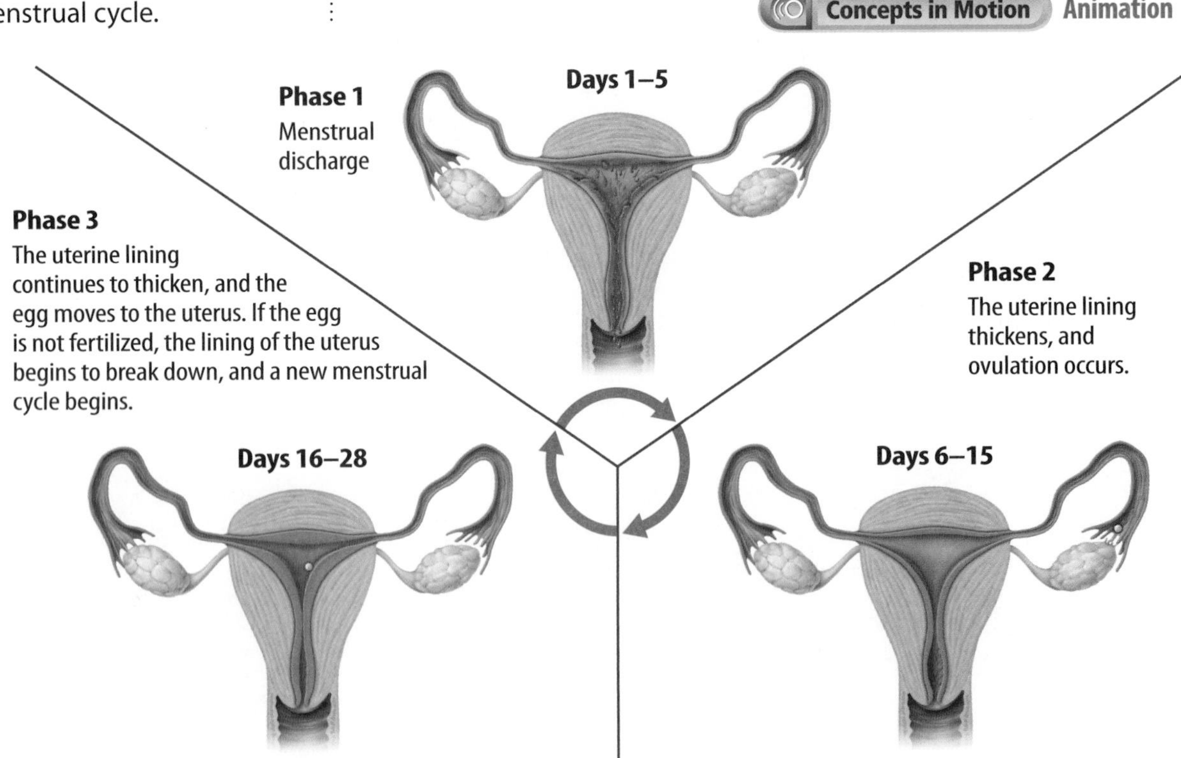

Visual Check Which are the days of menstrual flow?

Phase 3 During phase 3, the tissue that lines the uterus continues to thicken. *If sperm are present, the egg might join with a sperm in a process called* **fertilization.**

The tissue lining the uterus is called the endometrium (en doh MEE tree um). The endometrium provides a fertilized egg with nutrients and oxygen during its early development. If fertilization does not occur, the endometrium breaks down and the menstrual cycle repeats itself.

Key Concept Check How do the parts of the female reproductive system work together?

Menopause When females get older, the reproductive system stops releasing eggs. When this happens, a woman reaches menopause—a time when the menstrual cycle stops. Menopause occurs because a woman's ovaries produce fewer hormones. There are not enough hormones to cause oocyte maturation and ovulation. Menopause usually happens between the ages of 45 and 55.

The Reproductive System and Homeostasis

Although the reproductive system does not help the body maintain homeostasis, reproduction is essential to ensure the survival of the human species. The reproductive systems of both males and females are controlled by the endocrine system.

Recall that hormones produced by the endocrine system cause the menstrual cycle. These hormones act by positive and negative feedback mechanisms and control when oocytes mature. Hormones also control sperm maturation in the male reproductive system. The reproductive system and the endocrine system work together and control when sperm and eggs mature.

Key Concept Check How do body systems interact with the reproductive system?

Inquiry **MiniLab** 25 minutes

What occurs in menstrual cycle phases?

Because the function of the reproductive system is to produce a new human, an ovary releases an egg only when the uterus is prepared to nourish it.

1. Divide a sheet of **paper** into three columns. Label the three columns *Phase 1, Phase 2,* and *Phase 3.*

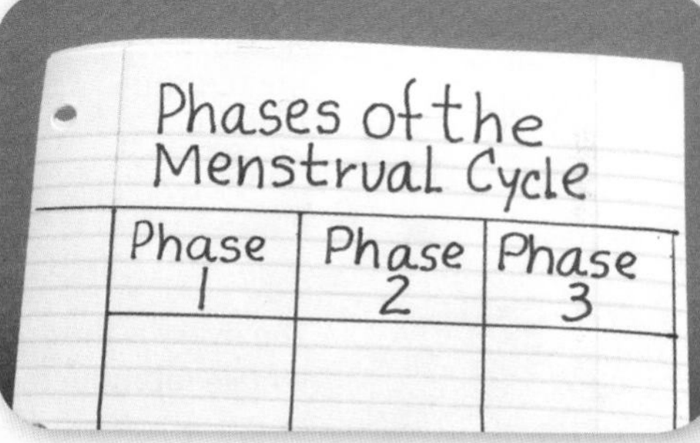

2. Begin the activity at the lab station assigned by your teacher. Read the description of one phase of the menstrual cycle at that lab station.
3. Decide whether the description is a part of the first, second, or third phase of the menstrual cycle. Use your textbook and your lesson Foldable if necessary to help you.
4. Write the lab station information in the appropriate column on your paper.

Analyze and Conclude

1. **Summarize** what happens in the first, second, and third phases of the menstrual cycle in your Science Journal.
2. **Compare and contrast** what happens in the menstrual cycle when the egg is fertilized versus when the egg is not fertilized.
3. **Key Concept** How do the parts of the female reproductive system work together so a baby can develop?

Lesson 1 Review

Assessment Online Quiz

Visual Summary

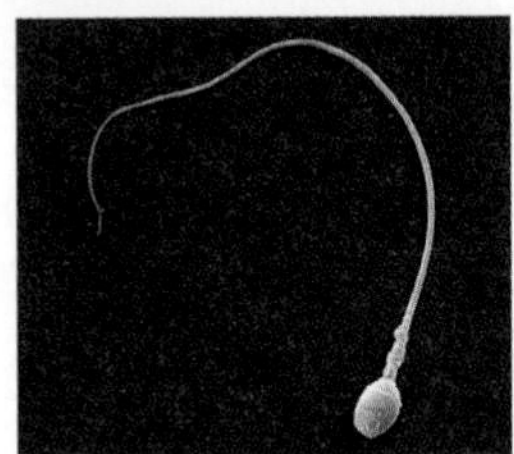

Sperm are male reproductive cells.

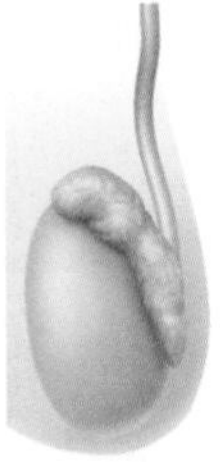

The testes produce sperm and are inside the scrotum.

The ovary is the female reproductive organ where oocytes are stored and mature into eggs.

FOLDABLES®

Use your lesson Foldable to review the lesson. Save your Foldable for the project at the end of the chapter.

What do you think NOW?

You first read the statements below at the beginning of the chapter.

1. Reproduction ensures that a species survives.
2. The male reproductive system has internal and external parts.
3. The menstrual cycle occurs in males and females.

Did you change your mind about whether you agree or disagree with the statements? Rewrite any false statements to make them true.

Use Vocabulary

1. **Use the terms** *testes* and *sperm* in a sentence.
2. **Distinguish** between ovulation and menstruation.
3. Cyclic changes in the uterus and ovaries of the female reproductive system are called the ________ ________.

Understand Key Concepts 

4. Which is a mixture of sperm, nutrients, and fluids?
   A. oocyte
   B. ovum
   C. penis
   D. semen
5. **Explain** why the scrotum is outside of the body.
6. **Compare** the structure of an egg to the structure of a sperm.

Interpret Graphics

7. **Summarize** Copy and fill in the table below to list the parts and functions of the male reproductive system.

Part	Function

8. **Identify** the phase of the menstrual cycle shown below. Explain your choice.

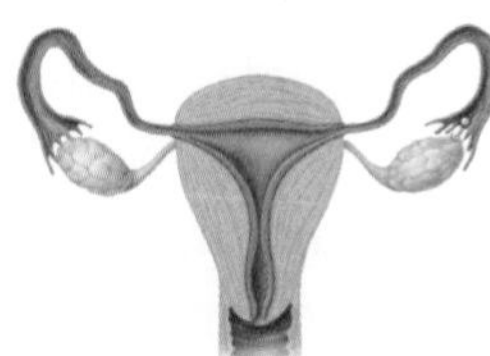

Critical Thinking

9. **Relate** the structures of the female reproductive system to their functions.
10. **Assess** the role of hormones in the function of the reproductive system.

SCIENCE & SOCIETY

A Medical Breakthrough

Scientific research provides help for infertile couples.

Have you heard the term "test-tube baby"? Can a baby really come from a test tube? Three decades ago, two British scientists answered that question. Patrick Steptoe, an obstetrician and gynecologist, and Robert Edwards, a biologist and physiologist, developed a procedure that helped a British woman overcome infertility, or the inability to become pregnant. As a result of this procedure, the world's first test-tube baby, Louise Joy Brown, was born on July 25, 1978.

Prior to meeting, Steptoe and Edwards worked separately on human infertility. Steptoe developed a process called laparoscopy. Laparoscopy uses a narrow, tubelike instrument fitted with a fiber-optic light and a lens. This enables a doctor to examine a woman's ovaries, fallopian tubes, and uterus through a small incision in her abdomen. Edwards researched the fertilization of human eggs outside the body under laboratory conditions. As a team, Steptoe and Edwards developed a procedure now known as in vitro fertilization (IVF). In IVF, a doctor uses laparoscopy to remove mature eggs from a woman's ovaries. The eggs are fertilized in a Petri dish (the test tube) and grow into zygotes. A doctor transfers the zygotes into a woman's uterus, where they can continue to develop.

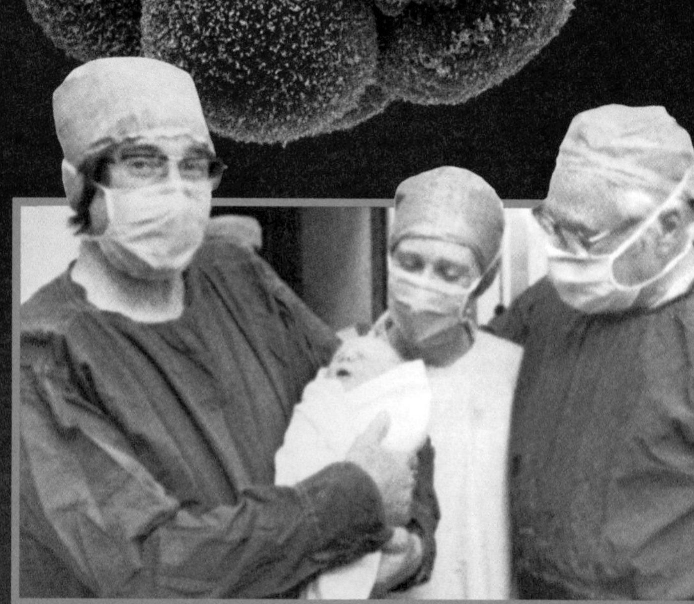

▲ **Dr. Steptoe (right) looks on as Dr. Edwards holds the world's first test-tube baby.**

Steptoe and Edwards worked for ten years to perfect their procedure. The successful implantation and pregnancy achieved through IVF opened new doors for infertile couples. By 2006, it was estimated that as many as 3 million babies had been born using IVF since the birth of Louise Joy Brown in 1978.

▲ **A doctor performs IVF by injecting a sperm cell into an egg cell.**

It's Your Turn

REPORT How long has in vitro fertilization been practiced in the United States? What is its success rate? Research these questions and write a short report.

Lesson 2

Human Growth and Development

Reading Guide

Key Concepts

ESSENTIAL QUESTIONS

- What happens during fertilization of a human egg?
- What are the major stages in the development of an embryo and a fetus?
- How do the life stages differ after birth?

Vocabulary

zygote p. 684

pregnancy p. 685

placenta p. 685

umbilical cord p. 685

embryo p. 686

fetus p. 686

cervix p. 688

puberty p. 690

 Multilingual eGlossary

 Video BrainPOP®

Inquiry When's my birthday?

This developing human is 12 weeks old. How did it develop from one cell? In this lesson, you will learn about development before and after birth.

Launch Lab

15 minutes

How does a fetus develop in the uterus?

Just as humans have several stages of development after birth, scientists have given names to the developmental stages from fertilization to birth.

1. Observe the copy of a **sonogram** provided by your teacher.
2. Match the numbered structures on the sonogram to the **list of body parts** that your teacher has given you. Record your answers in your Science Journal.

Think About This

1. What were the most difficult structures to identify? Why?
2. What questions do you have about fetal development?
3. **Key Concept** How do you think the stages of development after birth compare to the stages of development before birth?

Stages of Development

Have you ever been to a large family gathering? If you have, you might recall seeing people of every age. As shown in **Figure 7,** there probably were babies, children, teens, and adults.

As you might have noticed at the family gathering, people go through stages of development—infancy, childhood, adolescence (a duh LES unts), and adulthood. These stages are based on major developments that take place during each stage. In infancy there is rapid development of the nervous and muscular systems. During childhood, the abilities to speak, read, write, and reason develop. Adolescence is when a person becomes physically able to reproduce. Adulthood is the last stage—when growth of the muscular and skeletal systems stops.

Just as a human has several stages of development after he or she is born, the time before birth is divided into stages, beginning with fertilization.

Figure 7 People at this family reunion are in different stages of development. Humans grow and develop after birth.

Fertilization

If sperm enter the vagina, they can travel to the uterus and up into the fallopian tubes. The fallopian tubes are thin tubes that connect the uterus to the ovaries. Although millions of sperm are released into the vagina, most die before reaching the fallopian tubes.

A sperm contains substances that help its cell membrane join with the cell membrane of an egg. Once a sperm enters an egg, the egg's cell membrane changes rapidly. These changes ensure that only one male reproductive cell and one female reproductive cell combine to create a new human. When the nucleus of the sperm joins with the nucleus of the egg, fertilization is complete, as shown in **Figure 8.**

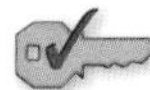

Key Concept Check What happens during fertilization?

Figure 8 Sperm travel through the uterus and fallopian tubes to fertilize an egg. Eggs are fertilized in the fallopian tube.

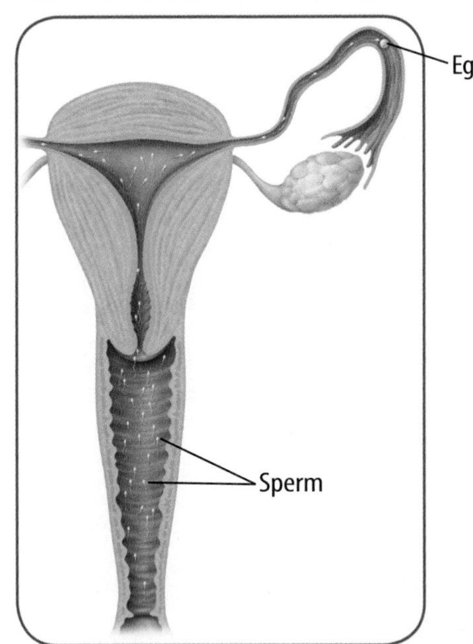

Fertilization

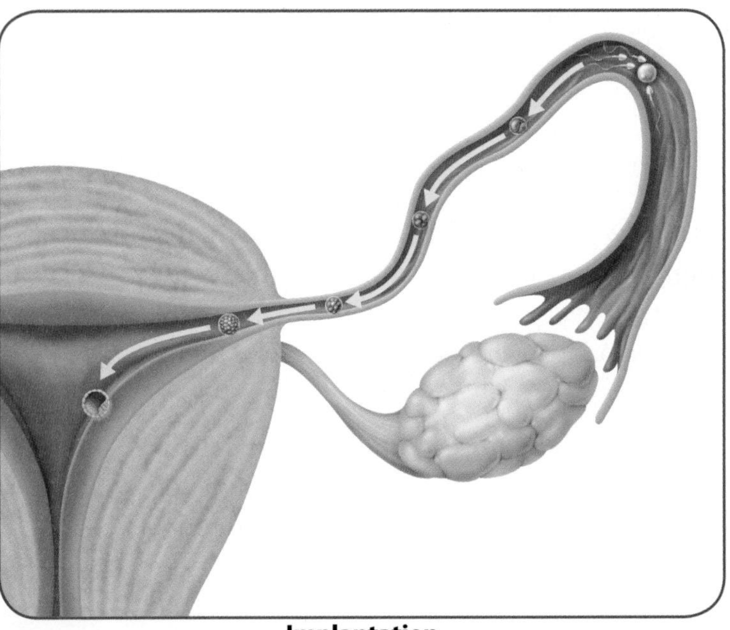

Implantation

Zygote Formation

A fertilized egg is called a **zygote** (ZI goht). Human zygotes contain 46 chromosomes of DNA—23 chromosomes from the sperm cell and 23 chromosomes from the egg cell. This means a zygote is a diploid cell. You might recall that reproductive cells form during **meiosis** and are haploid cells—they contain half the number of chromosomes of a diploid cell.

The zygote moves through the fallopian tube to the uterus. As it moves toward the uterus, the zygote undergoes mitosis and cell division many times, developing into a ball of cells.

REVIEW VOCABULARY

meiosis
the process in which one diploid cell divides to form four haploid cells

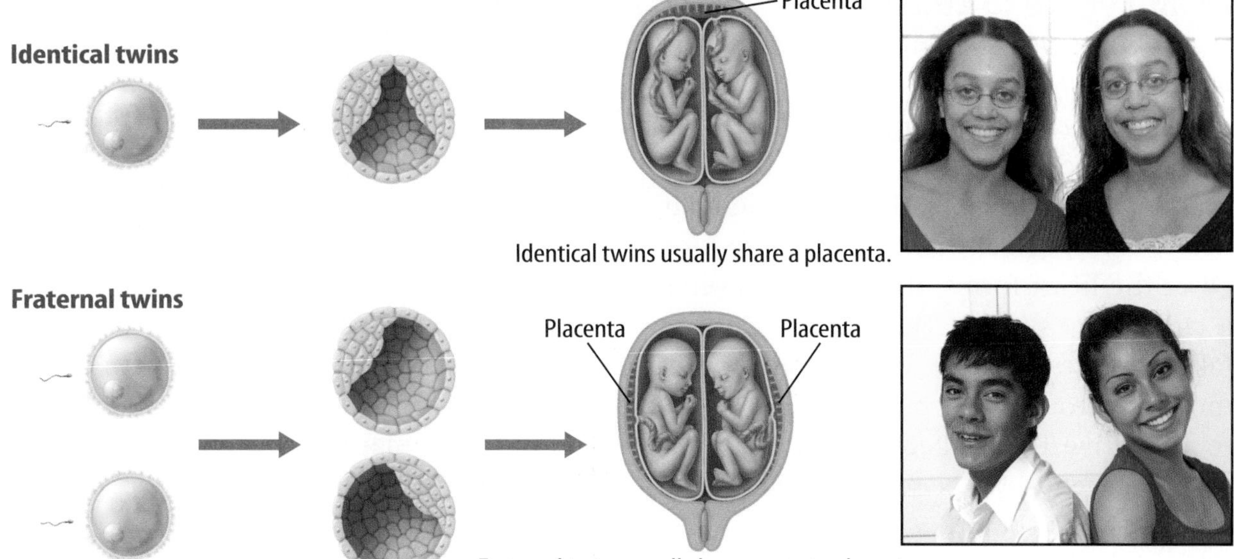

Identical twins usually share a placenta.

Fraternal twins usually have separate placentas.

Figure 9 Each inner cell mass can develop into a baby.

Multiple Births

A zygote is now a ball of cells with a group of cells on the inside. This group of cells, called the inner cell mass, can develop into a baby. Sometimes the inner cell mass divides in two. When one zygote contains two inner cell masses, as shown in **Figure 9,** identical twins can develop. Identical twins are always of the same gender and usually look very similar. Multiple births can also occur when more than one egg is released and each is fertilized by a different sperm. This results in fraternal twins, also shown in **Figure 9.** Unlike identical twins, fraternal twins can be different genders.

Development Before Birth

Recall that the tissue lining the uterus, the endometrium, thickens during the menstrual cycle. About seven days after fertilization, the zygote enters the uterus and attaches to the thickened endometrium. As shown in **Figure 8,** this process is called implantation. After implantation, a zygote can develop into a baby in about nine months. *The period of development from fertilized egg to birth is called* **pregnancy.**

After attaching to the uterus, *the outer cells of the zygote and cells from the uterus form an organ called the* **placenta** (pluh SEN tuh). *The outer zygote cells also form a rope-like structure, called the* **umbilical** (um BIH lih kul) **cord,** *which attaches the developing offspring to the placenta.* The developing offspring and the mother exchange materials through the umbilical cord. Nutrients and oxygen from the mother pass to the developing offspring and waste and carbon dioxide are removed from it.

Math Skills

Use Percentages

In 2005, 32.2 sets of twins were born for every 1,000 births in the United States. You can express this rate as 32.2/1,000. To convert this to a percentage, first convert the fraction to a decimal:

$$\frac{32.2}{1{,}000} = 0.0322$$

Multiply the decimal by 100 and add a percent sign.

$$0.0322 \times 100 = 3.22\%$$

Practice

In 2005, there were 161.8 multiple births per 100,000 live births. What percent of live births were multiples?

- Math Practice
- Personal Tutor

From Zygote to Embryo

From the time the zygote attaches to the uterus until the end of the eighth week of pregnancy, it is called an **embryo.** During this time, cells divide, grow, and gain unique functions. As shown in **Figure 10,** the brain, heart, limbs, fingers, and toes start to form. Once the heart forms, the embryo develops a circulatory system. The embryo can now take in more nutrients and oxygen from its mother through the placenta. Bones and reproductive tissues begin to develop. The ears and eyelids can be seen. By eight weeks, the embryo is about 2.5 cm long.

 **Reading Check** What changes take place in the embryo?

From Embryo to Fetus

During the time between nine weeks and birth, the developing offspring is called a **fetus.** Organ systems begin to function, and the fetus continues to grow in size, as shown in **Figure 10.** The fetus is now able to move its arms and legs. The heartbeat can be heard with a medical instrument called a stethoscope (STEH thuh skohp). During the remaining time until birth, the fetus grows rapidly. The bones fully develop but are still soft, and the lungs mature. The fetus can respond to sounds from outside the uterus, such as its mother's voice.

 **Key Concept Check** What are the major stages in the development of a fetus?

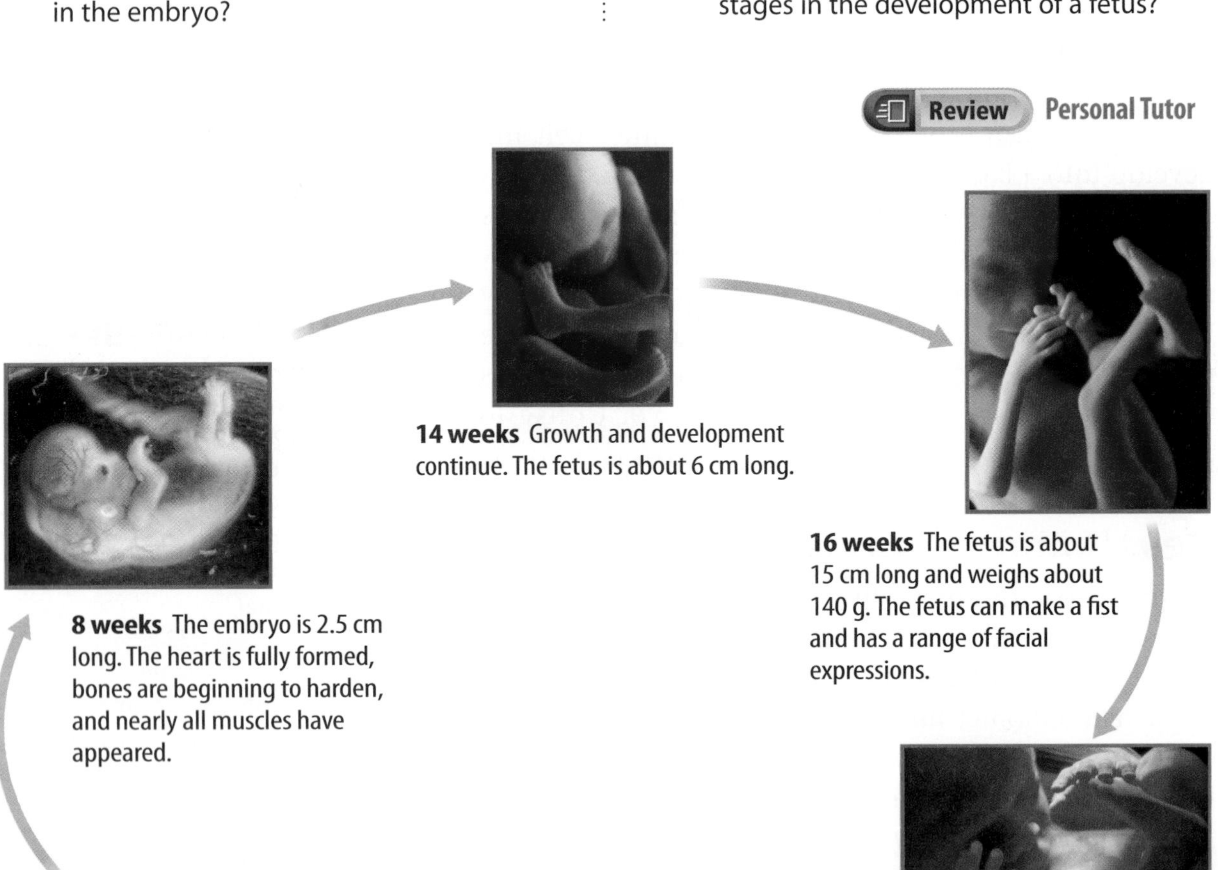

8 weeks The embryo is 2.5 cm long. The heart is fully formed, bones are beginning to harden, and nearly all muscles have appeared.

14 weeks Growth and development continue. The fetus is about 6 cm long.

16 weeks The fetus is about 15 cm long and weighs about 140 g. The fetus can make a fist and has a range of facial expressions.

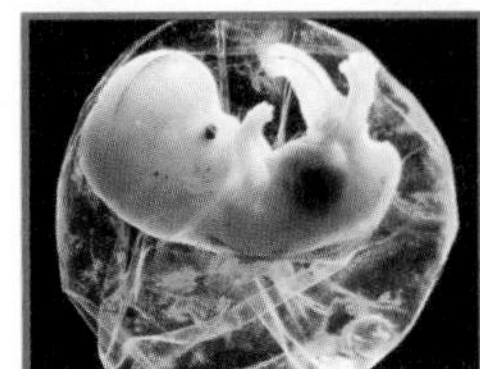

5 weeks The embryo is about 7 mm long. The heart and other organs have started to develop. The arms and legs are beginning to bud.

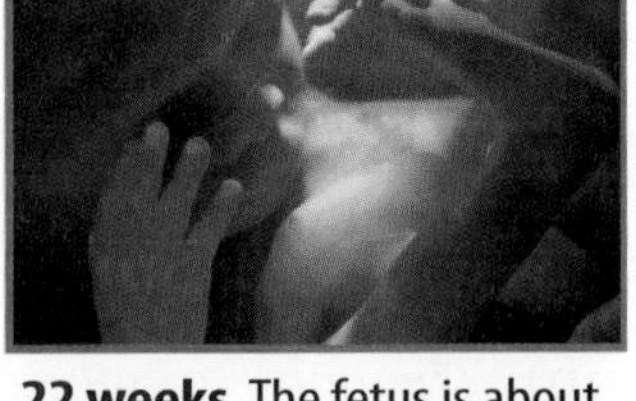

22 weeks The fetus is about 27 cm long and weighs about 430 g. Footprints and fingerprints are forming.

Figure 10 As an embryo develops into a fetus, it grows in size and organ systems form.

Visual Check What features do you recognize in the developing human?

Fetal Health

Because a fetus receives all nutrients from its mother, the fetus's growth and development depends on the food, water, and other nutrients that its mother eats and drinks. Other factors in a pregnant woman's environment, such as cigarette smoke or chemicals, can also **affect** the growth and development of her fetus.

Academic Vocabulary

affect
(verb) to influence or alter an outcome

Nutrition It is important that a pregnant woman takes in enough protein and vitamins to provide her fetus with the nutrients it needs to grow and develop. Nutrients such as vitamin D, folic acid, and zinc are needed for fetal development of bones and the nervous system. Protein is needed for making all of the new cells as the fetus grows.

Environmental Factors A fetus is protected from many environmental factors because it develops in a woman's uterus. However, the mother's exposure to substances such as chemicals and smoke can harm her fetus. Heavy metals such as lead and mercury can also affect its growth and development.

Drugs and Alcohol When a woman drinks alcohol during pregnancy, the developing fetus can be harmed. This is because her fetus also takes in the alcohol through the placenta. When a fetus is exposed to alcohol, the baby that develops can be born with fetal alcohol syndrome, as shown in **Figure 11.** Fetal alcohol syndrome is a group of lifelong problems that include growth problems, vision and hearing problems, and delayed mental development. Drugs such as cocaine and the nicotine in tobacco can also have harmful effects on a developing fetus if a woman uses them during pregnancy.

Figure 11 When a fetus is exposed to alcohol, many problems can occur. Some parts of the brain in the fetus with fetal alcohol syndrome are smaller than normal. This can cause learning and behavior problems.

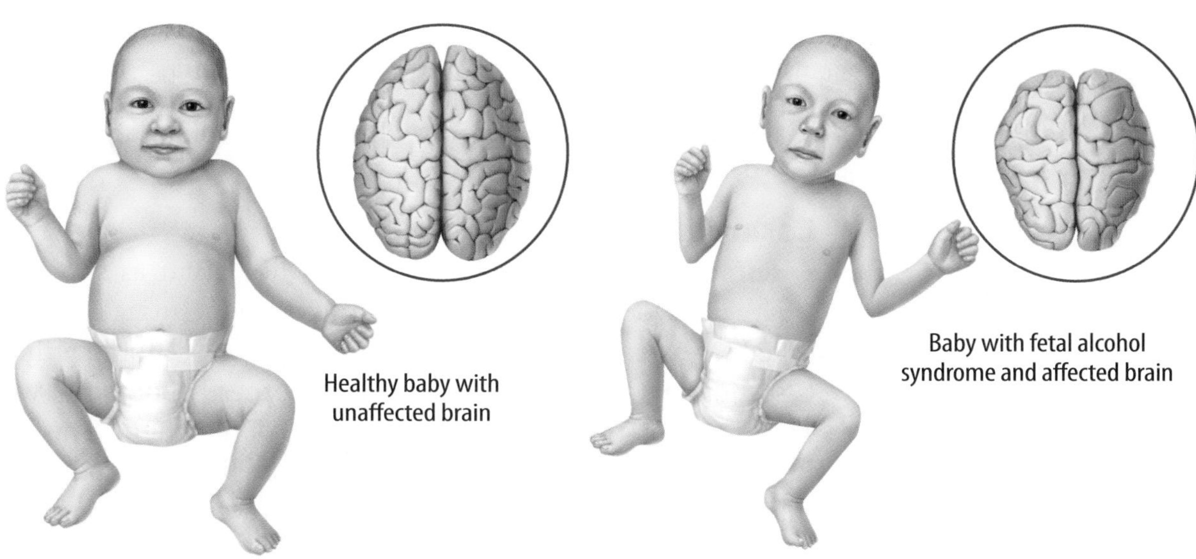

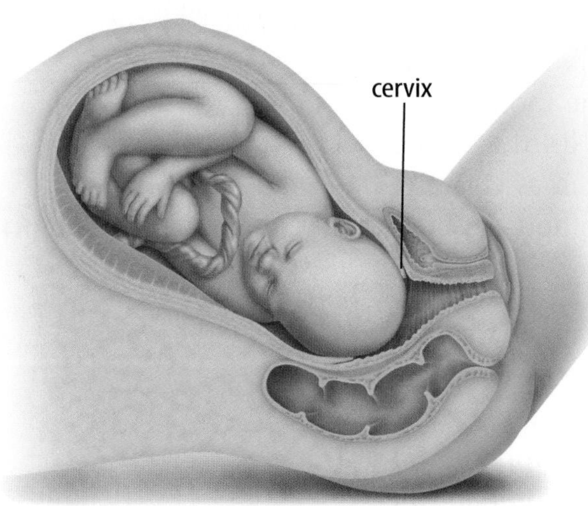

Stage 1 As the fetus moves into the birth canal, the opening to the uterus widens.

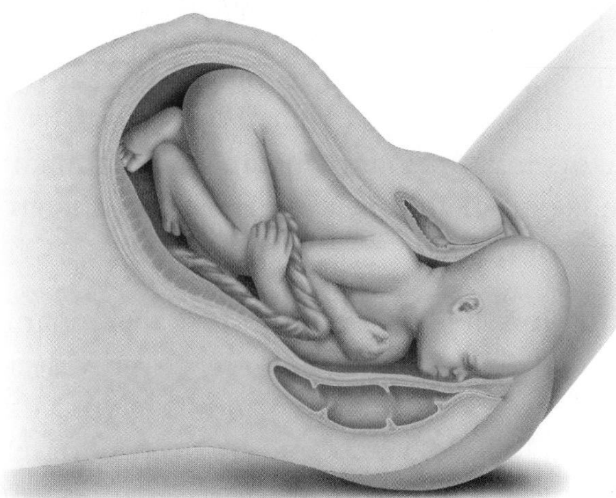

Stage 2 Muscle contractions help push the fetus out through the birth canal.

Figure 12 During birth, the head of the fetus moves toward the cervix. The fetus is delivered after the cervix opens.

WORD ORIGIN

cervix
Latin, means "neck"

Birth

A fetus leaves its mother's body and enters the world through a process called birth, as shown in **Figure 12.** Like the menstrual cycle, the birth of a fetus requires hormones. Hormones cause changes in the female reproductive system. These changes, called labor, help a fetus leave the uterus.

Labor and Delivery

Labor begins when hormones that are released by the endocrine system cause muscles in the uterus to contract. Also, *a small structure between the uterus and the vagina, called the* **cervix** (SUR vihks), begins to open. As more hormones are released, muscles in the uterus continue to contract faster and more strongly. As shown in **Figure 12,** the cervix opens wider to enable the fetus to leave the body. The contractions push a fetus into the vagina and out of the woman's body. After the fetus is delivered, the placenta breaks away from the uterus and also exits the woman's body through the vagina.

 **Reading Check** Summarize what happens during labor and delivery.

Cesarean Section

Sometimes delivery of a fetus does not occur as shown in **Figure 12.** Doctors can deliver the fetus by a surgical process called a cesarean (suh ZER ee un) section, or C-section. During a C-section, an incision is made in the mother's abdominal wall. Then another incision is made in the wall of the uterus. The baby is delivered through the openings in the uterine and abdominal walls. C-sections are often performed to prevent harm to a fetus and its mother.

Infancy

Once a baby is born, it stops depending directly on its mother for nutrients and oxygen and starts to function on its own. For the first time, the baby uses its own respiratory system and digestive system. The first two years of a newborn's life are called infancy. During infancy, the brain continues to develop, teeth form, and bones grow and get harder. As shown in **Figure 13,** an infant grows in size and learns to crawl, sit, walk, and speak. Organ systems continue to develop and mature, and the infant begins to eat solid food.

Childhood

The period following infancy is called childhood. During this time, the brain continues to grow and develop and thinking improves. Muscle strength increases, and arms and legs grow longer.

FOLDABLES®

Make a four-tab book from a sheet of notebook paper. Use it to organize your notes on the changes that occur during the stages of human development after birth.

Infancy
Childhood
Adolescence
Adulthood

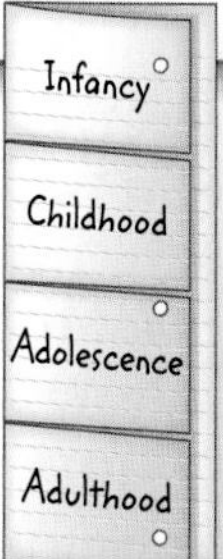

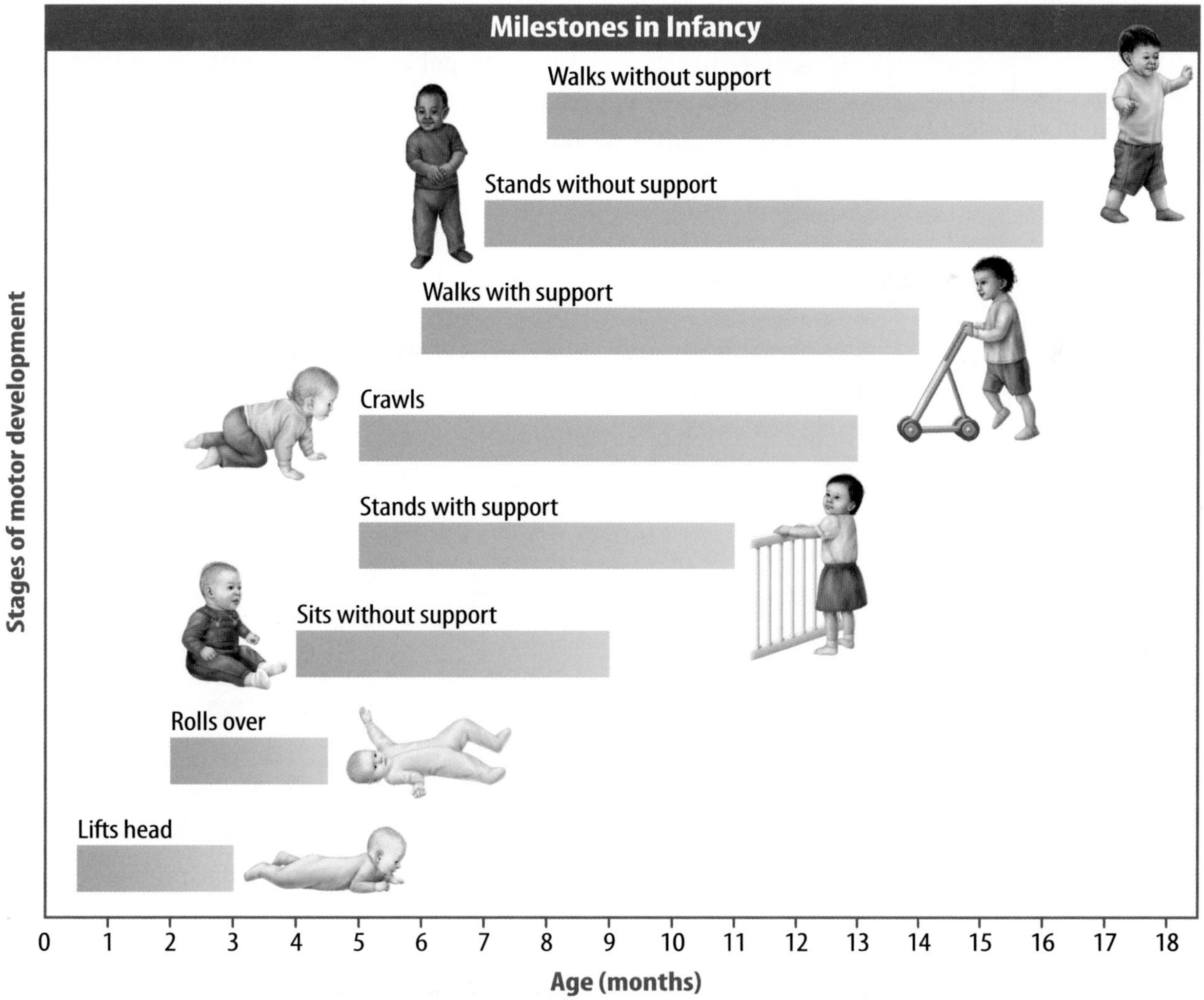

Milestones in Infancy
Walks without support
Stands without support
Walks with support
Crawls
Stands with support
Sits without support
Rolls over
Lifts head
Stages of motor development
0 1 2 3 4 5 6 7 8 9 10 11 12 13 14 15 16 17 18
Age (months)

Figure 13 During infancy, infants begin to move around independently.

Visual Check At what age do infants sit without support?

Figure 14 Many physical changes take place between adolescence and older adulthood.

Visual Check What changes can you identify in this photo?

Adolescence

Following childhood is a period of growth called adolescence. Both males and females grow taller as muscles and bones continue to grow. *During adolescence, the reproductive system matures in a process called* **puberty** (PYEW bur tee). Just as hormones are important for the menstrual cycle and for birth, hormones cause the changes that occur during puberty. In males, the voice deepens, muscles increase in size, and facial, pubic, and underarm hair grow. In females, breasts develop, pubic and underarm hair appear, and fatty tissue is added to the buttocks and thighs.

Adulthood and Aging

At the end of adolescence, a person enters adulthood, which continues through old age, as shown in **Figure 14.** Although adults will not grow taller, physical changes in body mass can still occur. Aging is the process of changes in the body over time. Hair can turn white or gray and stop growing, and the skin wrinkles. As humans get older, the sensory system and skeletal system decrease in function. Vision and hearing decline, bones become weaker, and the digestive system slows down.

Key Concept Check How do the life stages after birth differ?

Inquiry MiniLab

25 minutes

How do life stages after birth differ?

People go through several stages of development after birth—infancy, childhood, adolescence, and adulthood. These stages are based on major developments that take place during those years.

1. Read the information on the small **construction-paper shape** given to you by your teacher.
2. Discuss with other students in your group when exactly in a person's lifetime the event on your paper occurs. Decide where the paper should be placed on the "Human Life Continuum."
3. When your teacher calls upon your group, place the paper in the appropriate spot on the Human Life Continuum.

Analyze and Conclude

1. Where do the most events fall on the Human Life Continuum? Explain why this is so.
2. Where do the fewest events fall on the Human Life Continuum? Explain why this is so.
3. **Key Concept** Summarize how the life stages after birth differ from one another.

Lesson 2 Review

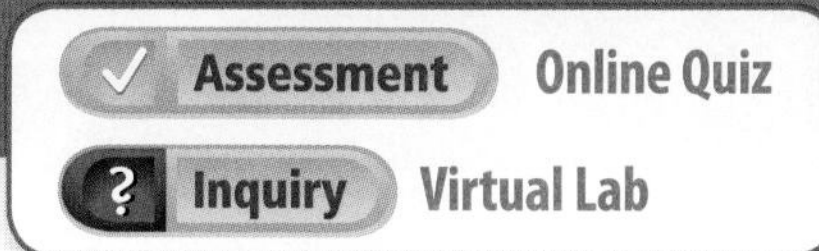

Visual Summary

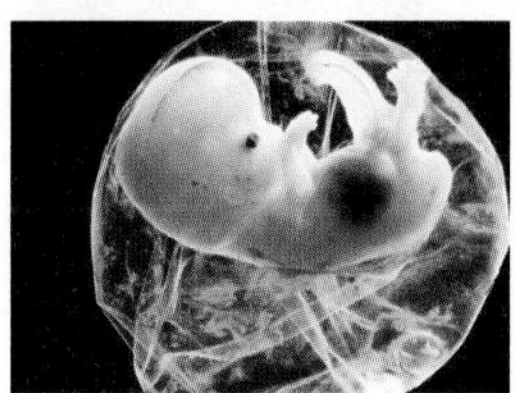

The developing zygote is called an embryo after it attaches to the uterus and until the end of the eighth week of pregnancy.

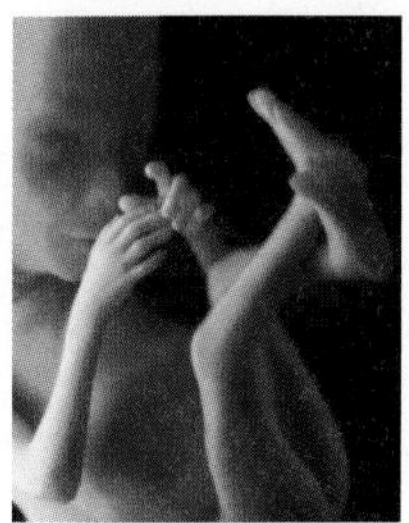

From the ninth week of pregnancy until birth, the developing embryo is called a fetus.

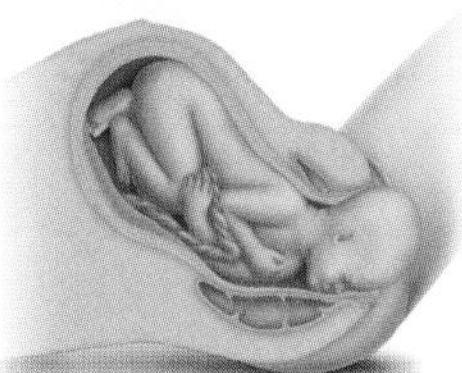

During labor, the fetus leaves the uterus through the cervix, a structure between the uterus and the vagina.

FOLDABLES®

Use your lesson Foldable to review the lesson. Save your Foldable for the project at the end of the chapter.

What do you think NOW?

You first read the statements below at the beginning of the chapter.

4. Eggs are fertilized in the ovary.

5. Lead is a nutrient that helps a fetus develop.

6. Puberty occurs during adolescence.

Did you change your mind about whether you agree or disagree with the statements? Rewrite any false statements to make them true.

Use Vocabulary

1. **Distinguish** between zygote and embryo.
2. **Define** *umbilical cord* in your own words.
3. **Use the term** *puberty* in a sentence.

Understand Key Concepts 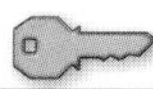

4. **Describe** the changes in the cervix and the uterus during birth.
5. **Compare** the functions of the placenta and the umbilical cord.
6. Which does not happen during aging?
   - **A.** Hair turns white.
   - **B.** Skin wrinkles.
   - **C.** An individual grows taller.
   - **D.** Vision and hearing decline.

Interpret Graphics

7. **Summarize** Copy and fill in the graphic organizer below to summarize the stages of development after an egg is fertilized and before birth.

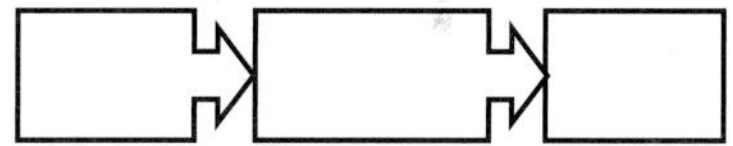

Critical Thinking

8. **Explain** why a developing human can be harmed during the stage of development shown in the figure below.

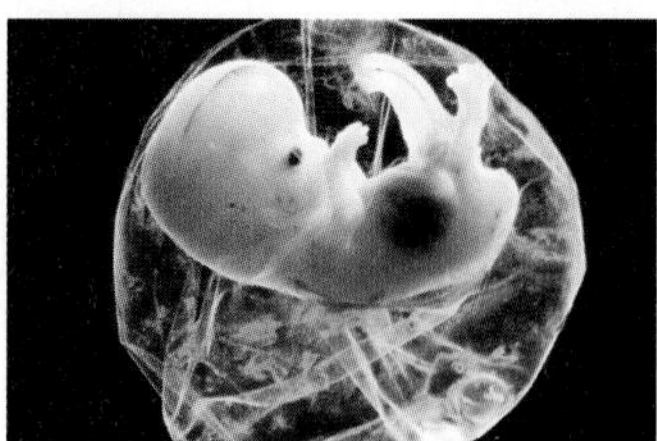

Math Skills

Math Practice

9. In 1990, 12 births out of 1,000 live births resulted in fraternal twins, and 4 births out of 1,000 live births resulted in identical twins. What percent of live births were fraternal twins? Identical twins?

45 minutes

Educating Extraterrestrials About Human Development

Materials

index cards

construction paper

small plastic bag of pictures

markers

Imagine you are an astronaut whose ship has landed on a faraway planet. Upon stepping out of your spaceship, you meet a friendly mob of aliens. They all look very similar. They are all approximately the same height, their arms and legs are all the same length, and their body shapes and proportions are all the same. You are surprised to find out that the aliens differ drastically in ages. They are very curious about you and your crew because you don't look alike. The aliens ask you and your crew to educate them on the human race, specifically how humans develop after birth.

Question

How do the bodies of humans change as they develop from infancy to adulthood?

Procedure

1. Using knowledge about yourself, as well as observations of other humans, make four sketches of yourself for the aliens to see. Sketch yourself in each of the following stages of development—infancy, childhood, adolescence, adulthood—making note of any obvious physical characteristics of that stage. Be sure to think about physical characteristics of each stage, such as length of the extremities, head size in relation to body size, muscle mass, body shape, and any other physical changes you have observed to happen in each stage of development (e.g., losing teeth, acne, gray hair).

2 Obtain the copies of pictures from your teacher. Place the photographs in the correct chronological order from infancy through childhood and adolescence and into adulthood. Compare and contrast the physical traits you observe in each of the photos.

Analyze and Conclude

3 **Compare and contrast** the physical characteristics of an infant and an adult.

4 **Compare and contrast** the physical characteristics of a child and an adolescent.

5 **The Big Idea** Explain how human bodies change as we go through the stages of development from infancy to adulthood. Describe the trends seen in body shape, extremity length, muscle mass, body proportions, and any other details you wish to include.

Communicate Your Results

Record a newscast or write a newspaper article detailing your expedition. Be sure to include information regarding where you went, who you encountered there, and what you taught them about your species.

Inquiry Extension

Think of a question you have regarding a specific stage of development. Design a controlled experiment or an observational study to investigate your question.

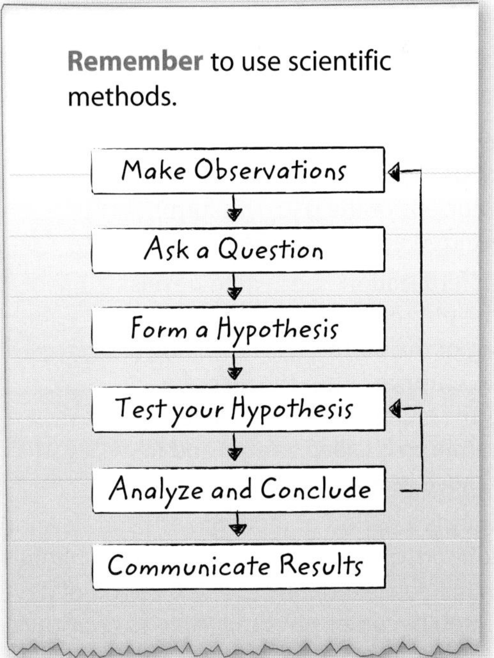

Chapter 19 Study Guide

Human reproduction and development begin when a sperm and an egg join. After fertilization, the zygote develops into an embryo and then a fetus. Human development continues after birth.

Key Concepts Summary

Lesson 1: The Reproductive System

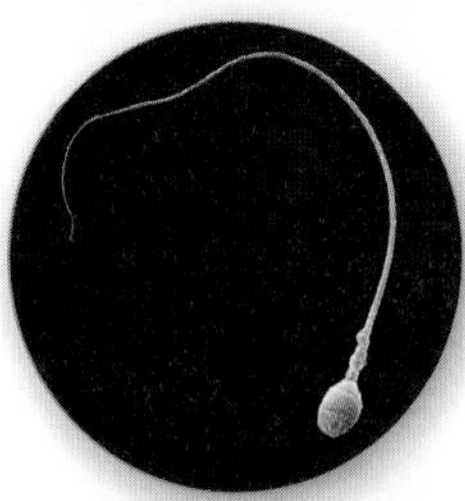

- The main function of the reproductive system is to ensure that the species survives. The male and female reproductive systems enable **sperm** and **egg** to join.
- The parts of the male reproductive system work together and produce and transport sperm to the female reproductive system.
- The parts of the female reproductive system work together and produce oocytes, provide a suitable environment for **fertilization** to occur, and nourish the developing offspring.
- The endocrine system interacts with the **testes** and **ovaries** and produces hormones that control and aid in sexual development.

Vocabulary

sperm p. 673
egg p. 673
testis p. 674
semen p. 674
penis p. 674
ovary p. 676
vagina p. 676
menstrual cycle p. 678
ovulation p. 678
fertilization p. 678

Lesson 2: Human Growth and Development

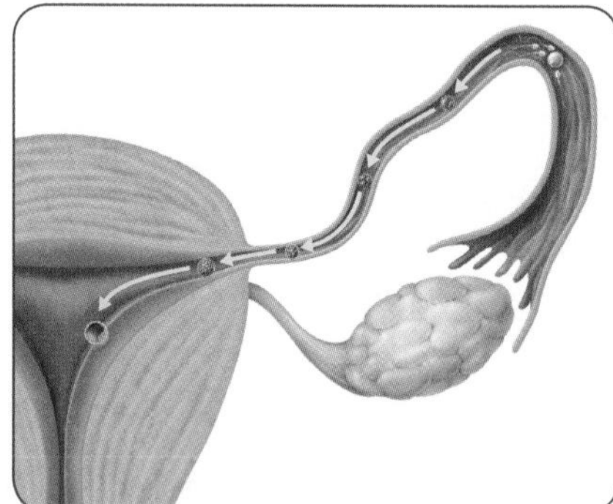

- After a sperm enters an egg, the nucleus of the sperm joins with the nucleus of the egg. This forms a fertilized egg, or **zygote.**
- After fertilization, the zygote travels to the uterus and attaches to the lining of the uterus. It is now known as an **embryo.** After the first two months of pregnancy, the developing embryo is called a **fetus** and it can move its arms and legs.
- Infancy is a period of rapid growth and development. During childhood, growth and development continue at a less rapid pace than in infancy. Adolescence is a period of development during which a person becomes physically able to reproduce. During adulthood, a person reaches his or her peak physical development, and then continues to change as aging occurs.

Vocabulary

zygote p. 684
pregnancy p. 685
placenta p. 685
umbilical cord p. 685
embryo p. 686
fetus p. 686
cervix p. 688
puberty p. 690

Review
- Personal Tutor
- Vocabulary eGames
- Vocabulary eFlashcards

FOLDABLES® Chapter Project

Assemble your lesson Foldables as shown to make a Chapter Project. Use the project to review what you have learned in this chapter.

Use Vocabulary

Distinguish between the vocabulary words in each pair.

1. sperm and egg
2. ovulation and menstruation
3. uterus and ovary

Determine if each statement is true or false.

4. The umbilical cord attaches the developing offspring to the uterus.
5. Puberty occurs during childhood.
6. A human zygote contains 23 chromosomes.

Link Vocabulary and Key Concepts

 Concepts in Motion Interactive Concept Map

Copy this concept map, and then use vocabulary terms from the previous page to complete the concept map.

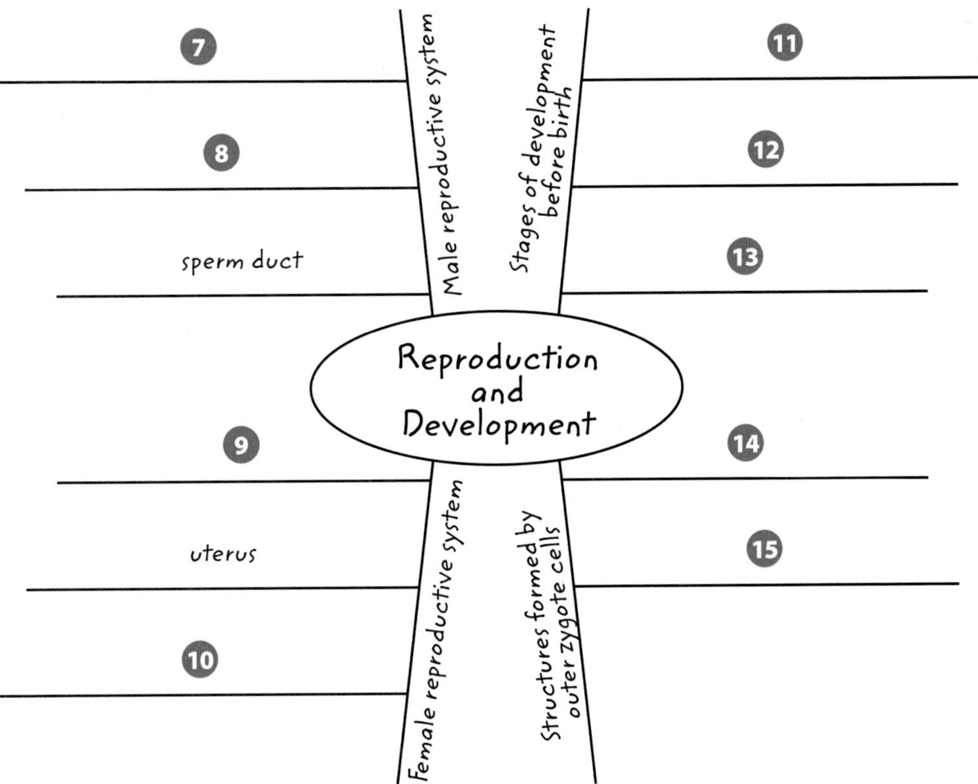

Chapter 19 Review

Understand Key Concepts 

1. Which part of the male reproductive system in the figure below produces sperm?

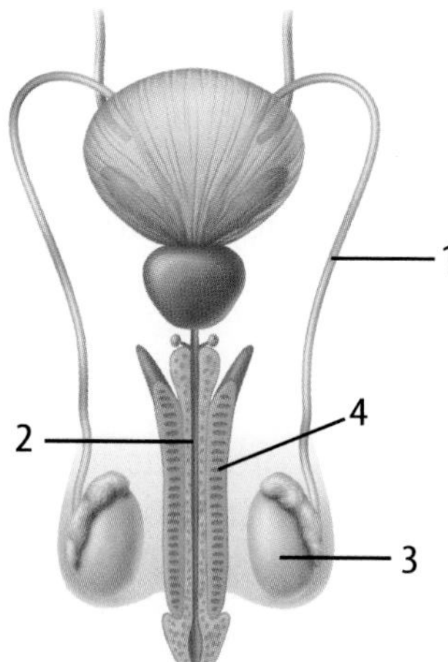

A. 1
B. 2
C. 3
D. 4

2. Which is NOT a part of sperm?
A. egg
B. head
C. midpiece
D. tail

3. Where do fertilized eggs develop?
A. ovary
B. scrotum
C. uterus
D. vagina

4. What happens to the menstrual cycle during menopause?
A. It begins.
B. It stops.
C. It gets faster.
D. It gets longer.

5. Which causes changes in male and female reproductive systems?
A. eggs
B. hormones
C. oocytes
D. sperm

6. How many diploid cells make up a zygote immediately following fertilization?
A. 0
B. 1
C. 2
D. 4

7. Fertilization of an egg takes place in which of the structures in the figure below?

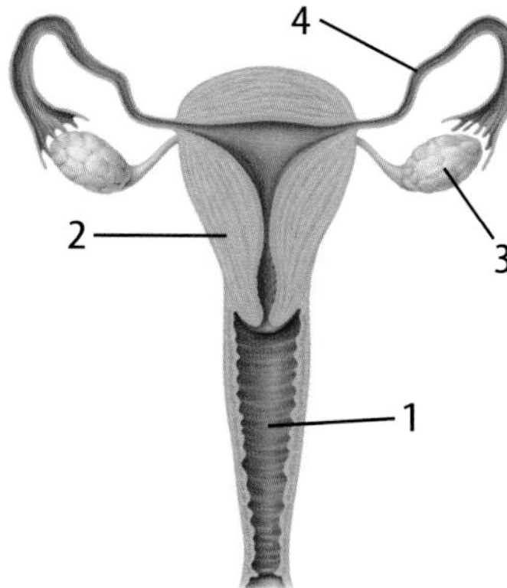

A. 1
B. 2
C. 3
D. 4

8. Which widens during delivery so the fetus can leave the uterus?
A. cervix
B. fallopian tube
C. ovary
D. uterus

9. Puberty occurs during which stage of development?
A. adolescence
B. adulthood
C. childhood
D. infancy

10. How is a fetus delivered during a cesarean section?
A. by hormones
B. by surgery
C. through the cervix
D. through the vagina

Assessment
Online Test Practice

Critical Thinking

11 **Infer** why semen contains millions of sperm.

12 **Compare** the parts of the male reproductive system to the parts of the female reproductive system.

13 **Infer** What would happen to sperm in a man whose testes were located inside his body instead of in the scrotum?

14 **Explain** why sperm and eggs contain only 23 chromosomes.

15 **Interpret** what is happening in the illustration below.

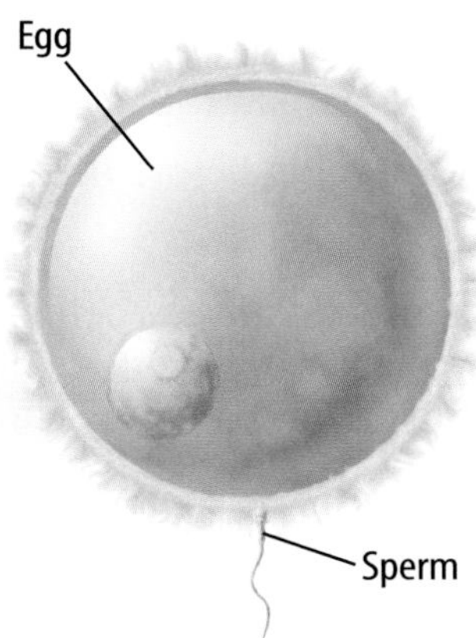

16 **Predict** how not having enough hormones affects puberty.

17 **Assess** the effect on the development of a fetus if a pregnant woman does not take in enough nutrients.

18 **Form** a hypothesis that explains why hormones are necessary for childbirth.

Writing in Science

19 **Write** a paragraph describing the changes in the uterus that lead to menstruation. Include a main idea, supporting details, and a concluding sentence.

REVIEW THE BIG IDEA

20 What are the stages in human reproduction and development? Explain the stages before and after birth.

21 The photo below shows an egg with many sperm on its surface. What changes occur in the egg to ensure that only one sperm fertilizes the egg? Why?

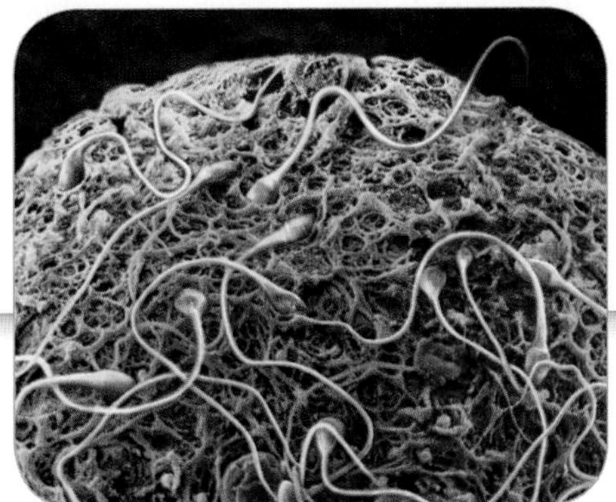

Math Skills

Review
Math Practice

Use Percentages

22 Women who are themselves fraternal twins give birth to about 200 sets of twins in every 2,000 births. What is the probability of a fraternal twin having twins?

23 Women who are themselves identical twins give birth to about 12 sets of twins in every 2,000 births. What is the probability of an identical twin having twins?

24 The Yoruba people in West Africa have the highest rate of twin births in the world. About 23 sets of twins are born per 500 live births. What is the probability of a Yoruba woman having twins?

Standardized Test Practice

Record your answers on the answer sheet provided by your teacher or on a sheet of paper.

Multiple Choice

1 Which body system works with the reproductive system to determine when sperm and eggs mature?

A circulatory

B endocrine

C lymphatic

D nervous

Use the diagram below to answer question 2.

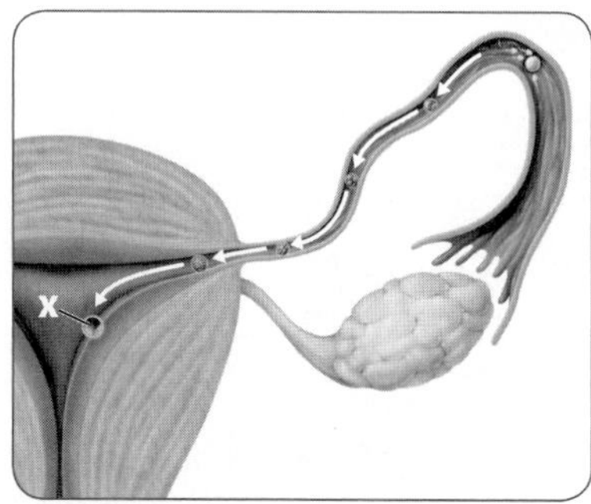

2 What takes place at the location marked *X* in the diagram above?

A fertilization

B implantation

C menstruation

D ovulation

3 How many chromosomes does each human zygote contain?

A 1

B 2

C 23

D 46

4 Which statement about fraternal twins is false?

A They can resemble each other.

B They must be different genders.

C They result from multiple eggs.

D They result from multiple sperm.

Use the table below to answer questions 5 and 6.

Milestone	Earliest Age	Latest Age
Rolls over	2 months	$4\frac{1}{2}$ months
Sits without support	4 months	9 months
Crawls	5 months	13 months
Stands alone	7 months	16 months
Walks alone	8 months	17 months

5 Which can be inferred from the above chart?

A Age brings more independence.

B All babies crawl before walking.

C Infants develop at the same rate.

D Muscles develop before the brain.

6 A parent wants to know when her baby might exhibit a certain skill. When is her baby most likely to begin crawling?

A 4–9 months

B 5–13 months

C 7–16 months

D 8–17 months

7 How are sperm and eggs similar?

A Both are large and round when they mature.

B Both are released one at a time.

C Both contain DNA.

D Both contain substances that nourish them.

8 Through which does nourishment pass directly to a fetus?

A intestine

B stomach

C fallopian tube

D umbilical cord

9 In which developmental stage do humans become physically capable of reproduction?

A adolescence

B adulthood

C childhood

D infancy

Use the diagram below to answer questions 10 and 11.

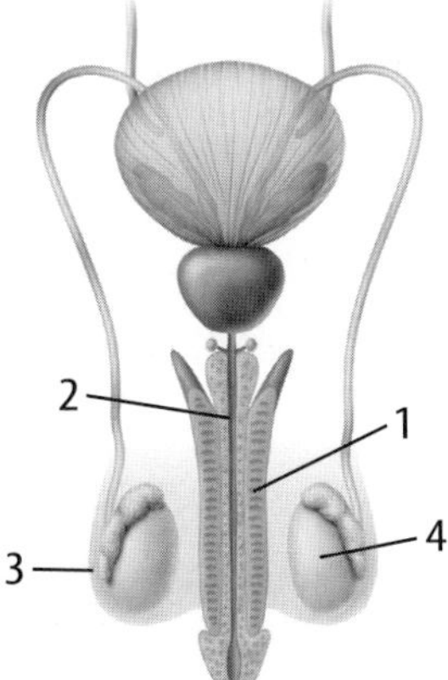

10 Which structure in the above diagram produces male reproductive cells?

A 1

B 2

C 3

D 4

11 Which structure in the diagram regulates temperature for sperm development?

A 1

B 2

C 3

D 4

Constructed Response

Use the diagram below to answer questions 12 and 13.

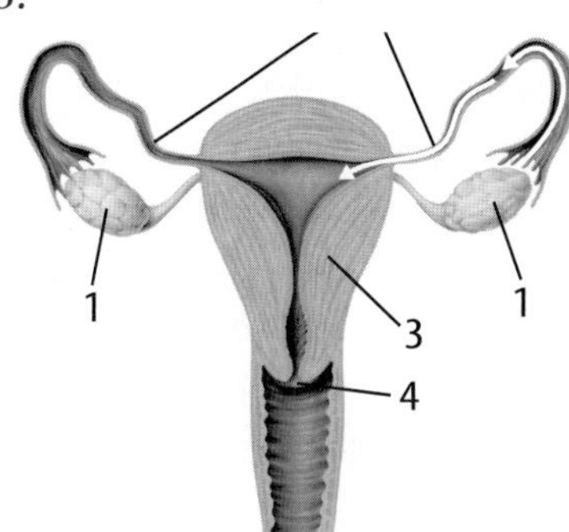

12 Name the numbered parts in the above diagram of the female reproductive system. Explain their functions.

13 What happens to a fertilized egg when it reaches the organ labeled 3 in the diagram above?

Use the table below to answer question 14.

Stage	Description
Fertilization	
Zygote	
Embryo	
Fetus	
Birth	

14 Complete the table above to describe the early stages of human development.

NEED EXTRA HELP?														
If You Missed Question...	1	2	3	4	5	6	7	8	9	10	11	12	13	14
Go to Lesson...	1	2	2	2	2	2	1	2	2	1	1	1	2	2

Student Resources

For Students and Parents/Guardians

These resources are designed to help you achieve success in science. You will find useful information on laboratory safety, math skills, and science skills. In addition, science reference materials are found in the Reference Handbook. You'll find the information you need to learn and sharpen your skills in these resources.

Table of Contents

Scientific Methods

Scientists use an orderly approach called the scientific method to solve problems. This includes organizing and recording data so others can understand them. Scientists use many variations in this method when they solve problems.

Identify a Question

The first step in a scientific investigation or experiment is to identify a question to be answered or a problem to be solved. For example, you might ask which gasoline is the most efficient.

Gather and Organize Information

After you have identified your question, begin gathering and organizing information. There are many ways to gather information, such as researching in a library, interviewing those knowledgeable about the subject, and testing and working in the laboratory and field. Fieldwork is investigations and observations done outside of a laboratory.

Researching Information Before moving in a new direction, it is important to gather the information that already is known about the subject. Start by asking yourself questions to determine exactly what you need to know. Then you will look for the information in various reference sources, like the student is doing in **Figure 1.** Some sources may include textbooks, encyclopedias, government documents, professional journals, science magazines, and the Internet. Always list the sources of your information.

Figure 1 The Internet can be a valuable research tool.

Evaluate Sources of Information Not all sources of information are reliable. You should evaluate all of your sources of information, and use only those you know to be dependable. For example, if you are researching ways to make homes more energy efficient, a site written by the U.S. Department of Energy would be more reliable than a site written by a company that is trying to sell a new type of weatherproofing material. Also, remember that research always is changing. Consult the most current resources available to you. For example, a 1985 resource about saving energy would not reflect the most recent findings.

Sometimes scientists use data that they did not collect themselves, or conclusions drawn by other researchers. This data must be evaluated carefully. Ask questions about how the data were obtained, if the investigation was carried out properly, and if it has been duplicated exactly with the same results. Would you reach the same conclusion from the data? Only when you have confidence in the data can you believe it is true and feel comfortable using it.

Interpret Scientific Illustrations As you research a topic in science, you will see drawings, diagrams, and photographs to help you understand what you read. Some illustrations are included to help you understand an idea that you can't see easily by yourself, like the tiny particles in an atom in **Figure 2.** A drawing helps many people to remember details more easily and provides examples that clarify difficult concepts or give additional information about the topic you are studying. Most illustrations have labels or a caption to identify or to provide more information.

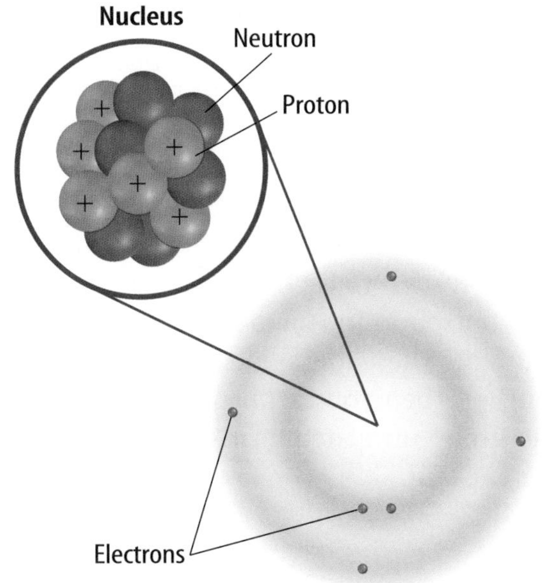

Figure 2 This drawing shows an atom of carbon with its six protons, six neutrons, and six electrons.

Concept Maps One way to organize data is to draw a diagram that shows relationships among ideas (or concepts). A concept map can help make the meanings of ideas and terms more clear, and help you understand and remember what you are studying. Concept maps are useful for breaking large concepts down into smaller parts, making learning easier.

Network Tree A type of concept map that not only shows a relationship, but how the concepts are related is a network tree, shown in **Figure 3.** In a network tree, the words are written in the ovals, while the description of the type of relationship is written across the connecting lines.

When constructing a network tree, write down the topic and all major topics on separate pieces of paper or notecards. Then arrange them in order from general to specific. Branch the related concepts from the major concept and describe the relationship on the connecting line. Continue to more specific concepts until finished.

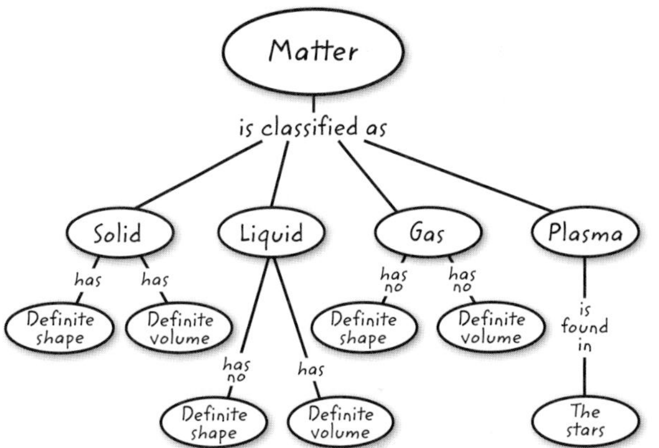

Figure 3 A network tree shows how concepts or objects are related.

Events Chain Another type of concept map is an events chain. Sometimes called a flow chart, it models the order or sequence of items. An events chain can be used to describe a sequence of events, the steps in a procedure, or the stages of a process.

When making an events chain, first find the one event that starts the chain. This event is called the initiating event. Then, find the next event and continue until the outcome is reached, as shown in **Figure 4** on the next page.

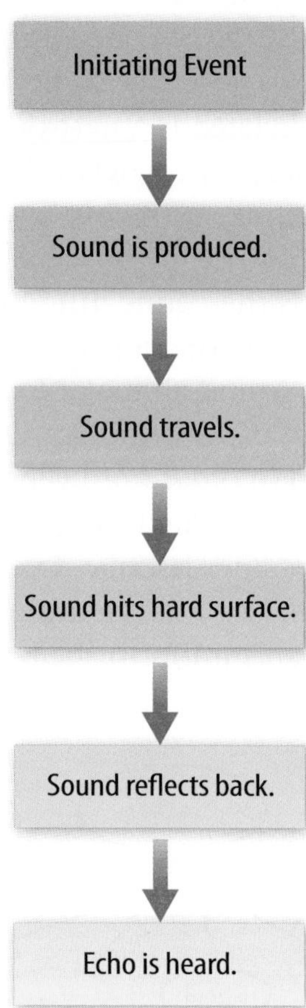

Figure 4 Events-chain concept maps show the order of steps in a process or event. This concept map shows how a sound makes an echo.

Cycle Map A specific type of events chain is a cycle map. It is used when the series of events do not produce a final outcome, but instead relate back to the beginning event, such as in **Figure 5.** Therefore, the cycle repeats itself.

To make a cycle map, first decide what event is the beginning event. This is also called the initiating event. Then list the next events in the order that they occur, with the last event relating back to the initiating event. Words can be written between the events that describe what happens from one event to the next. The number of events in a cycle map can vary, but usually contain three or more events.

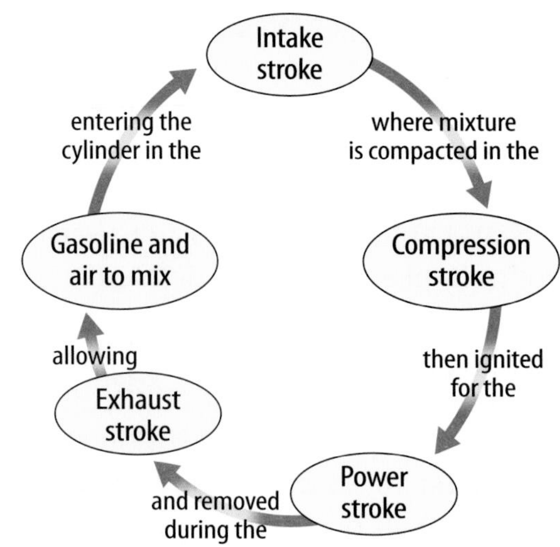

Figure 5 A cycle map shows events that occur in a cycle.

Spider Map A type of concept map that you can use for brainstorming is the spider map. When you have a central idea, you might find that you have a jumble of ideas that relate to it but are not necessarily clearly related to each other. The spider map on sound in **Figure 6** shows that if you write these ideas outside the main concept, then you can begin to separate and group unrelated terms so they become more useful.

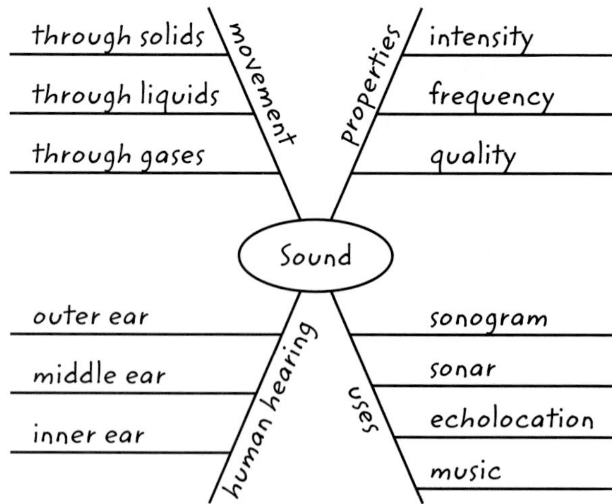

Figure 6 A spider map allows you to list ideas that relate to a central topic but not necessarily to one another.

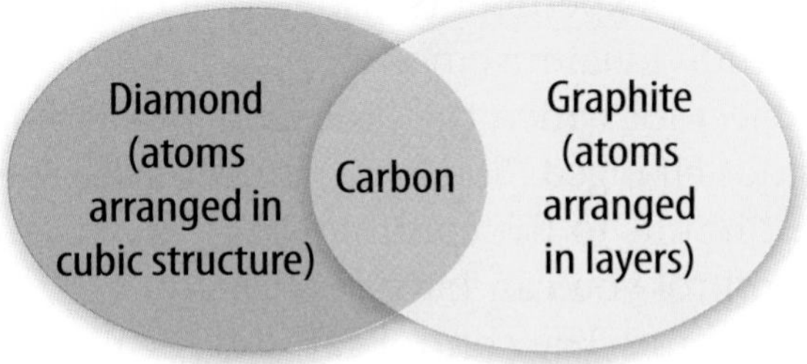

Diamond (atoms arranged in cubic structure) | Carbon | Graphite (atoms arranged in layers)

Figure 7 This Venn diagram compares and contrasts two substances made from carbon.

Venn Diagram To illustrate how two subjects compare and contrast you can use a Venn diagram. You can see the characteristics that the subjects have in common and those that they do not, shown in **Figure 7.**

To create a Venn diagram, draw two overlapping ovals that are big enough to write in. List the characteristics unique to one subject in one oval, and the characteristics of the other subject in the other oval. The characteristics in common are listed in the overlapping section.

Make and Use Tables One way to organize information so it is easier to understand is to use a table. Tables can contain numbers, words, or both.

To make a table, list the items to be compared in the first column and the characteristics to be compared in the first row. The title should clearly indicate the content of the table, and the column or row heads should be clear. Notice that in **Table 1** the units are included.

Table 1 Recyclables Collected During Week

Day of Week	Paper (kg)	Aluminum (kg)	Glass (kg)
Monday	5.0	4.0	12.0
Wednesday	4.0	1.0	10.0
Friday	2.5	2.0	10.0

Make a Model One way to help you better understand the parts of a structure, the way a process works, or to show things too large or small for viewing is to make a model. For example, an atomic model made of a plastic-ball nucleus and chenille stem electron shells can help you visualize how the parts of an atom relate to each other. Other types of models can be devised on a computer or represented by equations.

Form a Hypothesis

A possible explanation based on previous knowledge and observations is called a hypothesis. After researching gasoline types and recalling previous experiences in your family's car you form a hypothesis—our car runs more efficiently because we use premium gasoline. To be valid, a hypothesis has to be something you can test by using an investigation.

Predict When you apply a hypothesis to a specific situation, you predict something about that situation. A prediction makes a statement in advance, based on prior observation, experience, or scientific reasoning. People use predictions to make everyday decisions. Scientists test predictions by performing investigations. Based on previous observations and experiences, you might form a prediction that cars are more efficient with premium gasoline. The prediction can be tested in an investigation.

Design an Experiment A scientist needs to make many decisions before beginning an investigation. Some of these include: how to carry out the investigation, what steps to follow, how to record the data, and how the investigation will answer the question. It also is important to address any safety concerns.

SCIENCE SKILL HANDBOOK

MATH SKILL HANDBOOK

FOLDABLES HANDBOOK

REFERENCE HANDBOOK

GLOSSARY/GLOSARIO

INDEX

Test the Hypothesis

Now that you have formed your hypothesis, you need to test it. Using an investigation, you will make observations and collect data, or information. This data might either support or not support your hypothesis. Scientists collect and organize data as numbers and descriptions.

Follow a Procedure In order to know what materials to use, as well as how and in what order to use them, you must follow a procedure. **Figure 8** shows a procedure you might follow to test your hypothesis.

Procedure

Step 1	Use regular gasoline for two weeks.
Step 2	Record the number of kilometers between fill-ups and the amount of gasoline used.
Step 3	Switch to premium gasoline for two weeks.
Step 4	Record the number of kilometers between fill-ups and the amount of gasoline used.

Figure 8 A procedure tells you what to do step-by-step.

Identify and Manipulate Variables and Controls In any experiment, it is important to keep everything the same except for the item you are testing. The one factor you change is called the independent variable. The change that results is the dependent variable. Make sure you have only one independent variable, to assure yourself of the cause of the changes you observe in the dependent variable. For example, in your gasoline experiment the type of fuel is the independent variable. The dependent variable is the efficiency.

Many experiments also have a control—an individual instance or experimental subject for which the independent variable is not changed. You can then compare the test results to the control results. To design a control you can have two cars of the same type. The control car uses regular gasoline for four weeks. After you are done with the test, you can compare the experimental results to the control results.

Collect Data

Whether you are carrying out an investigation or a short observational experiment, you will collect data, as shown in **Figure 9.** Scientists collect data as numbers and descriptions and organize them in specific ways.

Observe Scientists observe items and events, then record what they see. When they use only words to describe an observation, it is called qualitative data. Scientists' observations also can describe how much there is of something. These observations use numbers, as well as words, in the description and are called quantitative data. For example, if a sample of the element gold is described as being "shiny and very dense" the data are qualitative. Quantitative data on this sample of gold might include "a mass of 30 g and a density of 19.3 g/cm^3."

Figure 9 Collecting data is one way to gather information directly.

Figure 10 Record data neatly and clearly so it is easy to understand.

When you make observations you should examine the entire object or situation first, and then look carefully for details. It is important to record observations accurately and completely. Always record your notes immediately as you make them, so you do not miss details or make a mistake when recording results from memory. Never put unidentified observations on scraps of paper. Instead they should be recorded in a notebook, like the one in **Figure 10.** Write your data neatly so you can easily read it later. At each point in the experiment, record your observations and label them. That way, you will not have to determine what the figures mean when you look at your notes later. Set up any tables that you will need to use ahead of time, so you can record any observations right away. Remember to avoid bias when collecting data by not including personal thoughts when you record observations. Record only what you observe.

Estimate Scientific work also involves estimating. To estimate is to make a judgment about the size or the number of something without measuring or counting. This is important when the number or size of an object or population is too large or too difficult to accurately count or measure.

Sample Scientists may use a sample or a portion of the total number as a type of estimation. To sample is to take a small, representative portion of the objects or organisms of a population for research. By making careful observations or manipulating variables within that portion of the group, information is discovered and conclusions are drawn that might apply to the whole population. A poorly chosen sample can be unrepresentative of the whole. If you were trying to determine the rainfall in an area, it would not be best to take a rainfall sample from under a tree.

Measure You use measurements every day. Scientists also take measurements when collecting data. When taking measurements, it is important to know how to use measuring tools properly. Accuracy also is important.

Length To measure length, the distance between two points, scientists use meters. Smaller measurements might be measured in centimeters or millimeters.

Length is measured using a metric ruler or meterstick. When using a metric ruler, line up the 0-cm mark with the end of the object being measured and read the number of the unit where the object ends. Look at the metric ruler shown in **Figure 11.** The centimeter lines are the long, numbered lines, and the shorter lines are millimeter lines. In this instance, the length would be 4.50 cm.

Figure 11 This metric ruler has centimeter and millimeter divisions.

Mass The SI unit for mass is the kilogram (kg). Scientists can measure mass using units formed by adding metric prefixes to the unit gram (g), such as milligram (mg). To measure mass, you might use a triple-beam balance similar to the one shown in **Figure 12.** The balance has a pan on one side and a set of beams on the other side. Each beam has a rider that slides on the beam.

When using a triple-beam balance, place an object on the pan. Slide the largest rider along its beam until the pointer drops below zero. Then move it back one notch. Repeat the process for each rider proceeding from the larger to smaller until the pointer swings an equal distance above and below the zero point. Sum the masses on each beam to find the mass of the object. Move all riders back to zero when finished.

Instead of putting materials directly on the balance, scientists often take a tare of a container. A tare is the mass of a container into which objects or substances are placed for measuring their masses. To find the mass of objects or substances, find the mass of a clean container. Remove the container from the pan, and place the object or substances in the container. Find the mass of the container with the materials in it. Subtract the mass of the empty container from the mass of the filled container to find the mass of the materials you are using.

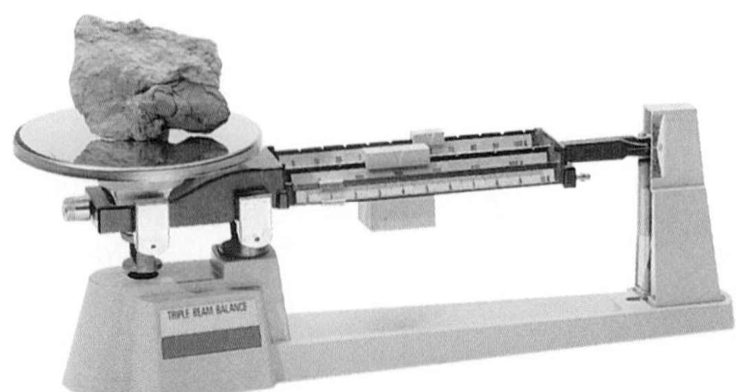

Figure 12 A triple-beam balance is used to determine the mass of an object.

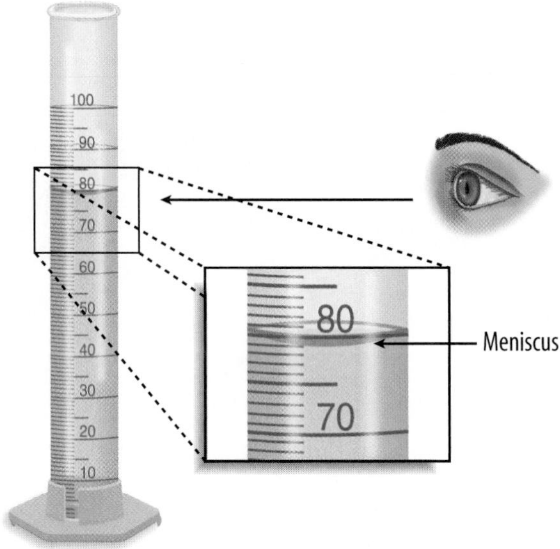

Figure 13 Graduated cylinders measure liquid volume.

Liquid Volume To measure liquids, the unit used is the liter. When a smaller unit is needed, scientists might use a milliliter. Because a milliliter takes up the volume of a cube measuring 1 cm on each side it also can be called a cubic centimeter ($cm^3 = cm \times cm \times cm$).

You can use beakers and graduated cylinders to measure liquid volume. A graduated cylinder, shown in **Figure 13,** is marked from bottom to top in milliliters. In lab, you might use a 10-mL graduated cylinder or a 100-mL graduated cylinder. When measuring liquids, notice that the liquid has a curved surface. Look at the surface at eye level, and measure the bottom of the curve. This is called the meniscus. The graduated cylinder in **Figure 13** contains 79.0 mL, or 79.0 cm^3, of a liquid.

Temperature Scientists often measure temperature using the Celsius scale. Pure water has a freezing point of 0°C and boiling point of 100°C. The unit of measurement is degrees Celsius. Two other scales often used are the Fahrenheit and Kelvin scales.

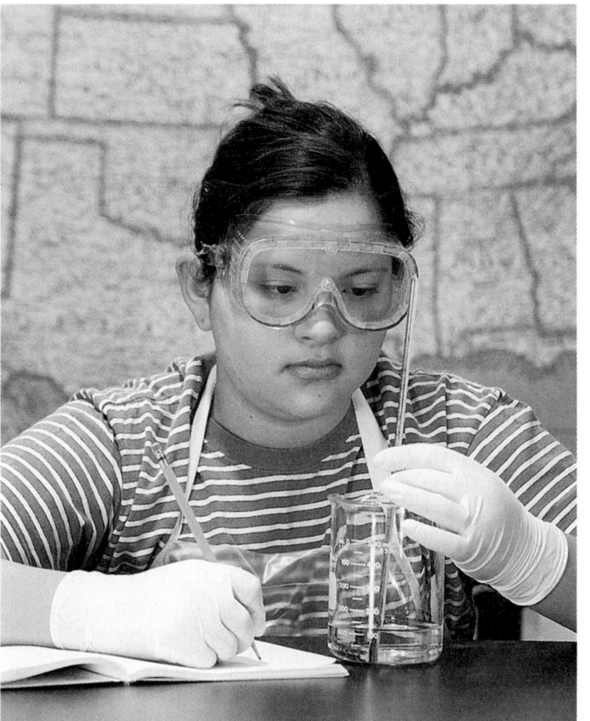

Figure 14 A thermometer measures the temperature of an object.

Scientists use a thermometer to measure temperature. Most thermometers in a laboratory are glass tubes with a bulb at the bottom end containing a liquid such as colored alcohol. The liquid rises or falls with a change in temperature. To read a glass thermometer like the thermometer in **Figure 14,** rotate it slowly until a red line appears. Read the temperature where the red line ends.

Form Operational Definitions An operational definition defines an object by how it functions, works, or behaves. For example, when you are playing hide and seek and a tree is home base, you have created an operational definition for a tree.

Objects can have more than one operational definition. For example, a ruler can be defined as a tool that measures the length of an object (how it is used). It can also be a tool with a series of marks used as a standard when measuring (how it works).

Analyze the Data

To determine the meaning of your observations and investigation results, you will need to look for patterns in the data. Then you must think critically to determine what the data mean. Scientists use several approaches when they analyze the data they have collected and recorded. Each approach is useful for identifying specific patterns.

Interpret Data The word *interpret* means "to explain the meaning of something." When analyzing data from an experiment, try to find out what the data show. Identify the control group and the test group to see whether changes in the independent variable have had an effect. Look for differences in the dependent variable between the control and test groups.

Classify Sorting objects or events into groups based on common features is called classifying. When classifying, first observe the objects or events to be classified. Then select one feature that is shared by some members in the group, but not by all. Place those members that share that feature in a subgroup. You can classify members into smaller and smaller subgroups based on characteristics. Remember that when you classify, you are grouping objects or events for a purpose. Keep your purpose in mind as you select the features to form groups and subgroups.

Compare and Contrast Observations can be analyzed by noting the similarities and differences between two or more objects or events that you observe. When you look at objects or events to see how they are similar, you are comparing them. Contrasting is looking for differences in objects or events.

Recognize Cause and Effect A cause is a reason for an action or condition. The effect is that action or condition. When two events happen together, it is not necessarily true that one event caused the other. Scientists must design a controlled investigation to recognize the exact cause and effect.

Draw Conclusions

When scientists have analyzed the data they collected, they proceed to draw conclusions about the data. These conclusions are sometimes stated in words similar to the hypothesis that you formed earlier. They may confirm a hypothesis, or lead you to a new hypothesis.

Infer Scientists often make inferences based on their observations. An inference is an attempt to explain observations or to indicate a cause. An inference is not a fact, but a logical conclusion that needs further investigation. For example, you may infer that a fire has caused smoke. Until you investigate, however, you do not know for sure.

Apply When you draw a conclusion, you must apply those conclusions to determine whether the data supports the hypothesis. If your data do not support your hypothesis, it does not mean that the hypothesis is wrong. It means only that the result of the investigation did not support the hypothesis. Maybe the experiment needs to be redesigned, or some of the initial observations on which the hypothesis was based were incomplete or biased. Perhaps more observation or research is needed to refine your hypothesis. A successful investigation does not always come out the way you originally predicted.

Avoid Bias Sometimes a scientific investigation involves making judgments. When you make a judgment, you form an opinion. It is important to be honest and not to allow any expectations of results to bias your judgments. This is important throughout the entire investigation, from researching to collecting data to drawing conclusions.

Communicate

The communication of ideas is an important part of the work of scientists. A discovery that is not reported will not advance the scientific community's understanding or knowledge. Communication among scientists also is important as a way of improving their investigations.

Scientists communicate in many ways, from writing articles in journals and magazines that explain their investigations and experiments, to announcing important discoveries on television and radio. Scientists also share ideas with colleagues on the Internet or present them as lectures, like the student is doing in **Figure 15.**

Figure 15 A student communicates to his peers about his investigation.

These safety symbols are used in laboratory and field investigations in this book to indicate possible hazards. Learn the meaning of each symbol and refer to this page often. *Remember to wash your hands thoroughly after completing lab procedures.*

PROTECTIVE EQUIPMENT

Do not begin any lab without the proper protection equipment.

GOGGLES

Proper eye protection must be worn when performing or observing science activities that involve items or conditions as listed below.

APRON

Wear an approved apron when using substances that could stain, wet, or destroy cloth.

SOAP

Wash hands with soap and water before removing goggles and after all lab activities.

GLOVES

Wear gloves when working with biological materials, chemicals, animals, or materials that can stain or irritate hands.

LABORATORY HAZARDS

Symbols	Potential Hazards	Precaution	Response
DISPOSAL	contamination of classroom or environment due to improper disposal of materials such as chemicals and live specimens	• DO NOT dispose of hazardous materials in the sink or trash can. • Dispose of wastes as directed by your teacher.	• If hazardous materials are disposed of improperly, notify your teacher immediately.
EXTREME TEMPERATURE	skin burns due to extremely hot or cold materials such as hot glass, liquids, or metals; liquid nitrogen; dry ice	• Use proper protective equipment, such as hot mitts and/or tongs, when handling objects with extreme temperatures.	• If injury occurs, notify your teacher immediately.
SHARP OBJECTS	punctures or cuts from sharp objects such as razor blades, pins, scalpels, and broken glass	• Handle glassware carefully to avoid breakage. • Walk with sharp objects pointed downward, away from you and others.	• If broken glass or injury occurs, notify your teacher immediately.
ELECTRICAL	electric shock or skin burn due to improper grounding, short circuits, liquid spills, or exposed wires	• Check condition of wires and apparatus for fraying or uninsulated wires, and broken or cracked equipment. • Use only GFCI-protected outlets	• DO NOT attempt to fix electrical problems. Notify your teacher immediately.
CHEMICAL	skin irritation or burns, breathing difficulty, and/or poisoning due to touching, swallowing, or inhalation of chemicals such as acids, bases, bleach, metal compounds, iodine, poinsettias, pollen, ammonia, acetone, nail polish remover, heated chemicals, mothballs, and any other chemicals labeled or known to be dangerous	• Wear proper protective equipment such as goggles, apron, and gloves when using chemicals. • Ensure proper room ventilation or use a fume hood when using materials that produce fumes. • NEVER smell fumes directly. • NEVER taste or eat any material in the laboratory.	• If contact occurs, immediately flush affected area with water and notify your teacher. • If a spill occurs, leave the area immediately and notify your teacher.
FLAMMABLE	unexpected fire due to liquids or gases that ignite easily such as rubbing alcohol	• Avoid open flames, sparks, or heat when flammable liquids are present.	• If a fire occurs, leave the area immediately and notify your teacher.
OPEN FLAME	burns or fire due to open flame from matches, Bunsen burners, or burning materials	• Tie back loose hair and clothing. • Keep flame away from all materials. • Follow teacher instructions when lighting and extinguishing flames. • Use proper protection, such as hot mitts or tongs, when handling hot objects.	• If a fire occurs, leave the area immediately and notify your teacher.
ANIMAL SAFETY	injury to or from laboratory animals	• Wear proper protective equipment such as gloves, apron, and goggles when working with animals. • Wash hands after handling animals.	• If injury occurs, notify your teacher immediately.
BIOLOGICAL	infection or adverse reaction due to contact with organisms such as bacteria, fungi, and biological materials such as blood, animal or plant materials	• Wear proper protective equipment such as gloves, goggles, and apron when working with biological materials. • Avoid skin contact with an organism or any part of the organism. • Wash hands after handling organisms.	• If contact occurs, wash the affected area and notify your teacher immediately.
FUME	breathing difficulties from inhalation of fumes from substances such as ammonia, acetone, nail polish remover, heated chemicals, and mothballs	• Wear goggles, apron, and gloves. • Ensure proper room ventilation or use a fume hood when using substances that produce fumes. • NEVER smell fumes directly.	• If a spill occurs, leave area and notify your teacher immediately.
IRRITANT	irritation of skin, mucous membranes, or respiratory tract due to materials such as acids, bases, bleach, pollen, mothballs, steel wool, and potassium permanganate	• Wear goggles, apron, and gloves. • Wear a dust mask to protect against fine particles.	• If skin contact occurs, immediately flush the affected area with water and notify your teacher.
RADIOACTIVE	excessive exposure from alpha, beta, and gamma particles	• Remove gloves and wash hands with soap and water before removing remainder of protective equipment.	• If cracks or holes are found in the container, notify your teacher immediately.

Safety in the Science Laboratory

Introduction to Science Safety

The science laboratory is a safe place to work if you follow standard safety procedures. Being responsible for your own safety helps to make the entire laboratory a safer place for everyone. When performing any lab, read and apply the caution statements and safety symbol listed at the beginning of the lab.

General Safety Rules

1. Complete the *Lab Safety Form* or other safety contract BEFORE starting any science lab.
2. Study the procedure. Ask your teacher any questions. Be sure you understand safety symbols shown on the page.
3. Notify your teacher about allergies or other health conditions that can affect your participation in a lab.
4. Learn and follow use and safety procedures for your equipment. If unsure, ask your teacher.

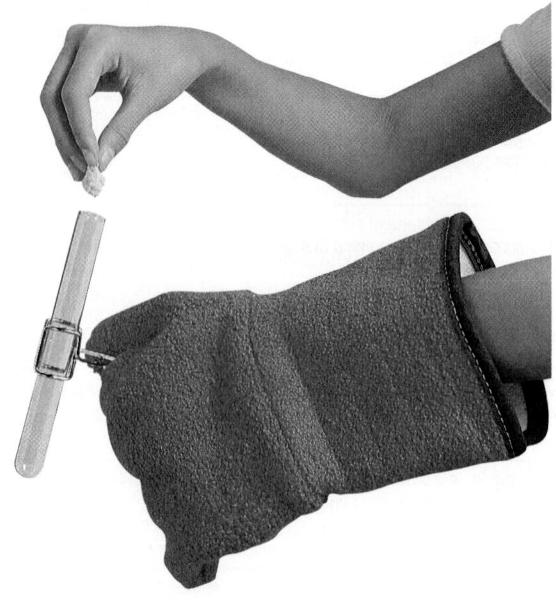

5. Never eat, drink, chew gum, apply cosmetics, or do any personal grooming in the lab. Never use lab glassware as food or drink containers. Keep your hands away from your face and mouth.
6. Know the location and proper use of the safety shower, eye wash, fire blanket, and fire alarm.

Prevent Accidents

1. Use the safety equipment provided to you. Goggles and a safety apron should be worn during investigations.
2. Do NOT use hair spray, mousse, or other flammable hair products. Tie back long hair and tie down loose clothing.
3. Do NOT wear sandals or other open-toed shoes in the lab.
4. Remove jewelry on hands and wrists. Loose jewelry, such as chains and long necklaces, should be removed to prevent them from getting caught in equipment.
5. Do not taste any substances or draw any material into a tube with your mouth.
6. Proper behavior is expected in the lab. Practical jokes and fooling around can lead to accidents and injury.
7. Keep your work area uncluttered.

Laboratory Work

1. Collect and carry all equipment and materials to your work area before beginning a lab.
2. Remain in your own work area unless given permission by your teacher to leave it.

3. Always slant test tubes away from yourself and others when heating them, adding substances to them, or rinsing them.
4. If instructed to smell a substance in a container, hold the container a short distance away and fan vapors toward your nose.
5. Do NOT substitute other chemicals/substances for those in the materials list unless instructed to do so by your teacher.
6. Do NOT take any materials or chemicals outside of the laboratory.
7. Stay out of storage areas unless instructed to be there and supervised by your teacher.

Laboratory Cleanup

1. Turn off all burners, water, and gas, and disconnect all electrical devices.
2. Clean all pieces of equipment and return all materials to their proper places.
3. Dispose of chemicals and other materials as directed by your teacher. Place broken glass and solid substances in the proper containers. Never discard materials in the sink.
4. Clean your work area.
5. Wash your hands with soap and water thoroughly BEFORE removing your goggles.

Emergencies

1. Report any fire, electrical shock, glassware breakage, spill, or injury, no matter how small, to your teacher immediately. Follow his or her instructions.
2. If your clothing should catch fire, STOP, DROP, and ROLL. If possible, smother it with the fire blanket or get under a safety shower. NEVER RUN.
3. If a fire should occur, turn off all gas and leave the room according to established procedures.
4. In most instances, your teacher will clean up spills. Do NOT attempt to clean up spills unless you are given permission and instructions to do so.
5. If chemicals come into contact with your eyes or skin, notify your teacher immediately. Use the eyewash, or flush your skin or eyes with large quantities of water.
6. The fire extinguisher and first-aid kit should only be used by your teacher unless it is an extreme emergency and you have been given permission.
7. If someone is injured or becomes ill, only a professional medical provider or someone certified in first aid should perform first-aid procedures.

Math Review

Use Fractions

A fraction compares a part to a whole. In the fraction $\frac{2}{3}$, the 2 represents the part and is the numerator. The 3 represents the whole and is the denominator.

Reduce Fractions To reduce a fraction, you must find the largest factor that is common to both the numerator and the denominator, the greatest common factor (GCF). Divide both numbers by the GCF. The fraction has then been reduced, or it is in its simplest form.

Example

Twelve of the 20 chemicals in the science lab are in powder form. What fraction of the chemicals used in the lab are in powder form?

Step 1 Write the fraction.

$$\frac{\text{part}}{\text{whole}} = \frac{12}{20}$$

Step 2 To find the GCF of the numerator and denominator, list all of the factors of each number.

Factors of 12: 1, 2, 3, 4, 6, 12
(the numbers that divide evenly into 12)

Factors of 20: 1, 2, 4, 5, 10, 20
(the numbers that divide evenly into 20)

Step 3 List the common factors.

1, 2, 4

Step 4 Choose the greatest factor in the list. The GCF of 12 and 20 is 4.

Step 5 Divide the numerator and denominator by the GCF.

$$\frac{12 \div 4}{20 \div 4} = \frac{3}{5}$$

In the lab, $\frac{3}{5}$ of the chemicals are in powder form.

Practice Problem At an amusement park, 66 of 90 rides have a height restriction. What fraction of the rides, in its simplest form, has a height restriction?

Add and Subtract Fractions with Like Denominators To add or subtract fractions with the same denominator, add or subtract the numerators and write the sum or difference over the denominator. After finding the sum or difference, find the simplest form for your fraction.

Example 1

In the forest outside your house, $\frac{1}{8}$ of the animals are rabbits, $\frac{3}{8}$ are squirrels, and the remainder are birds and insects. How many are mammals?

Step 1 Add the numerators.

$$\frac{1}{8} + \frac{3}{8} = \frac{(1 + 3)}{8} = \frac{4}{8}$$

Step 2 Find the GCF.

$\frac{4}{8}$ (GCF, 4)

Step 3 Divide the numerator and denominator by the GCF.

$$\frac{4 \div 4}{8 \div 4} = \frac{1}{2}$$

$\frac{1}{2}$ of the animals are mammals.

Example 2

If $\frac{7}{16}$ of the Earth is covered by freshwater, and $\frac{1}{16}$ of that is in glaciers, how much freshwater is not frozen?

Step 1 Subtract the numerators.

$$\frac{7}{16} - \frac{1}{16} = \frac{(7 - 1)}{16} = \frac{6}{16}$$

Step 2 Find the GCF.

$\frac{6}{16}$ (GCF, 2)

Step 3 Divide the numerator and denominator by the GCF.

$$\frac{6 \div 2}{16 \div 2} = \frac{3}{8}$$

$\frac{3}{8}$ of the freshwater is not frozen.

Practice Problem A bicycle rider is riding at a rate of 15 km/h for $\frac{4}{9}$ of his ride, 10 km/h for $\frac{2}{9}$ of his ride, and 8 km/h for the remainder of the ride. How much of his ride is he riding at a rate greater than 8 km/h?

Add and Subtract Fractions with Unlike Denominators To add or subtract fractions with unlike denominators, first find the least common denominator (LCD). This is the smallest number that is a common multiple of both denominators. Rename each fraction with the LCD, and then add or subtract. Find the simplest form if necessary.

Example 1

A chemist makes a paste that is $\frac{1}{2}$ table salt (NaCl), $\frac{1}{3}$ sugar ($C_6H_{12}O_6$), and the remainder is water (H_2O). How much of the paste is a solid?

Step 1 Find the LCD of the fractions.

$\frac{1}{2} + \frac{1}{3}$ (LCD, 6)

Step 2 Rename each numerator and each denominator with the LCD.

Step 3 Add the numerators.

$\frac{3}{6} + \frac{2}{6} = \frac{(3+2)}{6} = \frac{5}{6}$

$\frac{5}{6}$ of the paste is a solid.

Example 2

The average precipitation in Grand Junction, CO, is $\frac{7}{10}$ inch in November, and $\frac{3}{5}$ inch in December. What is the total average precipitation?

Step 1 Find the LCD of the fractions.

$\frac{7}{10} + \frac{3}{5}$ (LCD, 10)

Step 2 Rename each numerator and each denominator with the LCD.

Step 3 Add the numerators.

$\frac{7}{10} + \frac{6}{10} = \frac{(7+6)}{10} = \frac{13}{10}$

$\frac{13}{10}$ inches total precipitation, or $1\frac{3}{10}$ inches.

Practice Problem On an electric bill, about $\frac{1}{8}$ of the energy is from solar energy and about $\frac{1}{10}$ is from wind power. How much of the total bill is from solar energy and wind power combined?

Example 3

In your body, $\frac{7}{10}$ of your muscle contractions are involuntary (cardiac and smooth muscle tissue). Smooth muscle makes $\frac{3}{15}$ of your muscle contractions. How many of your muscle contractions are made by cardiac muscle?

Step 1 Find the LCD of the fractions.

$\frac{7}{10} - \frac{3}{15}$ (LCD, 30)

Step 2 Rename each numerator and each denominator with the LCD.

$\frac{7 \times 3}{10 \times 3} = \frac{21}{30}$

$\frac{3 \times 2}{15 \times 2} = \frac{6}{30}$

Step 3 Subtract the numerators.

$\frac{21}{30} - \frac{6}{30} = \frac{(21-6)}{30} = \frac{15}{30}$

Step 4 Find the GCF.

$\frac{15}{30}$ (GCF, 15)

$\frac{1}{2}$

$\frac{1}{2}$ of all muscle contractions are cardiac muscle.

Example 4

Tony wants to make cookies that call for $\frac{3}{4}$ of a cup of flour, but he only has $\frac{1}{3}$ of a cup. How much more flour does he need?

Step 1 Find the LCD of the fractions.

$\frac{3}{4} - \frac{1}{3}$ (LCD, 12)

Step 2 Rename each numerator and each denominator with the LCD.

$\frac{3 \times 3}{4 \times 3} = \frac{9}{12}$

$\frac{1 \times 4}{3 \times 4} = \frac{4}{12}$

Step 3 Subtract the numerators.

$\frac{9}{12} - \frac{4}{12} = \frac{(9-4)}{12} = \frac{5}{12}$

$\frac{5}{12}$ of a cup of flour

Practice Problem Using the information provided to you in Example 3 above, determine how many muscle contractions are voluntary (skeletal muscle).

SCIENCE SKILL HANDBOOK
MATH SKILL HANDBOOK
FOLDABLES HANDBOOK
REFERENCE HANDBOOK
GLOSSARY/GLOSARIO
INDEX

Multiply Fractions To multiply with fractions, multiply the numerators and multiply the denominators. Find the simplest form if necessary.

Example

Multiply $\frac{3}{5}$ by $\frac{1}{3}$.

Step 1 Multiply the numerators and denominators.

$\frac{3}{5} \times \frac{1}{3} = \frac{(3 \times 1)}{(5 \times 3)}\ \frac{3}{15}$

Step 2 Find the GCF.

$\frac{3}{15}$ (GCF, 3)

Step 3 Divide the numerator and denominator by the GCF.

$\frac{3 \div 3}{15 \div 3} = \frac{1}{5}$

$\frac{3}{5}$ multiplied by $\frac{1}{3}$ is $\frac{1}{5}$.

Practice Problem Multiply $\frac{3}{14}$ by $\frac{5}{16}$.

Find a Reciprocal Two numbers whose product is 1 are called multiplicative inverses, or reciprocals.

Example

Find the reciprocal of $\frac{3}{8}$.

Step 1 Inverse the fraction by putting the denominator on top and the numerator on the bottom.

$\frac{8}{3}$

The reciprocal of $\frac{3}{8}$ is $\frac{8}{3}$.

Practice Problem Find the reciprocal of $\frac{4}{9}$.

Divide Fractions To divide one fraction by another fraction, multiply the dividend by the reciprocal of the divisor. Find the simplest form if necessary.

Example 1

Divide $\frac{1}{9}$ by $\frac{1}{3}$.

Step 1 Find the reciprocal of the divisor.

The reciprocal of $\frac{1}{3}$ is $\frac{3}{1}$.

Step 2 Multiply the dividend by the reciprocal of the divisor.

$\frac{\frac{1}{9}}{\frac{1}{3}} = \frac{1}{9} \times \frac{3}{1} = \frac{(1 \times 3)}{(9 \times 1)} = \frac{3}{9}$

Step 3 Find the GCF.

$\frac{3}{9}$ (GCF, 3)

Step 4 Divide the numerator and denominator by the GCF.

$\frac{3 \div 3}{9 \div 3} = \frac{1}{3}$

$\frac{1}{9}$ divided by $\frac{1}{3}$ is $\frac{1}{3}$.

Example 2

Divide $\frac{3}{5}$ by $\frac{1}{4}$.

Step 1 Find the reciprocal of the divisor.

The reciprocal of $\frac{1}{4}$ is $\frac{4}{1}$.

Step 2 Multiply the dividend by the reciprocal of the divisor.

$\frac{\frac{3}{5}}{\frac{1}{4}} = \frac{3}{5} \times \frac{4}{1} = \frac{(3 \times 4)}{(5 \times 1)} = \frac{12}{5}$

$\frac{3}{5}$ divided by $\frac{1}{4}$ is $\frac{12}{5}$ or $2\frac{2}{5}$.

Practice Problem Divide $\frac{3}{11}$ by $\frac{7}{10}$.

Use Ratios

When you compare two numbers by division, you are using a ratio. Ratios can be written 3 to 5, 3:5, or $\frac{3}{5}$. Ratios, like fractions, also can be written in simplest form.

Ratios can represent one type of probability, called odds. This is a ratio that compares the number of ways a certain outcome occurs to the number of possible outcomes. For example, if you flip a coin 100 times, what are the odds that it will come up heads? There are two possible outcomes, heads or tails, so the odds of coming up heads are 50:100. Another way to say this is that 50 out of 100 times the coin will come up heads. In its simplest form, the ratio is 1:2.

Example 1

A chemical solution contains 40 g of salt and 64 g of baking soda. What is the ratio of salt to baking soda as a fraction in simplest form?

Step 1 Write the ratio as a fraction.

$$\frac{\text{salt}}{\text{baking soda}} = \frac{40}{64}$$

Step 2 Express the fraction in simplest form. The GCF of 40 and 64 is 8.

$$\frac{40}{64} = \frac{40 \div 8}{64 \div 8} = \frac{5}{8}$$

The ratio of salt to baking soda in the sample is 5:8.

Example 2

Sean rolls a 6-sided die 6 times. What are the odds that the side with a 3 will show?

Step 1 Write the ratio as a fraction.

$$\frac{\text{number of sides with a 3}}{\text{number of possible sides}} = \frac{1}{6}$$

Step 2 Multiply by the number of attempts.

$\frac{1}{6} \times 6$ attempts $= \frac{6}{6}$ attempts $= 1$ attempt

1 attempt out of 6 will show a 3.

Practice Problem Two metal rods measure 100 cm and 144 cm in length. What is the ratio of their lengths in simplest form?

Use Decimals

A fraction with a denominator that is a power of ten can be written as a decimal. For example, 0.27 means $\frac{27}{100}$. The decimal point separates the ones place from the tenths place.

Any fraction can be written as a decimal using division. For example, the fraction $\frac{5}{8}$ can be written as a decimal by dividing 5 by 8. Written as a decimal, it is 0.625.

Add or Subtract Decimals When adding and subtracting decimals, line up the decimal points before carrying out the operation.

Example 1

Find the sum of 47.68 and 7.80.

Step 1 Line up the decimal places when you write the numbers.

$$\begin{array}{r} 47.68 \\ +\ 7.80 \\ \hline \end{array}$$

Step 2 Add the decimals.

$$\begin{array}{r} \scriptstyle{1\,1} \\ 47.68 \\ +\ 7.80 \\ \hline 55.48 \end{array}$$

The sum of 47.68 and 7.80 is 55.48.

Example 2

Find the difference of 42.17 and 15.85.

Step 1 Line up the decimal places when you write the number.

$$\begin{array}{r} 42.17 \\ -15.85 \\ \hline \end{array}$$

Step 2 Subtract the decimals.

$$\begin{array}{r} \scriptstyle{3\,11\,1} \\ 42.17 \\ -15.85 \\ \hline 26.32 \end{array}$$

The difference of 42.17 and 15.85 is 26.32.

Practice Problem Find the sum of 1.245 and 3.842.

Multiply Decimals To multiply decimals, multiply the numbers like numbers without decimal points. Count the decimal places in each factor. The product will have the same number of decimal places as the sum of the decimal places in the factors.

Example

Multiply 2.4 by 5.9.

Step 1 Multiply the factors like two whole numbers.

$24 \times 59 = 1416$

Step 2 Find the sum of the number of decimal places in the factors. Each factor has one decimal place, for a sum of two decimal places.

Step 3 The product will have two decimal places.

14.16

The product of 2.4 and 5.9 is 14.16.

Practice Problem Multiply 4.6 by 2.2.

Divide Decimals When dividing decimals, change the divisor to a whole number. To do this, multiply both the divisor and the dividend by the same power of ten. Then place the decimal point in the quotient directly above the decimal point in the dividend. Then divide as you do with whole numbers.

Example

Divide 8.84 by 3.4.

Step 1 Multiply both factors by 10.

$3.4 \times 10 = 34, 8.84 \times 10 = 88.4$

Step 2 Divide 88.4 by 34.

```
   2.6
34)88.4
  -68
   204
  -204
     0
```

8.84 divided by 3.4 is 2.6.

Practice Problem Divide 75.6 by 3.6.

Use Proportions

An equation that shows that two ratios are equivalent is a proportion. The ratios $\frac{2}{4}$ and $\frac{5}{10}$ are equivalent, so they can be written as $\frac{2}{4} = \frac{5}{10}$. This equation is a proportion.

When two ratios form a proportion, the cross products are equal. To find the cross products in the proportion $\frac{2}{4} = \frac{5}{10}$, multiply the 2 and the 10, and the 4 and the 5. Therefore $2 \times 10 = 4 \times 5$, or $20 = 20$.

Because you know that both ratios are equal, you can use cross products to find a missing term in a proportion. This is known as solving the proportion.

Example

The heights of a tree and a pole are proportional to the lengths of their shadows. The tree casts a shadow of 24 m when a 6-m pole casts a shadow of 4 m. What is the height of the tree?

Step 1 Write a proportion.

$$\frac{\text{height of tree}}{\text{height of pole}} = \frac{\text{length of tree's shadow}}{\text{length of pole's shadow}}$$

Step 2 Substitute the known values into the proportion. Let h represent the unknown value, the height of the tree.

$\frac{h}{6} \times \frac{24}{4}$

Step 3 Find the cross products.

$h \times 4 = 6 \times 24$

Step 4 Simplify the equation.

$4h \times 144$

Step 5 Divide each side by 4.

$\frac{4h}{4} \times \frac{144}{4}$

$h = 36$

The height of the tree is 36 m.

Practice Problem The ratios of the weights of two objects on the Moon and on Earth are in proportion. A rock weighing 3 N on the Moon weighs 18 N on Earth. How much would a rock that weighs 5 N on the Moon weigh on Earth?

Use Percentages

The word *percent* means "out of one hundred." It is a ratio that compares a number to 100. Suppose you read that 77 percent of Earth's surface is covered by water. That is the same as reading that the fraction of Earth's surface covered by water is $\frac{77}{100}$. To express a fraction as a percent, first find the equivalent decimal for the fraction. Then, multiply the decimal by 100 and add the percent symbol.

Example 1

Express $\frac{13}{20}$ as a percent.

Step 1 Find the equivalent decimal for the fraction.

$$\begin{array}{r} 0.65 \\ 20\overline{)13.00} \\ \underline{12\,0} \\ 1\,00 \\ \underline{1\,00} \\ 0 \end{array}$$

Step 2 Rewrite the fraction $\frac{13}{20}$ as 0.65.

Step 3 Multiply 0.65 by 100 and add the % symbol.

$0.65 \times 100 = 65 = 65\%$

So, $\frac{13}{20} = 65\%$.

This also can be solved as a proportion.

Example 2

Express $\frac{13}{20}$ as a percent.

Step 1 Write a proportion.

$\frac{13}{20} = \frac{x}{100}$

Step 2 Find the cross products.

$1300 = 20x$

Step 3 Divide each side by 20.

$\frac{1300}{20} = \frac{20x}{20}$

$65\% = x$

Practice Problem In one year, 73 of 365 days were rainy in one city. What percent of the days in that city were rainy?

Solve One-Step Equations

A statement that two expressions are equal is an equation. For example, $A = B$ is an equation that states that A is equal to B.

An equation is solved when a variable is replaced with a value that makes both sides of the equation equal. To make both sides equal the inverse operation is used. Addition and subtraction are inverses, and multiplication and division are inverses.

Example 1

Solve the equation $x - 10 = 35$.

Step 1 Find the solution by adding 10 to each side of the equation.

$$x - 10 = 35$$
$$x - 10 + 10 = 35 - 10$$
$$x = 45$$

Step 2 Check the solution.

$$x - 10 = 35$$
$$45 - 10 = 35$$
$$35 = 35$$

Both sides of the equation are equal, so $x = 45$.

Example 2

In the formula $a = bc$, find the value of c if $a = 20$ and $b = 2$.

Step 1 Rearrange the formula so the unknown value is by itself on one side of the equation by dividing both sides by b.

$$a = bc$$
$$\frac{a}{b} = \frac{bc}{b}$$
$$\frac{a}{b} = c$$

Step 2 Replace the variables a and b with the values that are given.

$$\frac{a}{b} = c$$
$$\frac{20}{2} = c$$
$$10 = c$$

Step 3 Check the solution.

$$a = bc$$
$$20 = 2 \times 10$$
$$20 = 20$$

Both sides of the equation are equal, so $c = 10$ is the solution when $a = 20$ and $b = 2$.

Practice Problem In the formula $h = gd$, find the value of d if $g = 12.3$ and $h = 17.4$.

Science Skill Handbook
Math Skill Handbook
Foldables Handbook
Reference Handbook
Glossary/Glosario
Index

Use Statistics

The branch of mathematics that deals with collecting, analyzing, and presenting data is statistics. In statistics, there are three common ways to summarize data with a single number—the mean, the median, and the mode.

The **mean** of a set of data is the arithmetic average. It is found by adding the numbers in the data set and dividing by the number of items in the set.

The **median** is the middle number in a set of data when the data are arranged in numerical order. If there were an even number of data points, the median would be the mean of the two middle numbers.

The **mode** of a set of data is the number or item that appears most often.

Another number that often is used to describe a set of data is the range. The **range** is the difference between the largest number and the smallest number in a set of data.

Example

The speeds (in m/s) for a race car during five different time trials are 39, 37, 44, 36, and 44.

To find the mean:

Step 1 Find the sum of the numbers.

$39 + 37 + 44 + 36 + 44 = 200$

Step 2 Divide the sum by the number of items, which is 5.

$200 \div 5 = 40$

The mean is 40 m/s.

To find the median:

Step 1 Arrange the measures from least to greatest.

36, 37, 39, 44, 44

Step 2 Determine the middle measure.

36, 37, 39, 44, 44

The median is 39 m/s.

To find the mode:

Step 1 Group the numbers that are the same together.

44, 44, 36, 37, 39

Step 2 Determine the number that occurs most in the set.

44, 44, 36, 37, 39

The mode is 44 m/s.

To find the range:

Step 1 Arrange the measures from greatest to least.

44, 44, 39, 37, 36

Step 2 Determine the greatest and least measures in the set.

44, 44, 39, 37, 36

Step 3 Find the difference between the greatest and least measures.

$44 - 36 = 8$

The range is 8 m/s.

Practice Problem Find the mean, median, mode, and range for the data set 8, 4, 12, 8, 11, 14, 16.

A **frequency table** shows how many times each piece of data occurs, usually in a survey. **Table 1** below shows the results of a student survey on favorite color.

Table 1 Student Color Choice

Color	Tally	Frequency
red	IIII	4
blue	𝍸	5
black	II	2
green	III	3
purple	𝍸 II	7
yellow	𝍸 I	6

Based on the frequency table data, which color is the favorite?

Use Geometry

The branch of mathematics that deals with the measurement, properties, and relationships of points, lines, angles, surfaces, and solids is called geometry.

Perimeter The **perimeter** (P) is the distance around a geometric figure. To find the perimeter of a rectangle, add the length and width and multiply that sum by two, or $2(l + w)$. To find perimeters of irregular figures, add the length of the sides.

Example 1

Find the perimeter of a rectangle that is 3 m long and 5 m wide.

Step 1 You know that the perimeter is 2 times the sum of the width and length.

$P = 2(3 \text{ m} + 5 \text{ m})$

Step 2 Find the sum of the width and length.

$P = 2(8 \text{ m})$

Step 3 Multiply by 2.

$P = 16 \text{ m}$

The perimeter is 16 m.

Example 2

Find the perimeter of a shape with sides measuring 2 cm, 5 cm, 6 cm, 3 cm.

Step 1 You know that the perimeter is the sum of all the sides.

$P = 2 + 5 + 6 + 3$

Step 2 Find the sum of the sides.

$P = 2 + 5 + 6 + 3$

$P = 16$

The perimeter is 16 cm.

Practice Problem Find the perimeter of a rectangle with a length of 18 m and a width of 7 m.

Practice Problem Find the perimeter of a triangle measuring 1.6 cm by 2.4 cm by 2.4 cm.

Area of a Rectangle The **area** (A) is the number of square units needed to cover a surface. To find the area of a rectangle, multiply the length times the width, or $l \times w$. When finding area, the units also are multiplied. Area is given in square units.

Example

Find the area of a rectangle with a length of 1 cm and a width of 10 cm.

Step 1 You know that the area is the length multiplied by the width.

$A = (1 \text{ cm} \times 10 \text{ cm})$

Step 2 Multiply the length by the width. Also multiply the units.

$A = 10 \text{ cm}^2$

The area is 10 cm^2.

Practice Problem Find the area of a square whose sides measure 4 m.

Area of a Triangle To find the area of a triangle, use the formula:

$A = \frac{1}{2}(\text{base} \times \text{height})$

The base of a triangle can be any of its sides. The height is the perpendicular distance from a base to the opposite endpoint, or vertex.

Example

Find the area of a triangle with a base of 18 m and a height of 7 m.

Step 1 You know that the area is $\frac{1}{2}$ the base times the height.

$A = \frac{1}{2}(18 \text{ m} \times 7 \text{ m})$

Step 2 Multiply $\frac{1}{2}$ by the product of 18×7. Multiply the units.

$A = \frac{1}{2}(126 \text{ m}^2)$

$A = 63 \text{ m}^2$

The area is 63 m^2.

Practice Problem Find the area of a triangle with a base of 27 cm and a height of 17 cm.

Circumference of a Circle The **diameter** (d) of a circle is the distance across the circle through its center, and the **radius** (r) is the distance from the center to any point on the circle. The radius is half of the diameter. The distance around the circle is called the **circumference** (C). The formula for finding the circumference is:

$C = 2\pi r$ or $C = \pi d$

The circumference divided by the diameter is always equal to 3.1415926... This nonterminating and nonrepeating number is represented by the Greek letter π (pi). An approximation often used for π is 3.14.

Example 1

Find the circumference of a circle with a radius of 3 m.

Step 1 You know the formula for the circumference is 2 times the radius times π.

$C = 2\pi(3)$

Step 2 Multiply 2 times the radius.

$C = 6\pi$

Step 3 Multiply by π.

$C \approx 19$ m

The circumference is about 19 m.

Example 2

Find the circumference of a circle with a diameter of 24.0 cm.

Step 1 You know the formula for the circumference is the diameter times π.

$C = \pi(24.0)$

Step 2 Multiply the diameter by π.

$C \approx 75.4$ cm

The circumference is about 75.4 cm.

Practice Problem Find the circumference of a circle with a radius of 19 cm.

Area of a Circle The formula for the area of a circle is: $A = \pi r^2$

Example 1

Find the area of a circle with a radius of 4.0 cm.

Step 1 $A = \pi(4.0)^2$

Step 2 Find the square of the radius.

$A = 16\pi$

Step 3 Multiply the square of the radius by π.

$A \approx 50\ \text{cm}^2$

The area of the circle is about 50 cm^2.

Example 2

Find the area of a circle with a radius of 225 m.

Step 1 $A = \pi(225)^2$

Step 2 Find the square of the radius.

$A = 50625\pi$

Step 3 Multiply the square of the radius by π.

$A \approx 159043.1$

The area of the circle is about 159043.1 m^2.

Example 3

Find the area of a circle whose diameter is 20.0 mm.

Step 1 Remember that the radius is half of the diameter.

$A = \pi\left(\frac{20.0}{2}\right)^2$

Step 2 Find the radius.

$A = \pi(10.0)^2$

Step 3 Find the square of the radius.

$A = 100\pi$

Step 4 Multiply the square of the radius by π.

$A \approx 314\ \text{mm}^2$

The area of the circle is about 314 mm^2.

Practice Problem Find the area of a circle with a radius of 16 m.

Science Skill Handbook
Math Skill Handbook
Foldables Handbook
Reference Handbook
Glossary/Glosario
Index

Volume The measure of space occupied by a solid is the **volume** (V). To find the volume of a rectangular solid multiply the length times width times height, or $V = l \times w \times h$. It is measured in cubic units, such as cubic centimeters (cm^3).

Example

Find the volume of a rectangular solid with a length of 2.0 m, a width of 4.0 m, and a height of 3.0 m.

Step 1 You know the formula for volume is the length times the width times the height.

$V = 2.0 \text{ m} \times 4.0 \text{ m} \times 3.0 \text{ m}$

Step 2 Multiply the length times the width times the height.

$V = 24 \text{ m}^3$

The volume is 24 m^3.

Practice Problem Find the volume of a rectangular solid that is 8 m long, 4 m wide, and 4 m high.

To find the volume of other solids, multiply the area of the base times the height.

Example 1

Find the volume of a solid that has a triangular base with a length of 8.0 m and a height of 7.0 m. The height of the entire solid is 15.0 m.

Step 1 You know that the base is a triangle, and the area of a triangle is $\frac{1}{2}$ the base times the height, and the volume is the area of the base times the height.

$V = \left[\frac{1}{2}(b \times h)\right] \times 15$

Step 2 Find the area of the base.

$V = \left[\frac{1}{2}(8 \times 7)\right] \times 15$

$V = \left(\frac{1}{2} \times 56\right) \times 15$

Step 3 Multiply the area of the base by the height of the solid.

$V = 28 \times 15$

$V = 420 \text{ m}^3$

The volume is 420 m^3.

Example 2

Find the volume of a cylinder that has a base with a radius of 12.0 cm, and a height of 21.0 cm.

Step 1 You know that the base is a circle, and the area of a circle is the square of the radius times π, and the volume is the area of the base times the height.

$V = (\pi r^2) \times 21$

$V = (\pi 12^2) \times 21$

Step 2 Find the area of the base.

$V = 144\pi \times 21$

$V = 452 \times 21$

Step 3 Multiply the area of the base by the height of the solid.

$V \approx 9{,}500 \text{ cm}^3$

The volume is about 9,500 cm^3.

Example 3

Find the volume of a cylinder that has a diameter of 15 mm and a height of 4.8 mm.

Step 1 You know that the base is a circle with an area equal to the square of the radius times π. The radius is one-half the diameter. The volume is the area of the base times the height.

$V = (\pi r^2) \times 4.8$

$V = \left[\pi\left(\frac{1}{2} \times 15\right)^2\right] \times 4.8$

$V = (\pi 7.5^2) \times 4.8$

Step 2 Find the area of the base.

$V = 56.25\pi \times 4.8$

$V \approx 176.71 \times 4.8$

Step 3 Multiply the area of the base by the height of the solid.

$V \approx 848.2$

The volume is about 848.2 mm^3.

Practice Problem Find the volume of a cylinder with a diameter of 7 cm in the base and a height of 16 cm.

Science Applications

Measure in SI

The metric system of measurement was developed in 1795. A modern form of the metric system, called the International System (SI), was adopted in 1960 and provides the standard measurements that all scientists around the world can understand.

The SI system is convenient because unit sizes vary by powers of 10. Prefixes are used to name units. Look at **Table 2** for some common SI prefixes and their meanings.

Table 2 Common SI Prefixes

Prefix	Symbol	Meaning	
kilo–	k	1,000	thousandth
hecto–	h	100	hundred
deka–	da	10	ten
deci–	d	0.1	tenth
centi–	c	0.01	hundreth
milli–	m	0.001	thousandth

Example

How many grams equal one kilogram?

Step 1 Find the prefix *kilo–* in **Table 2.**

Step 2 Using **Table 2,** determine the meaning of *kilo–*. According to the table, it means 1,000. When the prefix *kilo–* is added to a unit, it means that there are 1,000 of the units in a "kilounit."

Step 3 Apply the prefix to the units in the question. The units in the question are grams. There are 1,000 grams in a kilogram.

Practice Problem Is a milligram larger or smaller than a gram? How many of the smaller units equal one larger unit? What fraction of the larger unit does one smaller unit represent?

Dimensional Analysis

Convert SI Units In science, quantities such as length, mass, and time sometimes are measured using different units. A process called dimensional analysis can be used to change one unit of measure to another. This process involves multiplying your starting quantity and units by one or more conversion factors. A conversion factor is a ratio equal to one and can be made from any two equal quantities with different units. If 1,000 mL equal 1 L then two ratios can be made.

$$\frac{1{,}000 \text{ mL}}{1 \text{ L}} = \frac{1 \text{ L}}{1{,}000 \text{ mL}} = 1$$

One can convert between units in the SI system by using the equivalents in **Table 2** to make conversion factors.

Example

How many cm are in 4 m?

Step 1 Write conversion factors for the units given. From **Table 2,** you know that 100 cm = 1 m. The conversion factors are

$$\frac{100 \text{ cm}}{1 \text{ m}} \text{ and } \frac{1 \text{ m}}{100 \text{ cm}}$$

Step 2 Decide which conversion factor to use. Select the factor that has the units you are converting from (m) in the denominator and the units you are converting to (cm) in the numerator.

$$\frac{100 \text{ cm}}{1 \text{ m}}$$

Step 3 Multiply the starting quantity and units by the conversion factor. Cancel the starting units with the units in the denominator. There are 400 cm in 4 m.

$$4 \cancel{\text{m}} = \frac{100 \text{ cm}}{1 \cancel{\text{m}}} = 400 \text{ cm}$$

Practice Problem How many milligrams are in one kilogram? (Hint: You will need to use two conversion factors from **Table 2.**)

Table 3 Unit System Equivalents

Type of Measurement	Equivalent
Length	1 in = 2.54 cm 1 yd = 0.91 m 1 mi = 1.61 km
Mass and weight*	1 oz = 28.35 g 1 lb = 0.45 kg 1 ton (short) = 0.91 tonnes (metric tons) 1 lb = 4.45 N
Volume	1 in^3 = 16.39 cm^3 1 qt = 0.95 L 1 gal = 3.78 L
Area	1 in^2 = 6.45 cm^2 1 yd^2 = 0.83 m^2 1 mi^2 = 2.59 km^2 1 acre = 0.40 hectares
Temperature	$°C = \frac{(°F - 32)}{1.8}$ K = °C + 273

*Weight is measured in standard Earth gravity.

Convert Between Unit Systems **Table 3** gives a list of equivalents that can be used to convert between English and SI units.

Example

If a meterstick has a length of 100 cm, how long is the meterstick in inches?

Step 1 Write the conversion factors for the units given. From **Table 3,** 1 in = 2.54 cm.

$$\frac{1 \text{ in}}{2.54 \text{ cm}} \textit{ and } \frac{2.54 \text{ cm}}{1 \text{ in}}$$

Step 2 Determine which conversion factor to use. You are converting from cm to in. Use the conversion factor with cm on the bottom.

$$\frac{1 \text{ in}}{2.54 \text{ cm}}$$

Step 3 Multiply the starting quantity and units by the conversion factor. Cancel the starting units with the units in the denominator. Round your answer to the nearest tenth.

$$100 \cancel{\text{cm}} \times \frac{1 \text{ in}}{2.54 \cancel{\text{cm}}} = 39.37 \text{ in}$$

The meterstick is about 39.4 in long.

Practice Problem 1 A book has a mass of 5 lb. What is the mass of the book in kg?

Practice Problem 2 Use the equivalent for in and cm (1 in = 2.54 cm) to show how 1 $in^3 \approx$ 16.39 cm^3.

Precision and Significant Digits

When you make a measurement, the value you record depends on the precision of the measuring instrument. This precision is represented by the number of significant digits recorded in the measurement. When counting the number of significant digits, all digits are counted except zeros at the end of a number with no decimal point such as 2,050, and zeros at the beginning of a decimal such as 0.03020. When adding or subtracting numbers with different precision, round the answer to the smallest number of decimal places of any number in the sum or difference. When multiplying or dividing, the answer is rounded to the smallest number of significant digits of any number being multiplied or divided.

Example

The lengths 5.28 and 5.2 are measured in meters. Find the sum of these lengths and record your answer using the correct number of significant digits.

Step 1 Find the sum.

5.28 m	2 digits after the decimal
+ 5.2 m	1 digit after the decimal
10.48 m	

Step 2 Round to one digit after the decimal because the least number of digits after the decimal of the numbers being added is 1.

The sum is 10.5 m.

Practice Problem 1 How many significant digits are in the measurement 7,071,301 m? How many significant digits are in the measurement 0.003010 g?

Practice Problem 2 Multiply 5.28 and 5.2 using the rule for multiplying and dividing. Record the answer using the correct number of significant digits.

Scientific Notation

Many times numbers used in science are very small or very large. Because these numbers are difficult to work with scientists use scientific notation. To write numbers in scientific notation, move the decimal point until only one non-zero digit remains on the left. Then count the number of places you moved the decimal point and use that number as a power of ten. For example, the average distance from the Sun to Mars is 227,800,000,000 m. In scientific notation, this distance is 2.278×10^{11} m. Because you moved the decimal point to the left, the number is a positive power of ten.

The mass of an electron is about 0.000 000 000 000 000 000 000 000 000 000 911 kg. Expressed in scientific notation, this mass is 9.11×10^{-31} kg. Because the decimal point was moved to the right, the number is a negative power of ten.

Example

Earth is 149,600,000 km from the Sun. Express this in scientific notation.

Step 1 Move the decimal point until one non-zero digit remains on the left.

1.496 000 00

Step 2 Count the number of decimal places you have moved. In this case, eight.

Step 2 Show that number as a power of ten, 10^8.

Earth is 1.496×10^8 km from the Sun.

Practice Problem 1 How many significant digits are in 149,600,000 km? How many significant digits are in 1.496×10^8 km?

Practice Problem 2 Parts used in a high performance car must be measured to 7×10^{-6} m. Express this number as a decimal.

Practice Problem 3 A CD is spinning at 539 revolutions per minute. Express this number in scientific notation.

Make and Use Graphs

Data in tables can be displayed in a graph—a visual representation of data. Common graph types include line graphs, bar graphs, and circle graphs.

Line Graph A line graph shows a relationship between two variables that change continuously. The independent variable is changed and is plotted on the x-axis. The dependent variable is observed, and is plotted on the *y*-axis.

Example

Draw a line graph of the data below from a cyclist in a long-distance race.

Table 4 Bicycle Race Data

Time (h)	Distance (km)
0	0
1	8
2	16
3	24
4	32
5	40

Step 1 Determine the *x*-axis and *y*-axis variables. Time varies independently of distance and is plotted on the *x*-axis. Distance is dependent on time and is plotted on the *y*-axis.

Step 2 Determine the scale of each axis. The *x*-axis data ranges from 0 to 5. The *y*-axis data ranges from 0 to 50.

Step 3 Using graph paper, draw and label the axes. Include units in the labels.

Step 4 Draw a point at the intersection of the time value on the *x*-axis and corresponding distance value on the *y*-axis. Connect the points and label the graph with a title, as shown in **Figure 8.**

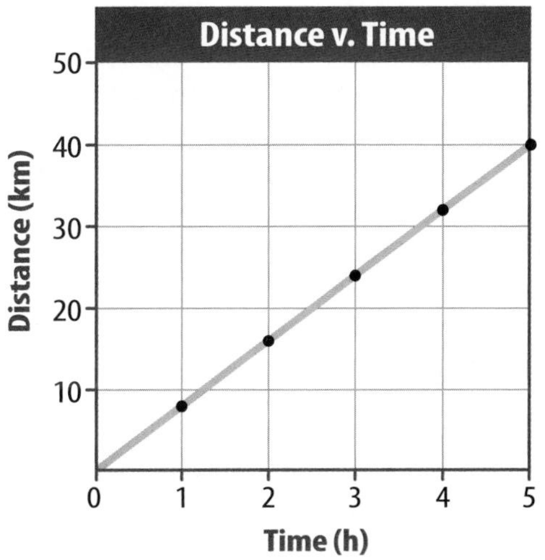

Figure 8 This line graph shows the relationship between distance and time during a bicycle ride.

Practice Problem A puppy's shoulder height is measured during the first year of her life. The following measurements were collected: (3 mo, 52 cm), (6 mo, 72 cm), (9 mo, 83 cm), (12 mo, 86 cm). Graph this data.

Find a Slope The slope of a straight line is the ratio of the vertical change, rise, to the horizontal change, run.

$$\text{Slope} = \frac{\text{vertical change (rise)}}{\text{horizontal change (run)}} = \frac{\text{change in } y}{\text{change in } x}$$

Example

Find the slope of the graph in **Figure 8**.

Step 1 You know that the slope is the change in *y* divided by the change in *x*.

$$\text{Slope} = \frac{\text{change in } y}{\text{change in } x}$$

Step 2 Determine the data points you will be using. For a straight line, choose the two sets of points that are the farthest apart.

$$\text{Slope} = \frac{(40 - 0)\text{ km}}{(5 - 0)\text{ h}}$$

Step 3 Find the change in *y* and *x*.

$$\text{Slope} = \frac{40\text{ km}}{5\text{ h}}$$

Step 4 Divide the change in *y* by the change in *x*.

$$\text{Slope} = \frac{8\text{ km}}{\text{h}}$$

The slope of the graph is 8 km/h.

Bar Graph To compare data that does not change continuously you might choose a bar graph. A bar graph uses bars to show the relationships between variables. The *x*-axis variable is divided into parts. The parts can be numbers such as years, or a category such as a type of animal. The *y*-axis is a number and increases continuously along the axis.

Example

A recycling center collects 4.0 kg of aluminum on Monday, 1.0 kg on Wednesday, and 2.0 kg on Friday. Create a bar graph of this data.

Step 1 Select the *x*-axis and *y*-axis variables. The measured numbers (the masses of aluminum) should be placed on the *y*-axis. The variable divided into parts (collection days) is placed on the *x*-axis.

Step 2 Create a graph grid like you would for a line graph. Include labels and units.

Step 3 For each measured number, draw a vertical bar above the *x*-axis value up to the *y*-axis value. For the first data point, draw a vertical bar above Monday up to 4.0 kg.

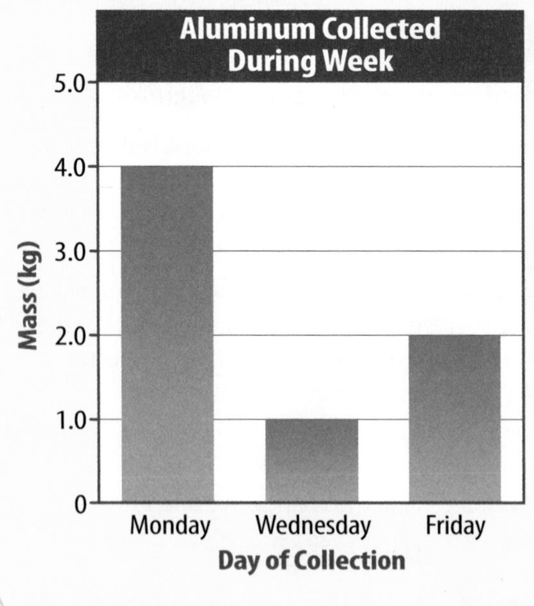

Practice Problem Draw a bar graph of the gases in air: 78% nitrogen, 21% oxygen, 1% other gases.

Circle Graph To display data as parts of a whole, you might use a circle graph. A circle graph is a circle divided into sections that represent the relative size of each piece of data. The entire circle represents 100%, half represents 50%, and so on.

Example

Air is made up of 78% nitrogen, 21% oxygen, and 1% other gases. Display the composition of air in a circle graph.

Step 1 Multiply each percent by 360° and divide by 100 to find the angle of each section in the circle.

$$78\% \times \frac{360^\circ}{100} = 280.8^\circ$$

$$21\% \times \frac{360^\circ}{100} = 75.6^\circ$$

$$1\% \times \frac{360^\circ}{100} = 3.6^\circ$$

Step 2 Use a compass to draw a circle and to mark the center of the circle. Draw a straight line from the center to the edge of the circle.

Step 3 Use a protractor and the angles you calculated to divide the circle into parts. Place the center of the protractor over the center of the circle and line the base of the protractor over the straight line.

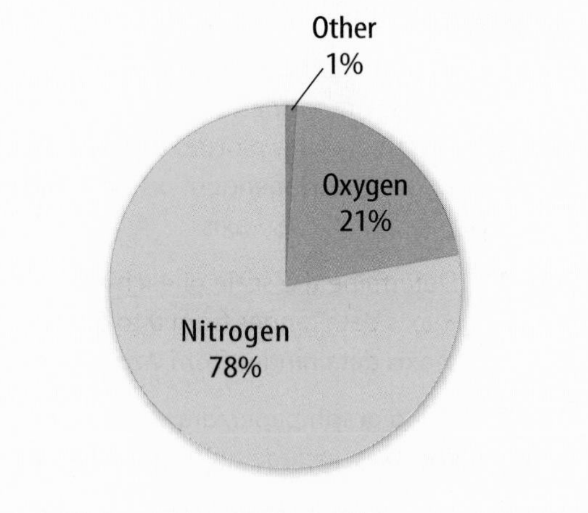

Practice Problem Draw a circle graph to represent the amount of aluminum collected during the week shown in the bar graph to the left.

Student Study Guides & Instructions
By Dinah Zike

1. You will find suggestions for Study Guides, also known as Foldables or books, in each chapter lesson and as a final project. Look at the end of the chapter to determine the project format and glue the Foldables in place as you progress through the chapter lessons.

2. Creating the Foldables or books is simple and easy to do by using copy paper, art paper, and internet printouts. Photocopies of maps, diagrams, or your own illustrations may also be used for some of the Foldables. Notebook paper is the most common source of material for study guides and 83% of all Foldables are created from it. When folded to make books, notebook paper Foldables easily fit into 11″ × 17″ or 12″ × 18″ chapter projects with space left over. Foldables made using photocopy paper are slightly larger and they fit into Projects, but snugly. Use the least amount of glue, tape, and staples needed to assemble the Foldables.

3. Seven of the Foldables can be made using either small or large paper. When 11″ × 17″ or 12″ × 18″ paper is used, these become projects for housing smaller Foldables. Project format boxes are located within the instructions to remind you of this option.

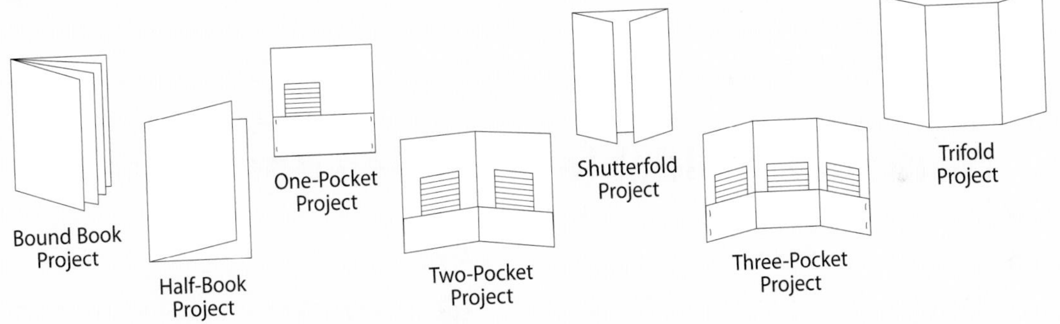

4. Use one-gallon self-locking plastic bags to store your projects. Place strips of two-inch clear tape along the left, long side of the bag and punch holes through the taped edge. Cut the bottom corners off the bag so it will not hold air. Store this Project Portfolio inside a three-hole binder. To store a large collection of project bags, use a giant laundry-soap box. Holes can be punched in some of the Foldable Projects so they can be stored in a three-hole binder without using a plastic bag. Punch holes in the pocket books before gluing or stapling the pocket.

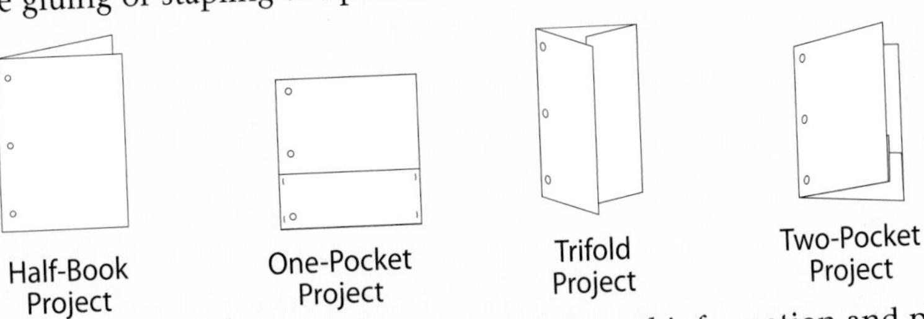

5. Maximize the use of the projects by collecting additional information and placing it on the back of the project and other unused spaces of the large Foldables.

Half-Book Foldable® By Dinah Zike

Step 1 Fold a sheet of notebook or copy paper in half.

Label the exterior tab and use the inside space to write information.

PROJECT FORMAT
Use 11″ × 17″ or 12″ × 18″ paper on the horizontal axis to make a large project book.

Variations

Paper can be folded horizontally, like a *hamburger* or vertically, like a *hot dog.*

A 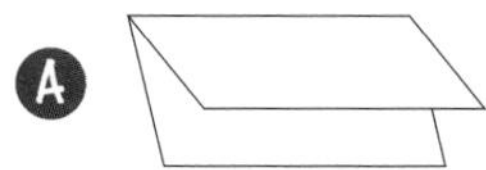

B 

C Half-books can be folded so that one side is ½ inch longer than the other side. A title or question can be written on the extended tab.

Worksheet Foldable or Folded Book® By Dinah Zike

Step 1 Make a half-book (see above) using work sheets, internet print-outs, diagrams, or maps.

Step 2 Fold it in half again.

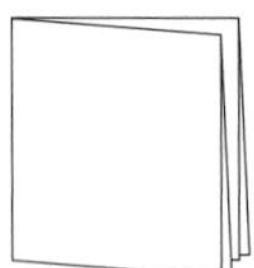

Variations

A This folded sheet as a small book with two pages can be used for comparing and contrasting, cause and effect, or other skills.

B When the sheet of paper is open, the four sections can be used separately or used collectively to show sequences or steps.

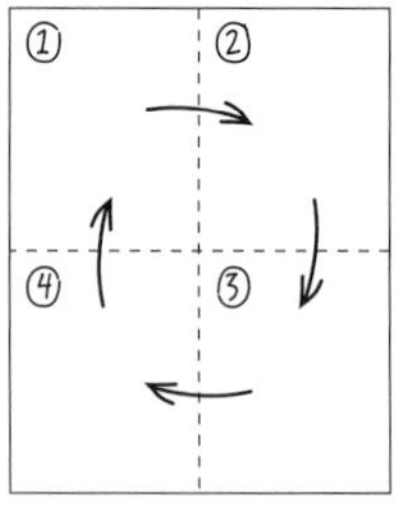

Two-Tab and Concept-Map Foldable® By Dinah Zike

Step 1 Fold a sheet of notebook or copy paper in half vertically or horizontally.

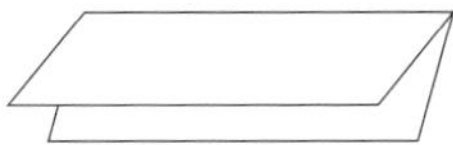

Step 2 Fold it in half again, as shown.

Step 3 Unfold once and cut along the fold line or valley of the top flap to make two flaps.

Variations

A Concept maps can be made by leaving a ½ inch tab at the top when folding the paper in half. Use arrows and labels to relate topics to the primary concept.

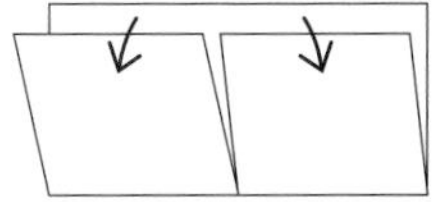

B Use two sheets of paper to make multiple page tab books. Glue or staple books together at the top fold.

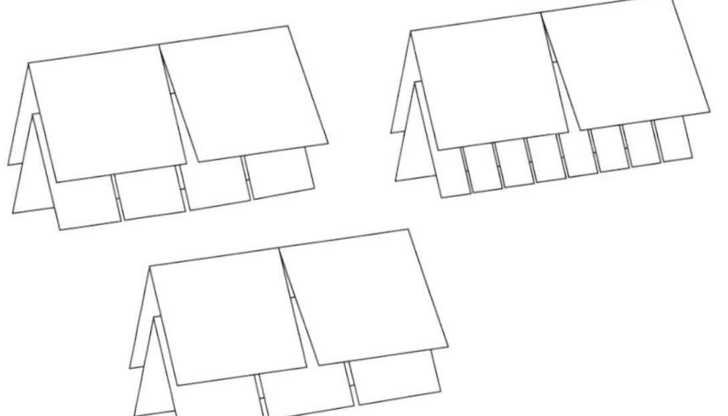

Three-Quarter Foldable® By Dinah Zike

Step 1 Make a two-tab book (see above) and cut the left tab off at the top of the fold line.

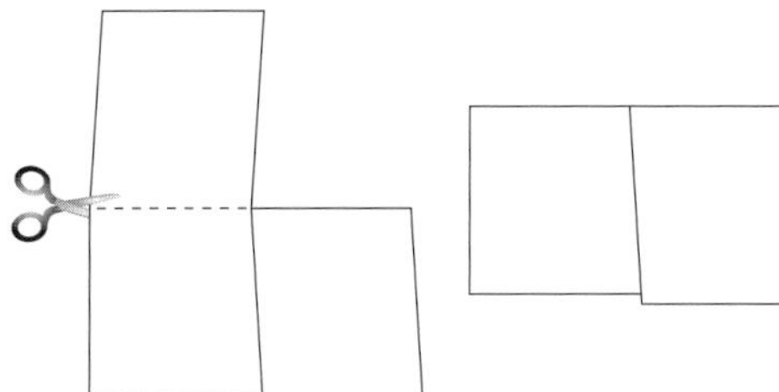

Variations

A Use this book to draw a diagram or a map on the exposed left tab. Write questions about the illustration on the top right tab and provide complete answers on the space under the tab.

B Compose a self-test using multiple choice answers for your questions. Include the correct answer with three wrong responses. The correct answers can be written on the back of the book or upside down on the bottom of the inside page.

Three-Tab Foldable® By Dinah Zike

Step 1 Fold a sheet of paper in half horizontally.

Step 2 Fold into thirds.

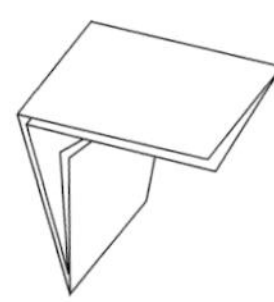

Step 3 Unfold and cut along the folds of the top flap to make three sections.

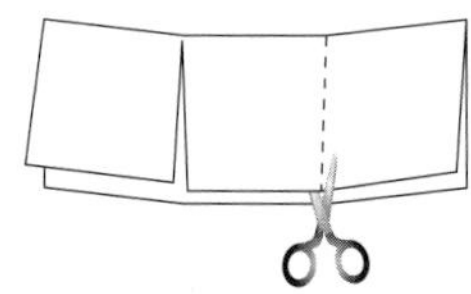

Variations

A Before cutting the three tabs draw a Venn diagram across the front of the book.

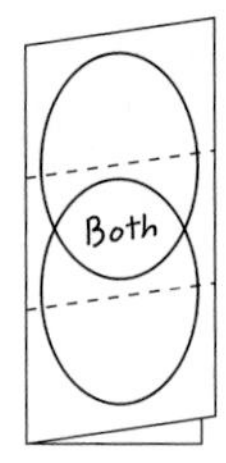

B Make a space to use for titles or concept maps by leaving a ½ inch tab at the top when folding the paper in half.

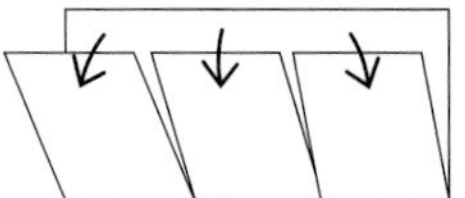

Four-Tab Foldable® By Dinah Zike

Step 1 Fold a sheet of paper in half horizontally.

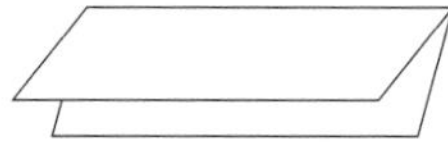

Step 2 Fold in half and then fold each half as shown below.

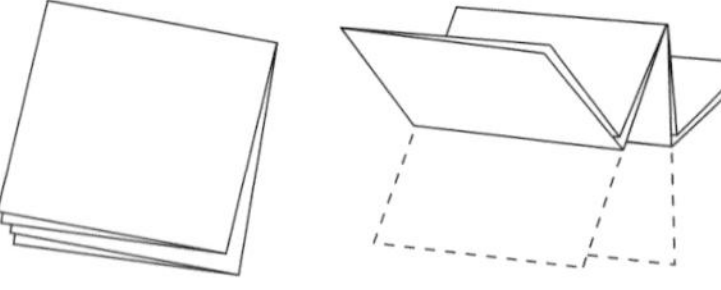

Step 3 Unfold and cut along the fold lines of the top flap to make four tabs.

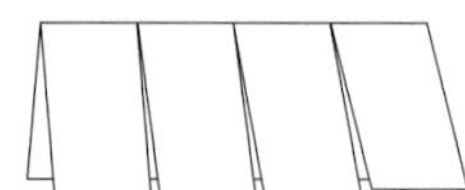

Variations

A Make a space to use for titles or concept maps by leaving a ½ inch tab at the top when folding the paper in half.

B Use the book on the vertical axis, with or without an extended tab.

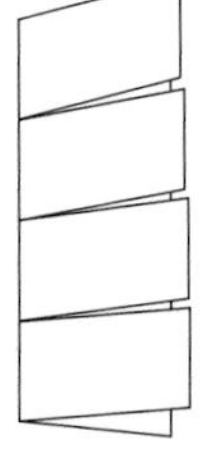

Folding Fifths for a Foldable® By Dinah Zike

Step 1 Fold a sheet of paper in half horizontally.

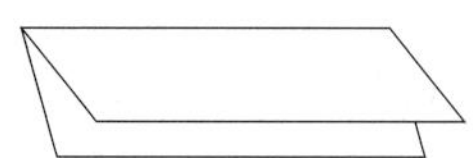

Step 2 Fold again so one-third of the paper is exposed and two-thirds are covered.

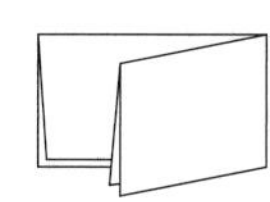

Step 3 Fold the two-thirds section in half.

Step 4 Fold the one-third section, a single thickness, backward to make a fold line.

Variations

A Unfold and cut along the fold lines to make five tabs.

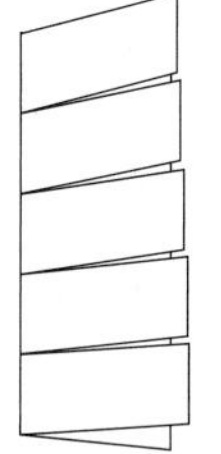

B Make a five-tab book with a ½ inch tab at the top (see two-tab instructions).

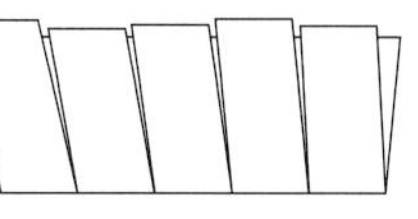

C Use 11″ × 17″ or 12″ × 18″ paper and fold into fifths for a five-column and/or row table or chart.

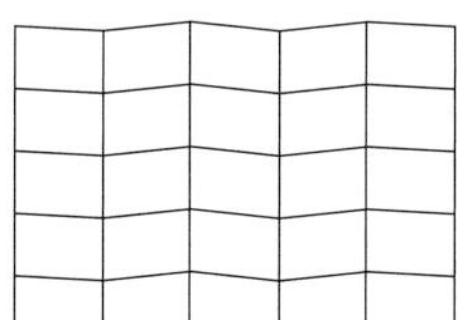

Folded Table or Chart, and Trifold Foldable® By Dinah Zike

Step 1 Fold a sheet of paper in the required number of vertical columns for the table or chart.

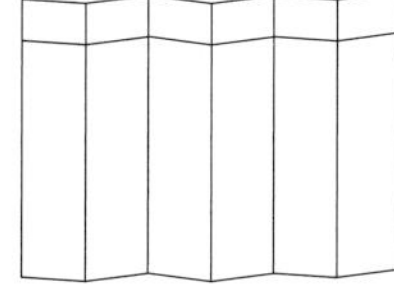

Step 2 Fold the horizontal rows needed for the table or chart.

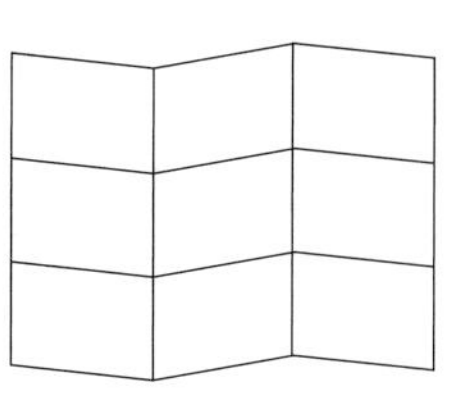

PROJECT FORMAT
Use 11″ × 17″ or 12″ × 18″ paper and fold it to make a large trifold project book or larger tables and charts.

Variations

A Make a trifold by folding the paper into thirds vertically or horizontally.

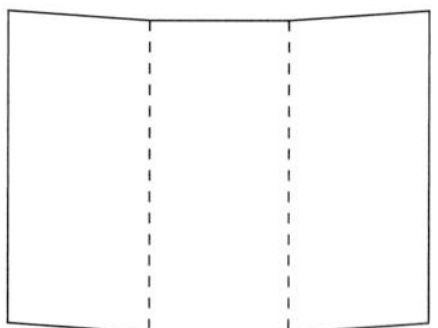

B Make a trifold book. Unfold it and draw a Venn diagram on the inside.

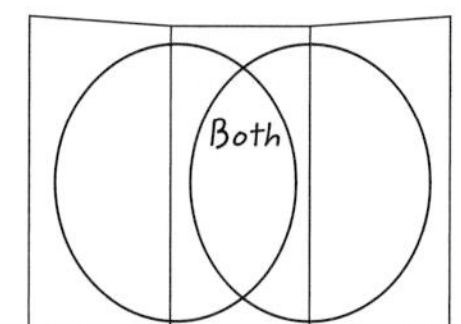

Two or Three-Pockets Foldable® By Dinah Zike

Step 1 Fold up the long side of a horizontal sheet of paper about 5 cm.

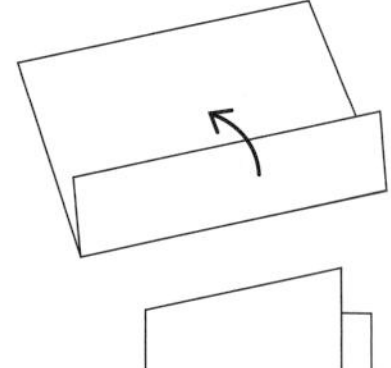

Step 2 Fold the paper in half.

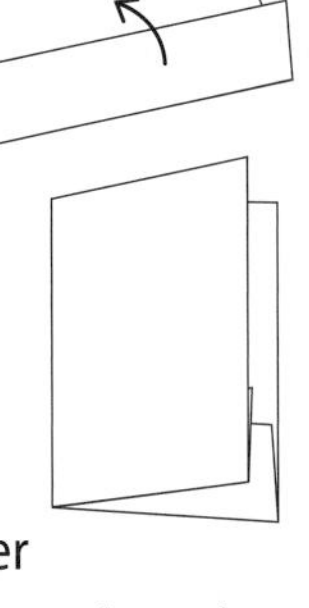

Step 3 Open the paper and glue or staple the outer edges to make two compartments.

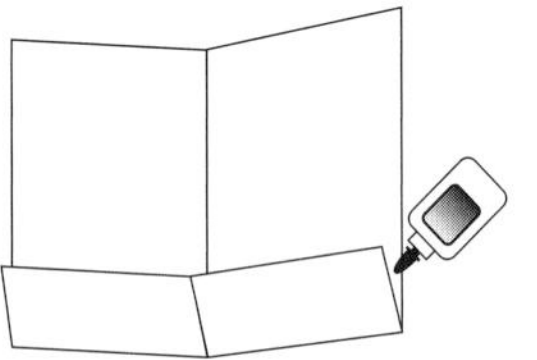

Variations

A Make a multi-page booklet by gluing several pocket books together.

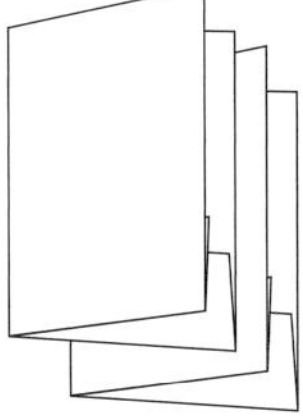

B Make a three-pocket book by using a trifold (see previous instructions).

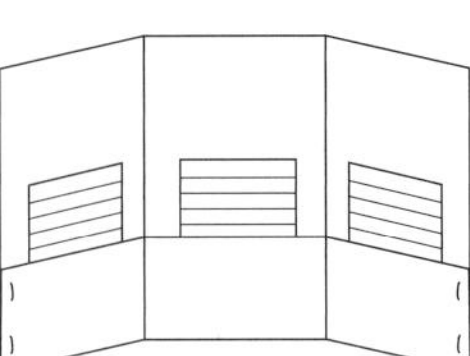

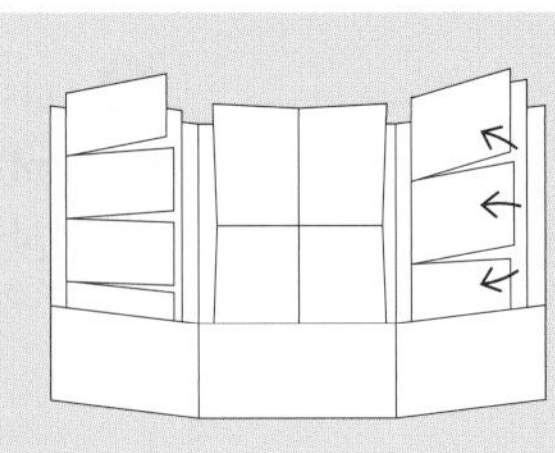

PROJECT FORMAT
Use 11″ × 17″ or 12″ × 18″ paper and fold it horizontally to make a large multi-pocket project.

Matchbook Foldable® By Dinah Zike

Step 1 Fold a sheet of paper almost in half and make the back edge about 1–2 cm longer than the front edge.

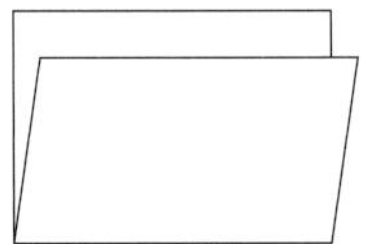

Step 2 Find the midpoint of the shorter flap.

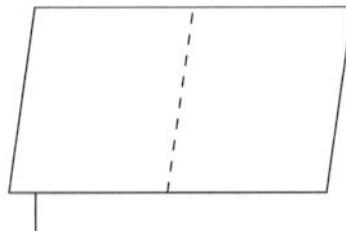

Step 3 Open the paper and cut the short side along the midpoint making two tabs.

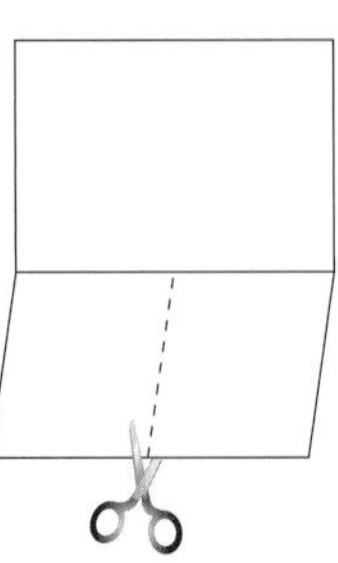

Step 4 Close the book and fold the tab over the short side.

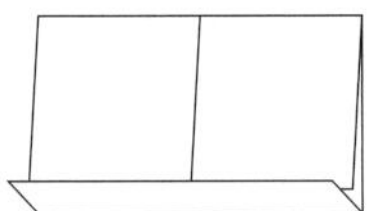

Variations

A Make a single-tab matchbook by skipping Steps 2 and 3.

B Make two smaller matchbooks by cutting the single-tab matchbook in half.

Shutterfold Foldable® By Dinah Zike

Step 1 Begin as if you were folding a vertical sheet of paper in half, but instead of creasing the paper, pinch it to show the midpoint.

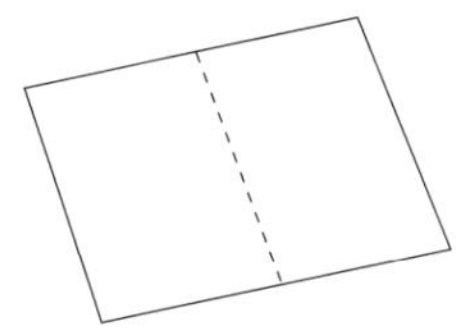

PROJECT FORMAT
Use 11″ × 17″ or 12″ × 18″ paper and fold it to make a large shutterfold project.

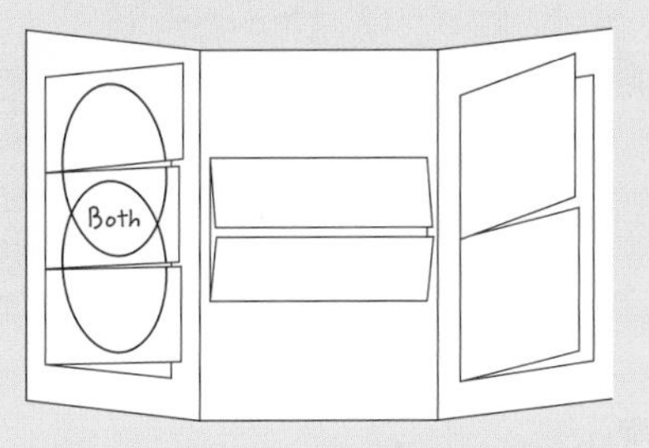

Step 2 Fold the top and bottom to the middle and crease the folds.

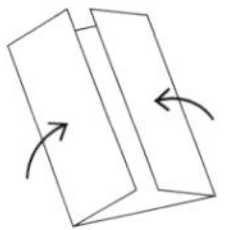

Variations

A Use the shutterfold on the horizontal axis.

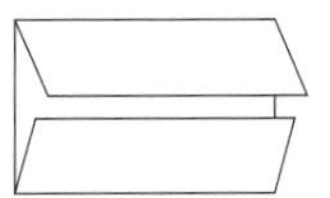

B Create a center tab by leaving .5–2 cm between the flaps in Step 2.

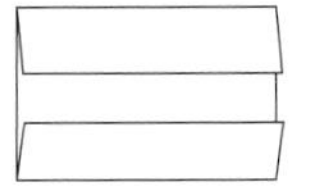

Four-Door Foldable® By Dinah Zike

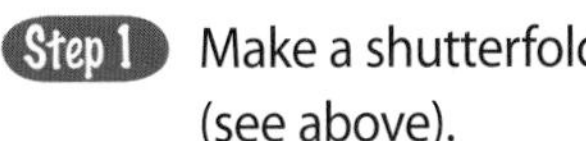

Step 1 Make a shutterfold (see above).

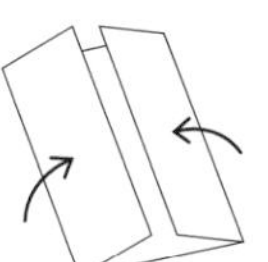

Step 2 Fold the sheet of paper in half.

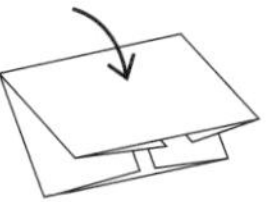

Step 3 Open the last fold and cut along the inside fold lines to make four tabs.

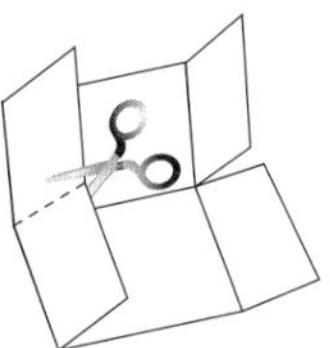

Variations

A Use the four-door book on the opposite axis.

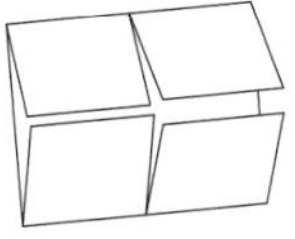

B Create a center tab by leaving .5–2 cm between the flaps in Step 1.

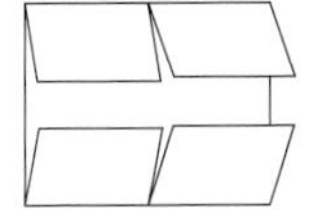

Bound Book Foldable® By Dinah Zike

Step 1 Fold three sheets of paper in half. Place the papers in a stack, leaving about .5 cm between each top fold. Mark all three sheets about 3 cm from the outer edges.

Step 2 Using two of the sheets, cut from the outer edges to the marked spots on each side. On the other sheet, cut between the marked spots.

Step 3 Take the two sheets from Step 1 and slide them through the cut in the third sheet to make a 12-page book.

Step 4 Fold the bound pages in half to form a book.

Variation

A Use two sheets of paper to make an eight-page book, or increase the number of pages by using more than three sheets.

PROJECT FORMAT
Use two or more sheets of 11″ × 17″ or 12″ × 18″ paper and fold it to make a large bound book project.

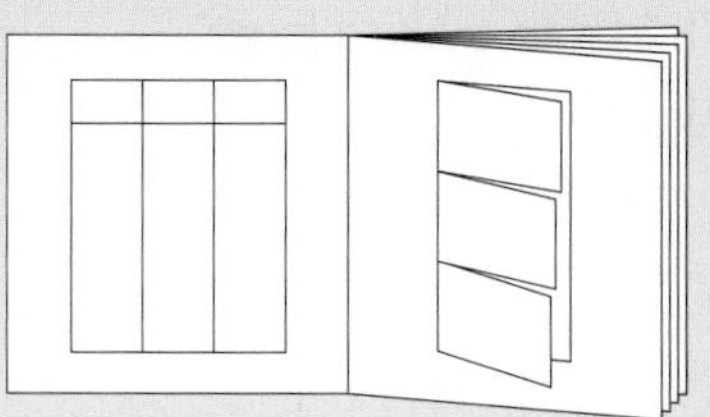

Accordian Foldable® By Dinah Zike

Step 1 Fold the selected paper in half vertically, like a *hamburger*.

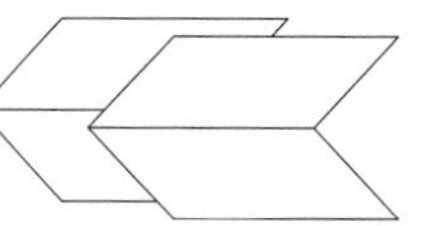

Step 2 Cut each sheet of folded paper in half along the fold lines.

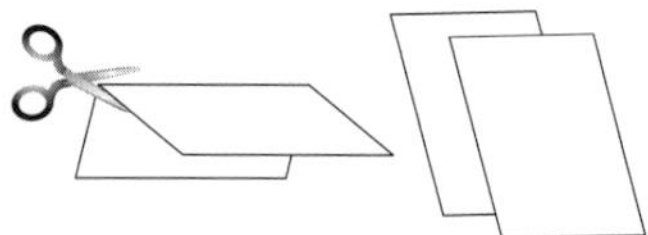

Step 3 Fold each half-sheet almost in half, leaving a 2 cm tab at the top.

Step 4 Fold the top tab over the short side, then fold it in the opposite direction.

Variations

A Glue the straight edge of one paper inside the tab of another sheet. Leave a tab at the end of the book to add more pages.

B Tape the straight edge of one paper to the tab of another sheet, or just tape the straight edges of nonfolded paper end to end to make an accordian.

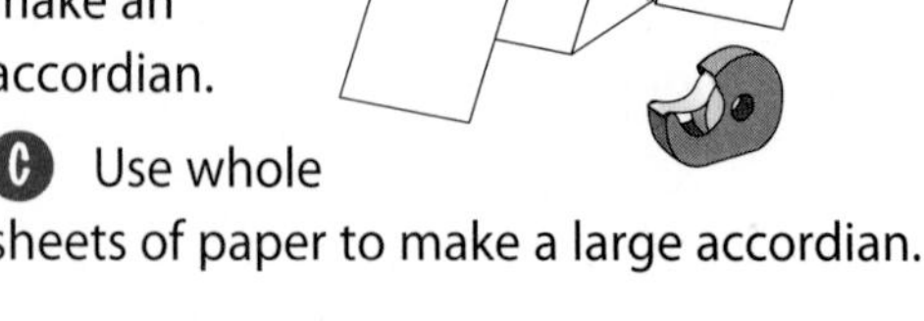

C Use whole sheets of paper to make a large accordian.

Layered Foldable® By Dinah Zike

Step 1 Stack two sheets of paper about 1–2 cm apart. Keep the right and left edges even.

Step 2 Fold up the bottom edges to form four tabs. Crease the fold to hold the tabs in place.

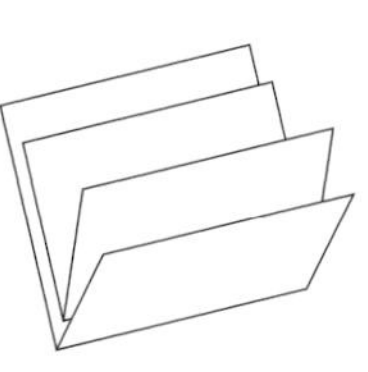

Step 3 Staple along the folded edge, or open and glue the papers together at the fold line.

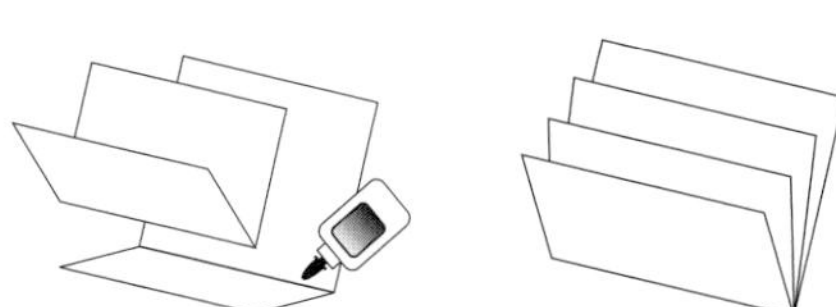

Variations

A Rotate the book so the fold is at the top or to the side.

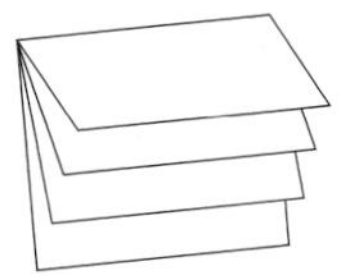

B Extend the book by using more than two sheets of paper.

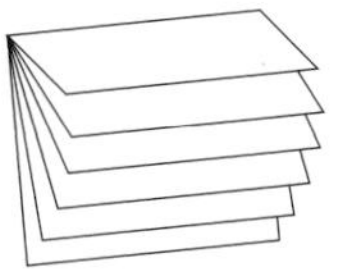

Envelope Foldable® By Dinah Zike

Step 1 Fold a sheet of paper into a *taco*. Cut off the tab at the top.

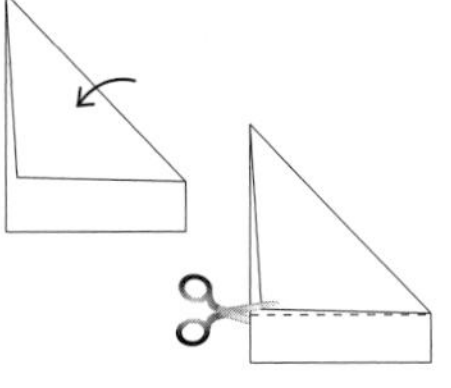

Step 2 Open the *taco* and fold it the opposite way making another *taco* and an X-fold pattern on the sheet of paper.

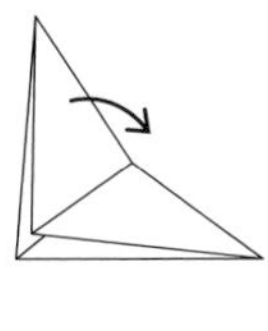

Step 3 Cut a map, illustration, or diagram to fit the inside of the envelope.

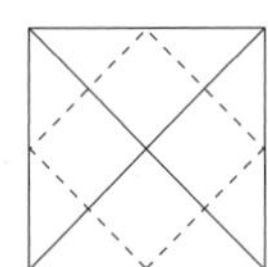

Step 4 Use the outside tabs for labels and inside tabs for writing information.

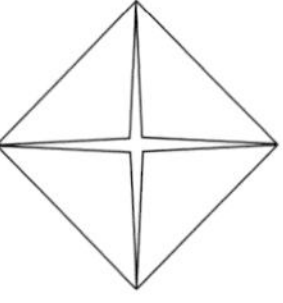

Variations

A Use 11″ × 17″ or 12″ × 18″ paper to make a large envelope.

B Cut off the points of the four tabs to make a window in the middle of the book.

Sentence Strip Foldable® By Dinah Zike

Step 1 Fold two sheets of paper in half vertically, like a *hamburger*.

Step 2 Unfold and cut along fold lines making four half sheets.

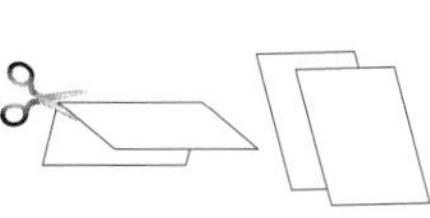

Step 3 Fold each half sheet in half horizontally, like a *hot dog*.

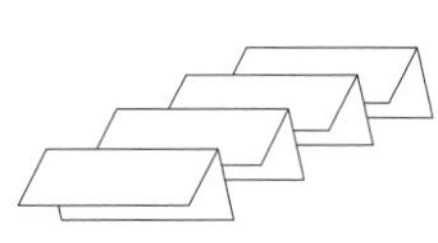

Step 4 Stack folded horizontal sheets evenly and staple together on the left side.

Step 5 Open the top flap of the first sentence strip and make a cut about 2 cm from the stapled edge to the fold line. This forms a flap that can be raisied and lowered. Repeat this step for each sentence strip.

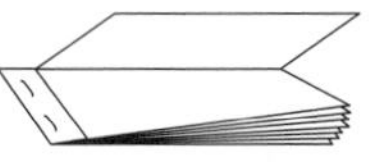

Variations

A Expand this book by using more than two sheets of paper.

B Use whole sheets of paper to make large books.

Pyramid Foldable® By Dinah Zike

Step 1 Fold a sheet of paper into a *taco*. Crease the fold line, but do not cut it off.

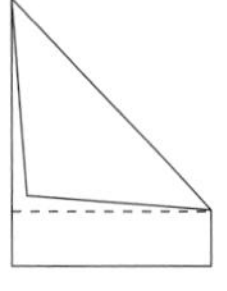

Step 2 Open the folded sheet and refold it like a *taco* in the opposite direction to create an X-fold pattern.

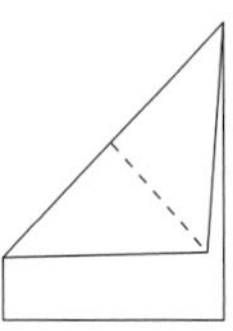

Step 3 Cut one fold line as shown, stopping at the center of the X-fold to make a flap.

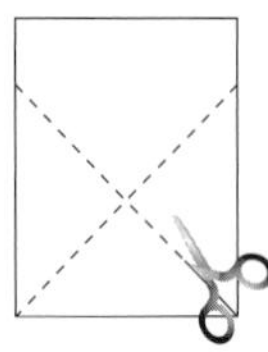

Step 4 Outline the fold lines of the X-fold. Label the three front sections and use the inside spaces for notes. Use the tab for the title.

Step 5 Glue the tab into a project book or notebook. Use the space under the pyramid for other information.

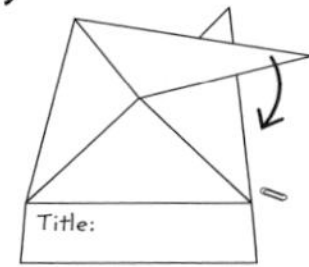

Step 6 To display the pyramid, fold the flap under and secure with a paper clip, if needed.

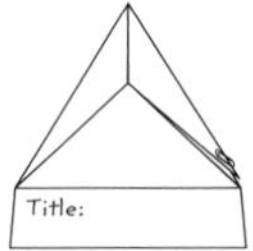

Single-Pocket or One-Pocket Foldable® By Dinah Zike

Step 1 Using a large piece of paper on a vertical axis, fold the bottom edge of the paper upwards, about 5 cm.

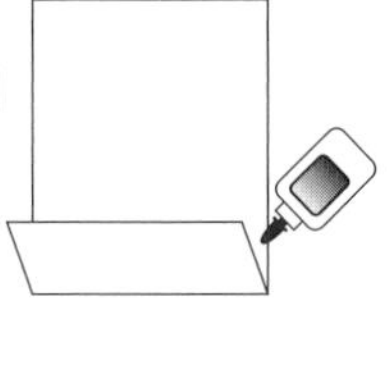

Step 2 Glue or staple the outer edges to make a large pocket.

PROJECT FORMAT
Use 11″ × 17″ or 12″ × 18″ paper and fold it vertically or horizontally to make a large pocket project.

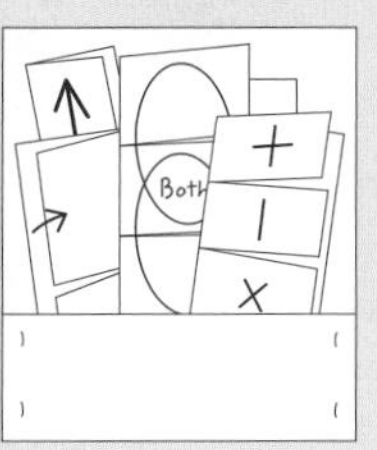

Variations

A Make the one-pocket project using the paper on the horizontal axis.

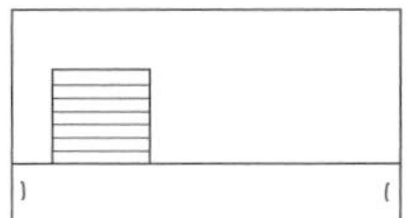

B To store materials securely inside, fold the top of the paper almost to the center, leaving about 2–4 cm between the paper edges. Slip the Foldables through the opening and under the top and bottom pockets.

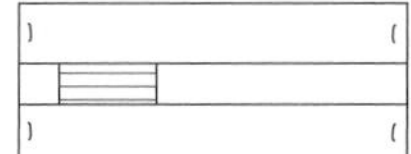

Multi-Tab Foldable® By Dinah Zike

Step 1 Fold a sheet of notebook paper in half like a *hot dog*.

Step 2 Open the paper and on one side cut every third line. This makes ten tabs on wide ruled notebook paper and twelve tabs on college ruled.

Step 3 Label the tabs on the front side and use the inside space for definitions or other information.

Variation

A Make a tab for a title by folding the paper so the holes remain uncovered. This allows the notebook Foldable to be stored in a three-hole binder.

PERIODIC TABLE OF THE ELEMENTS

Element — Hydrogen
Atomic number — 1
Symbol — H
Atomic mass — 1.01
State of matter

Gas
Liquid
Solid
Synthetic

A column in the periodic table is called a **group.**

A row in the periodic table is called a **period.**

Period	1	2	3	4	5	6	7	8	9
1	Hydrogen 1 **H** 1.01								
2	Lithium 3 **Li** 6.94	Beryllium 4 **Be** 9.01							
3	Sodium 11 **Na** 22.99	Magnesium 12 **Mg** 24.31							
4	Potassium 19 **K** 39.10	Calcium 20 **Ca** 40.08	Scandium 21 **Sc** 44.96	Titanium 22 **Ti** 47.87	Vanadium 23 **V** 50.94	Chromium 24 **Cr** 52.00	Manganese 25 **Mn** 54.94	Iron 26 **Fe** 55.85	Cobalt 27 **Co** 58.93
5	Rubidium 37 **Rb** 85.47	Strontium 38 **Sr** 87.62	Yttrium 39 **Y** 88.91	Zirconium 40 **Zr** 91.22	Niobium 41 **Nb** 92.91	Molybdenum 42 **Mo** 95.96	Technetium 43 **Tc** (98)	Ruthenium 44 **Ru** 101.07	Rhodium 45 **Rh** 102.91
6	Cesium 55 **Cs** 132.91	Barium 56 **Ba** 137.33	Lanthanum 57 **La** 138.91	Hafnium 72 **Hf** 178.49	Tantalum 73 **Ta** 180.95	Tungsten 74 **W** 183.84	Rhenium 75 **Re** 186.21	Osmium 76 **Os** 190.23	Iridium 77 **Ir** 192.22
7	Francium 87 **Fr** (223)	Radium 88 **Ra** (226)	Actinium 89 **Ac** (227)	Rutherfordium 104 **Rf** (267)	Dubnium 105 **Db** (268)	Seaborgium 106 **Sg** (271)	Bohrium 107 **Bh** (272)	Hassium 108 **Hs** (270)	Meitnerium 109 **Mt** (276)

The number in parentheses is the mass number of the longest lived isotope for that element.

Lanthanide series	Cerium 58 **Ce** 140.12	Praseodymium 59 **Pr** 140.91	Neodymium 60 **Nd** 144.24	Promethium 61 **Pm** (145)	Samarium 62 **Sm** 150.36	Europium 63 **Eu** 151.96
Actinide series	Thorium 90 **Th** 232.04	Protactinium 91 **Pa** 231.04	Uranium 92 **U** 238.03	Neptunium 93 **Np** (237)	Plutonium 94 **Pu** (244)	Americium 95 **Am** (243)

Metal

Metalloid

Nonmetal

Recently discovered

10	11	12	13	14	15	16	17	18
								Helium 2 He 4.00
			Boron 5 B 10.81	Carbon 6 C 12.01	Nitrogen 7 N 14.01	Oxygen 8 O 16.00	Fluorine 9 F 19.00	Neon 10 Ne 20.18
			Aluminum 13 Al 26.98	Silicon 14 Si 28.09	Phosphorus 15 P 30.97	Sulfur 16 S 32.07	Chlorine 17 Cl 35.45	Argon 18 Ar 39.95
Nickel 28 Ni 58.69	Copper 29 Cu 63.55	Zinc 30 Zn 65.38	Gallium 31 Ga 69.72	Germanium 32 Ge 72.64	Arsenic 33 As 74.92	Selenium 34 Se 78.96	Bromine 35 Br 79.90	Krypton 36 Kr 83.80
Palladium 46 Pd 106.42	Silver 47 Ag 107.87	Cadmium 48 Cd 112.41	Indium 49 In 114.82	Tin 50 Sn 118.71	Antimony 51 Sb 121.76	Tellurium 52 Te 127.60	Iodine 53 I 126.90	Xenon 54 Xe 131.29
Platinum 78 Pt 195.08	Gold 79 Au 196.97	Mercury 80 Hg 200.59	Thallium 81 Tl 204.38	Lead 82 Pb 207.20	Bismuth 83 Bi 208.98	Polonium 84 Po (209)	Astatine 85 At (210)	Radon 86 Rn (222)
Darmstadtium 110 Ds (281)	Roentgenium 111 Rg (280)	Copernicium 112 Cn (285)	Ununtrium * 113 Uut (284)	Ununquadium * 114 Uuq (289)	Ununpentium * 115 Uup (288)	Ununhexium * 116 Uuh (293)		Ununoctium * 118 Uuo (294)

* The names and symbols for elements 113-116 and 118 are temporary. Final names will be selected when the elements' discoveries are verified.

Gadolinium 64 Gd 157.25	Terbium 65 Tb 158.93	Dysprosium 66 Dy 162.50	Holmium 67 Ho 164.93	Erbium 68 Er 167.26	Thulium 69 Tm 168.93	Ytterbium 70 Yb 173.05	Lutetium 71 Lu 174.97
Curium 96 Cm (247)	Berkelium 97 Bk (247)	Californium 98 Cf (251)	Einsteinium 99 Es (252)	Fermium 100 Fm (257)	Mendelevium 101 Md (258)	Nobelium 102 No (259)	Lawrencium 103 Lr (262)

Diversity of Life: Classification of Living Organisms

A six-kingdom system of classification of organisms is used today. Two kingdoms—Kingdom Archaebacteria and Kingdom Eubacteria—contain organisms that do not have a nucleus and that lack membrane-bound structures in the cytoplasm of their cells. The members of the other four kingdoms have a cell or cells that contain a nucleus and structures in the cytoplasm, some of which are surrounded by membranes. These kingdoms are Kingdom Protista, Kingdom Fungi, Kingdom Plantae, and Kingdom Animalia.

Kingdom Archaebacteria

one-celled; some absorb food from their surroundings; some are photosynthetic; some are chemosynthetic; many are found in extremely harsh environments including salt ponds, hot springs, swamps, and deep-sea hydrothermal vents

Kingdom Eubacteria

one-celled; most absorb food from their surroundings; some are photosynthetic; some are chemosynthetic; many are parasites; many are round, spiral, or rod-shaped; some form colonies

Kingdom Protista

Phylum Euglenophyta one-celled; photosynthetic or take in food; most have one flagellum; euglenoids

Kingdom Eubacteria
Bacillus anthracis

Phylum Chlorophyta
Desmids

Phylum Bacillariophyta one-celled; photosynthetic; have unique double shells made of silica; diatoms

Phylum Dinoflagellata one-celled; photosynthetic; contain red pigments; have two flagella; dinoflagellates

Phylum Chlorophyta one-celled, many-celled, or colonies; photosynthetic; contain chlorophyll; live on land, in freshwater, or salt water; green algae

Phylum Rhodophyta most are many-celled; photosynthetic; contain red pigments; most live in deep, saltwater environments; red algae

Phylum Phaeophyta most are many-celled; photosynthetic; contain brown pigments; most live in saltwater environments; brown algae

Phylum Rhizopoda one-celled; take in food; are free-living or parasitic; move by means of pseudopods; amoebas

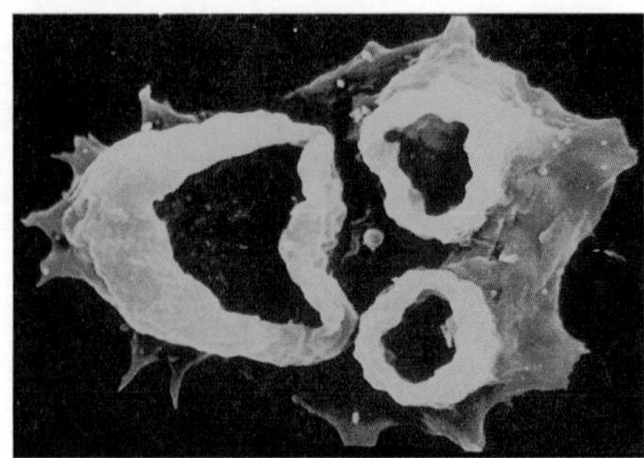

Amoeba

Phylum Zoomastigina one-celled; take in food; free-living or parasitic; have one or more flagella; zoomastigotes

Phylum Ciliophora one-celled; take in food; have large numbers of cilia; ciliates

Phylum Sporozoa one-celled; take in food; have no means of movement; are parasites in animals; sporozoans

Phylum Myxomycota
Slime mold

Phylum Oomycota
Phytophthora infestans

Phyla Myxomycota and Acrasiomycota one- or many-celled; absorb food; change form during life cycle; cellular and plasmodial slime molds

Phylum Oomycota many-celled; are either parasites or decomposers; live in freshwater or salt water; water molds, rusts and downy mildews

Kingdom Fungi

Phylum Zygomycota many-celled; absorb food; spores are produced in sporangia; zygote fungi; bread mold

Phylum Ascomycota one- and many-celled; absorb food; spores produced in asci; sac fungi; yeast

Phylum Basidiomycota many-celled; absorb food; spores produced in basidia; club fungi; mushrooms

Phylum Deuteromycota members with unknown reproductive structures; imperfect fungi; *Penicillium*

Phylum Mycophycota organisms formed by symbiotic relationship between an ascomycote or a basidiomycote and green alga or cyanobacterium; lichens

Lichens

Kingdom Plantae

Divisions Bryophyta (mosses), **Anthocerophyta** (hornworts), **Hepaticophyta** (liverworts), **Psilophyta** (whisk ferns) many-celled nonvascular plants; reproduce by spores produced in capsules; green; grow in moist, land environments

Division Lycophyta many-celled vascular plants; spores are produced in conelike structures; live on land; are photosynthetic; club mosses

Division Arthrophyta vascular plants; ribbed and jointed stems; scalelike leaves; spores produced in conelike structures; horsetails

Division Pterophyta vascular plants; leaves called fronds; spores produced in clusters of sporangia called sori; live on land or in water; ferns

Division Ginkgophyta deciduous trees; only one living species; have fan-shaped leaves with branching veins and fleshy cones with seeds; ginkgoes

Division Cycadophyta palmlike plants; have large, featherlike leaves; produces seeds in cones; cycads

Division Coniferophyta deciduous or evergreen; trees or shrubs; have needlelike or scalelike leaves; seeds produced in cones; conifers

Division Gnetophyta shrubs or woody vines; seeds are produced in cones; division contains only three genera; gnetum

Division Anthophyta dominant group of plants; flowering plants; have fruits with seeds

Kingdom Animalia

Phylum Porifera aquatic organisms that lack true tissues and organs; are asymmetrical and sessile; sponges

Phylum Cnidaria radially symmetrical organisms; have a digestive cavity with one opening; most have tentacles armed with stinging cells; live in aquatic environments singly or in colonies; includes jellyfish, corals, hydra, and sea anemones

Phylum Platyhelminthes bilaterally symmetrical worms; have flattened bodies; digestive system has one opening; parasitic and free-living species; flatworms

Division Anthophyta
Tomato plant

Division Bryophyta
Liverwort

Phylum Platyhelminthes
Flatworm

Phylum Chordata

Phylum Nematoda round, bilaterally symmetrical body; have digestive system with two openings; free-living forms and parasitic forms; roundworms

Phylum Mollusca soft-bodied animals, many with a hard shell and soft foot or footlike appendage; a mantle covers the soft body; aquatic and terrestrial species; includes clams, snails, squid, and octopuses

Phylum Annelida bilaterally symmetrical worms; have round, segmented bodies; terrestrial and aquatic species; includes earthworms, leeches, and marine polychaetes

Phylum Arthropoda largest animal group; have hard exoskeletons, segmented bodies, and pairs of jointed appendages; land and aquatic species; includes insects, crustaceans, and spiders

Phylum Echinodermata marine organisms; have spiny or leathery skin and a water-vascular system with tube feet; are radially symmetrical; includes sea stars, sand dollars, and sea urchins

Phylum Chordata organisms with internal skeletons and specialized body systems; most have paired appendages; all at some time have a notochord, nerve cord, gill slits, and a post-anal tail; include fish, amphibians, reptiles, birds, and mammals

Use and Care of a Microscope

Eyepiece Contains magnifying lenses you look through.

Arm Supports the body tube.

Low-power objective Contains the lens with the lowest power magnification.

Stage clips Hold the microscope slide in place.

Coarse adjustment Focuses the image under low power.

Fine adjustment Sharpens the image under high magnification.

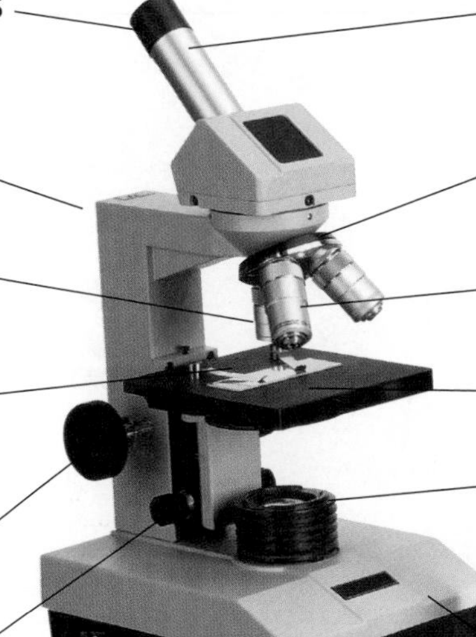

Body tube Connects the eyepiece to the revolving nosepiece.

Revolving nosepiece Holds and turns the objectives into viewing position.

High-power objective Contains the lens with the highest magnification.

Stage Supports the microscope slide.

Light source Provides light that passes upward through the diaphragm, the specimen, and the lenses.

Base Provides support for the microscope.

Caring for a Microscope

1. Always carry the microscope holding the arm with one hand and supporting the base with the other hand.
2. Don't touch the lenses with your fingers.
3. The coarse adjustment knob is used only when looking through the lowest-power objective lens. The fine adjustment knob is used when the high-power objective is in place.
4. Cover the microscope when you store it.

Using a Microscope

1. Place the microscope on a flat surface that is clear of objects. The arm should be toward you.
2. Look through the eyepiece. Adjust the diaphragm so light comes through the opening in the stage.
3. Place a slide on the stage so the specimen is in the field of view. Hold it firmly in place by using the stage clips.
4. Always focus with the coarse adjustment and the low-power objective lens first. After the object is in focus on low power, turn the nosepiece until the high-power objective is in place. Use ONLY the fine adjustment to focus with the high-power objective lens.

Making a Wet-Mount Slide

1. Carefully place the item you want to look at in the center of a clean, glass slide. Make sure the sample is thin enough for light to pass through.
2. Use a dropper to place one or two drops of water on the sample.
3. Hold a clean coverslip by the edges and place it at one edge of the water. Slowly lower the coverslip onto the water until it lies flat.
4. If you have too much water or a lot of air bubbles, touch the edge of a paper towel to the edge of the coverslip to draw off extra water and draw out unwanted air.

Glossary/Glosario

Multilingual eGlossary

A science multilingual glossary is available on the science Web site. The glossary includes the following languages.

Arabic	Hmong	Tagalog
Bengali	Korean	Urdu
Chinese	Portuguese	Vietnamese
English	Russian	
Haitian Creole	Spanish	

Cómo usar el glosario en español:

1. Busca el término en inglés que desees encontrar.
2. El término en español, junto con la definición, se encuentran en la columna de la derecha.

Pronunciation Key

Use the following key to help you sound out words in the glossary.

a	b**a**ck (BAK)	**ew**	f**oo**d (FEWD)
ay	d**ay** (DAY)	**yoo**	p**u**re (PYOOR)
ah	f**ather** (FAH thur)	**yew**	f**ew** (FYEW)
ow	fl**ow**er (FLOW ur)	**uh**	comm**a** (CAH muh)
ar	**car** (CAR)	**u (+ con)**	r**u**b (RUB)
e	**less** (LES)	**sh**	**sh**elf (SHELF)
ee	**leaf** (LEEF)	**ch**	na**t**ure (NAY chur)
ih	**trip** (TRIHP)	**g**	**g**ift (GIHFT)
i (i + com + e)	**i**dea (i DEE uh)	**j**	**g**em (JEM)
oh	**go** (GOH)	**ing**	s**ing** (SING)
aw	**so**ft (SAWFT)	**zh**	vi**si**on (VIH zhun)
or	**orb**it (OR buht)	**k**	ca**k**e (KAYK)
oy	**coi**n (COYN)	**s**	**s**eed, **c**ent (SEED, SENT)
oo	f**oo**t (FOOT)	**z**	**z**one, rai**s**e (ZOHN, RAYZ)

English — **A** — **Español**

active immunity/antigen — **inmunidad activa/antígeno**

active immunity: the process by which the human body produces antibodies in response to an antigen. (p. 617)

allergy: an overly sensitive immune response to common antigens. (p. 616)

alveolus (al VEE uh lus; plural, alveoli): microscopic sacs or pouches at the end of the bronchioles where gas exchange occurs. (p. 561)

antibiotic: a medicine that stops the growth and reproduction of bacteria. (p. 624)

antibody: a protein that can attach to a pathogen and makes it useless. (p. 616)

antigen: a substance that causes an immune response. (p. 616)

inmunidad activa: proceso por el cual el cuerpo humano produce anticuerpos en respuesta a un antígeno. (pág. 617)

alergia: respuesta inmune demasiado sensible a los antígenos comunes. (pág. 616)

alveolo (plural, alveolos): bolsas o sacos microscópicos en los extremos de los bronquiolos donde ocurre el intercambio de gas. (pág. 561)

antibiótico: medicina que detiene el crecimiento y reproducción de bacterias. (pág. 624)

anticuerpo: proteína que se adhiere a un patógeno y lo hace inútil. (pág. 616)

antígeno: sustancia que causa una respuesta inmune. (pág. 616)

artery: a vessel that carries blood away from the heart. (p. 570)

arthritis (ar THRI tus): a disease in which joints become irritated or inflamed, such as when cartilage in joints is damaged or wears away. (p. 492)

atherosclerosis (a thuh roh skluh ROH sus): the buildup of fatty material within the walls of arteries. (p. 572)

atria (AY tree uh; singular, atrium): the upper two chambers of the heart. (p. 568)

arteria: vaso que lleva sangre fuera del corazón. (pág. 570)

artritis: enfermedad en la que las articulaciones se irritan o inflaman, como cuando el cartílago en las articulaciones se lastima o desgasta. (pág. 492)

arteriosclerosis: acumulación de material graso en el interior de las paredes de las arterias. (pág. 572)

atrios (singular, atrio): las dos cámaras superiores del corazón. (pág. 568)

B

B cell: a type of white blood cell that forms and matures in the bone marrow and secretes antibodies into the blood. (p. 616)

bladder: a muscular sac that holds urine until the urine is excreted. (p. 545)

brain stem: the area of the brain that controls involuntary functions. (p. 641)

breathing: the movement of air into and out of the lungs. (p. 559)

bronchus (BRAHN kus; plural, bronchi): one of two narrow tubes that carry air into the lungs from the trachea. (p. 561)

bruise: an injury where blood vessels in the skin are broken, but the skin is not cut or opened. (p. 508)

célula B: tipo de glóbulo blanco que se forma y madura en la médula ósea y secreta anticuerpos a la sangre. (pág. 616)

vejiga: bolsa muscular que contiene la orina hasta que se excreta. (pág. 545)

tallo cerebral: área del cerebro que controla funciones involuntarias. (pág. 641)

respiración: movimiento de aire hacia adentro y hacia afuera de los pulmones. (pág. 559)

bronquio (plural, bronquios): uno de los dos tubos delgados que llevan aire de la tráquea a los pulmones. (pág. 561)

moretón: herida en la cual los vasos sanguíneos de la piel se rompen, pero la piel no se corta ni abre. (pág. 508)

C

Calorie: the amount of energy it takes to raise the temperature of 1 kg of water by 1°C. (p. 523)

cancer: a disease in which cells reproduce uncontrollably without the usual signals to stop. (p. 607)

capillary: a tiny blood vessel that delivers supplies to an individual cell and takes away waste materials. (p. 570)

carbohydrate (kar boh HI drayt): a macromolecule made up of one or more sugar molecules, which are composed of carbon, hydrogen, and oxygen; usually the body's major source of energy. (p. 524)

cardiac (KAR dee ak) muscle: muscle found only in the heart. (p. 500)

caloría: cantidad de energía necesaria p... aumentar la temperatura de 1 kg d... 1°C. (pág. 523)

cáncer: enfermedad en la cual la... reproducen sin control sin l... para detenerse. (pág. 607)

capilar: vaso sanguíneo di... suministros a una célu... los materiales de des...

carbohidrato: macrom... una o más moléc... están compuest... oxígeno; usua... energía del c...

músculo cardí... encuentr...

Science Skill Handbook
Math Skill Handbook
Reference Handbook

cartilage (KAR tuh lihj): a strong, flexible tissue that covers the ends of bones. (p. 490)

central nervous system (CNS): system made up of the brain and the spinal cord. (p. 640)

cerebellum (ser uh BEH lum): the part of the brain that coordinates voluntary muscle movement and regulates balance and posture. (p. 640)

cerebrum (suh REE brum): the part of the brain that controls memory, language, and thought. (p. 640)

cervix (SUR vihks): a small structure between the uterus and the vagina. (p. 688)

chemical digestion: a process in which chemical reactions break down pieces of food into small molecules. (p. 532)

chemotherapy: a type of cancer treatment in which chemicals are used to kill the cells that are reproducing uncontrollably. (p. 624)

chyme (KIME): a thin, watery liquid made of broken down food molecules and gastric juice. (p. 535)

coronary circulation: the network of arteries and veins that supplies blood to all the cells of the heart. (p. 571)

cartílago: tejido fuerte y flexible que reviste los extremos de los huesos. (pág. 490)

sistema nervioso central (SNC): sistema constituido por el cerebro y la médula espinal. (pág. 640)

cerebelo: parte del cerebro que coordina el movimiento muscular voluntario y regula el equilibrio y la postura. (pág. 640)

cerebrum: parte del cerebro que controla la memoria, el lenguaje y el pensamiento. (pág. 640)

cérvix: estructura pequeña entre el útero y la vagina. (pág. 688)

digestión química: proceso por el cual las reacciones químicas descomponen partes del alimento en moléculas pequeñas. (pág. 532)

quimioterapia: tipo de tratamiento para el cáncer, en el cual se usan químicos para matar las células que se están reproduciendo sin control. (pág. 624)

quimo: líquido diluido y acuoso constituido de moléculas de alimento descompuestas y jugos gástricos. (pág. 535)

circulación coronaria: red de arterias y venas que suministran sangre a todas las células del corazón. (pág. 571)

D

[illegible]ermis: a thick layer of skin that gives skin strength, nourishment, and flexibility. [illegible] 507)

dia[illegible]agm (DI uh fram): a large muscle below th[illegible] [illegible]ngs that contracts and relaxes as air mo[illegible] [illegible]nto and out of your lungs. (p. 562)

[illegible]sti[illegible] [illegible] e mechanical and chemical break- [illegible]d into small particles and mole- [illegible]r body can absorb and use.

dermis: capa gruesa de piel que le proporciona a la piel fuerza, nutrimento y flexibilidad. (pág. 507)

diafragma: músculo grande debajo de los pulmones que se contrae y relaja a medida que el aire entra y sale a los pulmones. (pág. 562)

digestión: descomposición mecánica y química del alimento en partículas y moléculas pequeñas que el cuerpo absorbe y usa. (pág. 531)

E

eardr[illegible] ear a[illegible] m[illegible] [illegible]er [illegible]ne between the outer [illegible]rod[illegible] 650)

egg: the [illegible] forms i[illegible] (p. 6[illegible] [illegible] or sex, cell;

embryo: a [illegible] huma[illegible] attaches t[illegible] [illegible]s until [illegible]m the time it [illegible] of pregnan[illegible] eighth week

tímpano: membrana delgada en medio del oído externo y del oído interno. (pág. 650)

óvulo: célula reproductiva femenina o sexual; forma en un ovario. (pág. 673)

embrión: ser humano en desarrollo desde el momento en que se adhiere al útero hasta la octava semana de embarazo. (pág. 686)

endocrine (EN duh krun) system: system consisting of groups of organs and tissues that release chemical messages into the bloodstream. (p. 657)

enzyme (EN zime): a protein that helps break down larger molecules into smaller molecules and speeds up, or catalyzes, the rate of chemical reactions. (p. 532)

epidermis (eh puh DUR mus): the outermost layer of skin and the only layer in direct contact with the outside environment. (p. 507)

esophagus (ih SAH fuh gus): a muscular tube that connects the mouth to the stomach. (p. 534)

excretory system: the system that collects and eliminates wastes from the body and regulates the level of fluid in the body. (p. 541)

sistema endocrino: sistema que consta de grupos de órganos y tejidos que liberan mensajes químicos en la corriente sanguínea. (pág. 657)

enzima: proteína que descompone moléculas más grandes en moléculas más pequeñas y acelera, o cataliza, la velocidad de las reacciones químicas. (pág. 532)

epidermis: capa más externa de la piel y la única capa que está en contacto directo con el medioambiente externo. (pág. 507)

esófago: tubo muscular que conecta la boca al estómago. (pág. 534)

sistema excretor: sistema que recolecta y elimina los desperdicios del cuerpo y regula el nivel de fluidos en el cuerpo. (pág. 541)

F

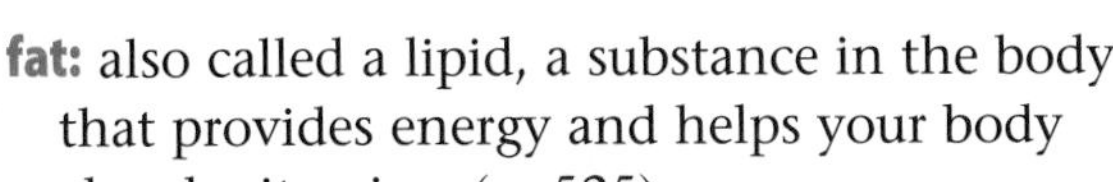

fat: also called a lipid, a substance in the body that provides energy and helps your body absorb vitamins. (p. 525)

fertilization: a reproductive process in which a sperm joins with an egg. (p. 679)

fetus: term used to describe a developing human from the ninth week of the pregnancy until birth. (p. 686)

grasa: también llamada lípido, sustancia en el cuerpo que proporciona energía y ayuda al cuerpo a absorber vitaminas. (pág. 525)

fertilización: proceso reproductivo en el cual un espermatozoide se une con un óvulo. (pág. 679)

feto: término usado para describir al ser humano en desarrollo desde la novena semana de embarazo hasta el nacimiento. (pág. 686)

H

hormone: a chemical signal that is produced by an endocrine gland in one part of an organism and is carried in the bloodstream to another part of the organism. (p. 657)

hormona: señal química producido por una glándula endocrina en una parte de un organismo y llevado en la corriente sanguínea a otra parte del organismo. (pág. 657)

I

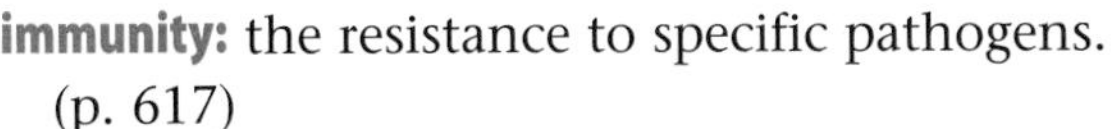

immunity: the resistance to specific pathogens. (p. 617)

infectious disease: a disease caused by a pathogen that can be transmitted from one person to another. (p. 605)

inflammation: a process that causes a bodily area to become red and swollen. (p. 615)

inmunidad: resistencia a patógeno específicos. (pág. 617)

enfermedad infecciosa: enfermedad causada por un patógeno que se puede transmitir de una persona a otra. (pág. 605)

inflamación: proceso que causa que un área del cuerpo se vuelva roja e hinchada. (pág. 615)

integumentary (ihn teg gyuh MEN tuh ree) system: the body system that includes all the external coverings of the body, including the skin, nails, and hair. (p. 505)

involuntary muscle: muscle you cannot consciously control. (p. 500)

sistema tegumentario: sistema corporal que comprende todas las coberturas externas del cuerpo, incluidos la piel, las uñas y el cabello. (pág. 505)

músculos involuntarios: músculo que se controla conscientemente. (pág. 500)

J

joint: where two or more bones meet. (p. 491)

articulación: donde dos o más huesos se unen. (pág. 491)

K

kidney: a bean-shaped organ that filters, or removes, wastes from blood. (p. 543)

riñón: órgano con forma de frijol que filtra, o extrae, los desechos de la sangre. (pág. 543)

L

larynx (LER ingks): a triangle-shaped area into which air passes from the pharynx; also called the voice box. (p. 560)

ligament (LIH guh munt): the tissue that connects bones to other bones. (p. 491)

lungs: the main organs of the respiratory system. (p. 561)

lymph node: a small, spongy structure that filters particles from lymph. (p. 586)

lymph: tissue fluid that has diffused into lymph vessels. (p. 586)

lymphatic system: part of the immune system that helps destroy microorganisms that enter the body. (p. 585)

laringe: área en forma de triángulo dentro de la cual pasa el aire proveniente de la faringe; también se le llama caja sonora de voz. (pág. 560)

ligamento: tejido que conecta los huesos con otros huesos. (pág. 491)

pulmones: órganos principales del sistema respiratorio. (pág. 561)

nódulo linfático: estructura pequeña y esponjosa que filtra partículas de la linfa. (pág. 586)

linfa: fluido de los tejidos que se esparce en los vasos linfáticos. (pág. 586)

sistema linfático: parte del sistema inmune que destruye los microorganismos que entran al cuerpo. (pág. 585)

M

mechanical digestion: a process in which food is physically broken into smaller pieces. (p. 532)

melanin (MEH luh nun): a pigment that protects the body by absorbing some of the Sun's damaging ultraviolet rays. (p. 507)

menstrual (MEN stroo ul) cycle: a process of reproductive-related changes involving the ovaries and uterus. (p. 678)

mineral: any of several inorganic nutrients that help the body regulate many chemical reactions. (p. 525)

digestión mecánica: proceso por el cual el alimento se descompone físicamente en pedazos más pequeños. (pág. 532)

melanina: pigmento que protege el cuerpo absorbiendo parte de los rayos ultravioleta dañinos del sol. (pág. 507)

ciclo menstrual: proceso de cambios relacionados con la reproducción que involucra los ovarios y el útero. (pág. 678)

mineral: cualquiera de los varios nutrientes inorgánicos que ayudan al cuerpo a regular muchas reacciones químicas. (pág. 525)

Science Skill Handbook
Math Skill Handbook
Reference Handbook
Glossary/Glosario
Index

muscle: strong body tissue that can contract in an orderly way. (p. 497)

músculo: tejido corporal fuerte que se contrae de manera sistemática. (pág. 497)

negative feedback: a control system in which the effect of a hormone inhibits further release of the hormone; sends a signal to stop a response. (p. 660)

nephron (NEH frahn): a network of capillaries and small tubes, or tubules, where filtration of blood occurs. (p. 543)

nervous system: the part of an organism that gathers, processes, and responds to information. (p. 637)

neuron (NOO rahn): the basic functioning unit of the nervous system; a nerve cell. (p. 639)

noninfectious disease: a disease that cannot pass from person to person. (p. 606)

retroalimentación negativa: sistema de control en el cual el efecto de una hormona inhibe más liberación de la hormona; envía una señal para detener una respuesta. (pág. 660)

nefrona: red de capilares y tubos pequeños, o túbulos, donde ocurre la filtración de la sangre. (pág. 543)

sistema nervioso: parte de un organismo que recoge, procesa y responde a la información. (pág. 637)

neurona: unidad básica de funcionamiento del sistema nervioso; célula nerviosa. (pág. 639)

enfermedad no infecciosa: enfermedad que no se puede pasar de una persona a otra. (pág. 606)

osteoporosis (ahs tee oh puh ROH sus): a bone disease that causes bones to weaken and become brittle. (p. 492)

ovary (OH vah ree): the female reproductive organ that produces egg cells; stores oocytes which mature into ova. (p. 676)

ovulation: a process occurring near the end of phase 2 of the menstrual cycle in which hormones cause an egg to be released from the ovary. (p. 678)

osteoporosis: enfermedad de los huesos que los debilita y los vuelve quebradizos. (pág. 492)

ovario: el órgano reproductivo femenino que produce óvulos; tiendas de ovocitos que maduran en los óvulos. (pág. 676)

ovulación: proceso que ocurre cerca de la finalización de la segunda fase del ciclo menstrual en el cual las hormonas causan la liberación de un óvulo del ovario. (pág. 678)

passive immunity: the introduction of antibodies that were produced outside the body. (p. 617)

pasteurization: a process of heating food or liquid to a temperature that kills most harmful bacteria. (p. 603)

pathogen: an agent that causes disease. (p. 601)

penis: a tubelike structure that delivers sperm to the female reproductive system. (p. 674)

periosteum (per ee AHS tee um): a membrane that surrounds bone. (p. 490)

inmunidad pasiva: introducción de anticuerpos producidos fuera del cuerpo. (pág. 617)

pasteurización: proceso en el cual se calientan los alimentos o líquidos para matar la mayoría de bacterias dañinas. (pág. 603)

patógeno: agente que causa enfermedad. (pág. 601)

pene: estructura en forma de tubo que deposita esperma en el sistema reproductor femenino. (pág. 674)

periostio: membrana que recubre los huesos. (pág. 490)

peripheral nervous system (PNS): system made of sensory and motor neurons that transmit information between the central nervous system (CNS) and the rest of the body. (p. 641)

peristalsis (per uh STAHL sus): waves of muscle contractions that move food through the digestive tract. (p. 534)

pharynx (FER ingks): a tubelike passageway at the top of the throat that receives air, food, and liquids from the mouth or nose. (p. 560)

placenta (pluh SEN tuh): an organ formed by the outer cells of the zygote and cells from the uterus. (p. 685)

plasma: the yellowish, liquid part of blood that transports blood cells. (p. 579)

platelet: a small, irregularly shaped piece of a cell that plugs wounds to stop bleeding. (p. 579)

positive feedback: a control system in which the effect of a hormone causes more of the hormone to be released; sends a signal to increase a response. (p. 660)

pregnancy: the period of human development from fertilized egg to birth. (p. 685)

protein: a long chain of amino acid molecules; contains carbon, hydrogen, oxygen, nitrogen, and sometimes sulfur. (p. 524)

puberty (PYEW bur tee): the process by which the reproductive system matures during adolescence. (p. 690)

pulmonary circulation: the network of vessels that carries blood to and from the lungs. (p. 571)

sistema nervioso periférico (SNP): sistema formado por neuronas sensoriales y motoras que transmiten información entre el sistema nervioso central (SNC) y el resto del cuerpo. (pág. 641)

peristalsis: ondas de contracciones musculares que mueven el alimento por el tracto digestivo. (pág. 534)

faringe: pasadizo parecido a un tubo en la parte superior de la garganta que recibe el aire, el alimento y los líquidos provenientes de la boca o de la nariz. (pág. 560)

placenta: órgano formado por las células externas del cigoto y células del útero. (pág. 685)

plasma: parte líquida y amarillenta de la sangre que transporta las células sanguíneas. (pág. 579)

plaqueta: fragmento de una célula, pequeño y de forma irregular, que tapona las heridas para detener el sangrado. (pág. 579)

retroalimentación positiva: sistema de control en el cual el efecto de una hormona causa más liberación de la hormona; envía una señal para aumentar la respuesta. (pág. 660)

embarazo: período del desarrollo del ser humano desde que se fertiliza el óvulo hasta el nacimiento. (pág. 685)

proteína: larga cadena de aminoácidos; contiene carbono, hidrógeno, oxígeno, nitrógeno y, algunas veces, sulfuro. (pág. 524)

pubertad: proceso por el cual el sistema reproductor madura durante la adolescencia. (pág. 690)

circulación pulmonar: red de vasos que lleva sangre hacia y desde los pulmones. (pág. 571)

R

receptor: special structures in all parts of the sensory system that detect stimuli. (p. 647)

reflex: an automatic movement in response to a stimulus. (p. 642)

retina (RET nuh): an area at the back of the eye that includes special light-sensitive cells—rod cells and cone cells. (p. 648)

receptor: estructuras especiales en todas partes del sistema sensorial que detectan los estímulos. (pág. 647)

reflejo: movimiento automático en respuesta a un estímulo. (pág. 642)

retina: área en la parte posterior del ojo que incluye especiales sensibles a la luz—bastones y conos. (pág. 648)

Rh factor: a protein found on red blood cells; a chemical marker. (p. 581)

factor Rh: proteína que se encuentra en los glóbulos rojos; es un marcador químico. (pág. 581)

S

semen (SEE mun): a mixture of sperm and fluids produced by several glands. (p. 674)

sensory system: the part of your nervous system that detects or senses the environment. (p. 647)

skeletal muscle: a type of muscle that attaches to bones. (p. 499)

skeletal system: body system that contains bones as well as other structures that connect and protect the bones and that support other functions in the body. (p. 487)

smooth muscle: involuntary muscle named for its smooth appearance. (p. 500)

sperm: a male reproductive, or sex, cell; forms in a testis. (p. 673)

spinal cord: a tubelike structure of neurons that sends signals to and from the brain. (p. 641)

spleen: an organ of the lymphatic system that recycles worn-out red blood cells and produces and stores lymphocytes. (p. 587)

stimulus (STIHM yuh lus): a change in an organism's environment that causes a response. (p. 638)

synapse (SIH naps): the gap between two neurons. (p. 639)

systemic circulation: the network of vessels that carry blood from the heart to the body and from the body back to the heart. (p. 571)

semen: mezcla de esperma y fluidos producidos por varias glándulas. (pág. 674)

sistema sensorial: parte del sistema nervioso que detecta o siente el medioambiente. (pág. 647)

músculo esquelético: tipo de músculo que se adhiere a los huesos. (pág. 499)

sistema esquelético: sistema corporal que comprende los huesos al igual que otras estructuras que conectan y protegen los huesos y que apoyan otras funciones en el cuerpo. (pág. 487)

músculo liso: músculo involuntario llamado así por su apariencia lisa. (pág. 500)

esperma: célula reproductora masculina o sexual; forma en un testículo. (pág. 673)

médula espinal: estructura de neuronas en forma de tubo que envía señales hacia y del cerebro. (pág. 641)

bazo: órgano del sistema linfático que recicla los glóbulos rojos muertos y produce y almacena linfocitos. (pág. 587)

estímulo: cualquier cambio en el medioambiente de un organismo que causa una respuesta. (pág. 638)

sinapsis: espacio en medio de dos neuronas. (pág. 639)

circulación sistémica: red de vasos que llevan sangre del corazón al cuerpo y de regreso del cuerpo al corazón. (pág. 571)

T

T cell: a type of white blood cell that forms in the bone marrow and matures in the thymus gland; produces a protein antibody that becomes part of the cell membrane. (p. 616)

testis (TES tihs): the male reproductive organ that produces sperm. (p. 674)

thymus: the organ of the lymphatic system in which T cells complete their development. (p. 587)

célula T: tipo de glóbulo blanco que se forma en la médula ósea y madura en la glándula del timo; produce un anticuerpo de proteína que se vuelve parte de la membrana celular. (pág. 616)

testículos: órgano reproductivo masculino que produce espermatozoides. (pág. 674)

timo: órgano del sistema linfático en el cual las células T completan su desarrollo. (pág. 587)

trachea (TRAY kee uh): a tube that is held open by C-shaped rings of cartilage; connects the larynx and the bronchi. (p. 560)

tráquea: tubo que los anillos en forma de C del cartílago mantienen abierto; este conecta la laringe y los bronquios. (pág. 560)

U

umbilical (um BIH lih kul) cord: a rope-like structure formed by the outer zygote cells that attaches the developing offspring to the placenta. (p. 685)

cordón umbilical: estructura parecida a una cuerda formada por las células externas del cigoto que unen el hijo a la placenta. (pág. 685)

ureter (YOO ruh tur): a tube through which urine leaves each kidney. (p. 545)

uréter: tubo por el cual la orina sale de cada riñón. (pág. 545)

urethra (yoo REE thruh): a tube through which urine leaves the bladder. (p. 545)

uretra: tubo por el cual la orina sale de la vejiga. (pág. 545)

urine: the fluid produced when blood is filtered by the kidneys. (p. 543)

orina: fluido que se produce cuando los riñones filtran la sangre. (pág. 543)

V

vaccination: weakened or dead pathogens placed in the body, usually by injection or by mouth. (p. 617)

vacunación: patógenos debilitados o muertos introducidos en el cuerpo, generalmente por medio de una inyección o por la boca. (pág. 617)

vagina: the part of the female reproductive system that connects the uterus to the outside of the body. (p. 676)

vagina: parte del sistema reproductor femenino que une el útero con el exterior del cuerpo. (p. 676)

vector: a disease-carrying organism that does not develop the disease. (p. 605)

vector: organismo portador de una enfermedad pero que no la desarrolla. (pág. 605)

vein: a vessel that carries blood toward the heart. (p. 570)

vena: vaso que lleva sangre hacia el corazón. (pág. 570)

ventricles (VEN trih kulz): the lower two chambers of the heart. (p. 568)

ventrículos: las dos cámaras inferiores del corazón. (pág. 568)

villus (VIH luhs): a fingerlike projection, many of which cover the folds of the small intestine. (p. 536)

vellosidad: proyección parecida a un dedo, muchas de las cuales cubren los pliegues del intestino delgado. (pág. 536)

vitamin: any of several nutrients that are needed in small amounts for growth, regulating body functions, and preventing some diseases. (p. 525)

vitamina: cualquiera de los varios nutrientes que se necesitan en cantidades pequeñas para el crecimiento, para regular las funciones del cuerpo y para prevenir algunas enfermedades. (pág. 525)

voluntary muscle: muscle that you can consciously control. (p. 499)

musculares voluntarios: músculo que controlas conscientemente. (pág. 499)

Z

zygote (ZI goht): the new cell that forms when a sperm cell fertilizes an egg cell (p. 684)

zigoto: célula nueva que se forma cuando un espermatozoide fertiliza un óvulo. (pág. 684)

Index

C

E

F

G

H

I

K

S

T

U

Y

Z

Credits

Photo Credits

Front Cover Spine Photodisc/Getty Images; **Back Cover** Thinkstock/Getty Images; **Inside front,back cove**r Thinkstock/Getty Images; **Connect Ed** (t)Richard Hutchings, (c)Getty Images, (b)Jupiter Images/Thinkstock/Alamy; **i** Thinkstock/Getty Images; viii–ix The McGraw-Hill Companies; **ix** (b)Fancy Photography/Veer; **482** (l)Nick Laham/Getty Images, (r)Nick Laham/Getty Images; AFP/Getty Images; **484–485** Charlie Schuck/Getty Images; **486** Gustoimages/Photo Researchers; **487** (t)Hutchings Photography/Digital Light Source, (b)CMCD/Getty Images; **488** Hutchings Photography/Digital Light Source; **492** Mark Steinmetz; **493** (l)Dennis MacDonald/PhotoEdit, (c) Daniel Dempster Photography/Alamy, (r)Bob Daemmrich/PhotoEdit; **494** (b)CMCD/Getty Images; **495** (bkgd)Tomi/PhotoLink/Getty Images; **496** Phil Loftus/Capital Pictures; **497** Hutchings Photography/Digital Light Source; **498** CMCD/Getty Images; **499** Innerspace Imaging/Photo Researchers; **500** (t)Dr, Gladden Willis/Visuals Unlimited/Getty Images, (b)Dr, Gladden Willis/Visuals Unlimited/Getty Images; **503** Hutchings Photography/Digital Light Source; **504** Steve Gschmeissner/Photo Researchers; **505** (t)Hutchings Photography/Digital Light Source, (b)Comstock/Jupiterimages; **506** (t) Thomas Barwick/Getty Images, (b)Hutchings Photography/Digital Light Source; **508** (l)Stephen J. Krasemann/Photo Researchers, (c)Scientifica/Visuals Unlimited, (r)Ian Leonard/Alamy; **510** Stockbyte/Getty Images; **511** (tl)Comstock/Jupiterimages, (r)Scientifica/Visuals Unlimited, (bl)Steve Gschmeissner/Photo Researchers; **512** Hutchings Photography/Digital Light Source; **513** Hutchings Photography/Digital Light Source; **517** Charlie Schuck/Getty Images; **520–521** MedicalRF.com/Getty Images; **522** Peter Menzel; **523** Hutchings Photography/Digital Light Source; **524** (t)Dorling Kindersley/Getty Images, (b)Hutchings Photography/Digital Light Source; **525** (t)The McGraw-Hill Companies, (b)Michael Rosenfeld/Getty Images; **526** Hutchings Photography/Digital Light Source; **527** (l)David Young-Wolff/PhotoEdit, (r)Mark Steinmetz; **528** (t)Peter Menzel, (tc)Dorling Kindersley/Getty Images, (b)Mark Steinmetz, (bc)David Young-Wolff / PhotoEdit; **529** (tl.tr)Macmillan/McGraw-Hill, (others)Hutchings Photography/Digital Light Source; **530** Steve Gschmeissner/Photo Researchers; **531,532,534** Hutchings Photography/Digital Light Source; **537** (l)Scimat/Photo Researchers, (r)CNRI/Photo Researchers; **538** Hutchings Photography/Digital Light Source; **539** (bkgd)Dr. Dennis Kunkel/Visuals Unlimited/Getty Images; **540** Wolf Fahrenbach/Visuals Unlimited/Getty Images; **541** Hutchings Photography/Digital Light Source; **543** Photodisc/Getty Images; **545** Hutchings Photography/Digital Light Source; **547** Wolf Fahrenbach/Visuals Unlimited/Getty Images; **548** (tl,cr,br) Hutchings Photography/Digital Light Source, (others)Macmillan/McGraw-Hill; **550** (t,b)Hutchings Photography/Digital Light Source, (c)Photodisc/Getty Images; **552** Hutchings Photography/Digital Light Source; **553** MedicalRF.com/Getty Images; **556–557** Bo Tornvig/Getty Images; **558** Juergen Berger/Photo Researchers; **559** Hutchings Photography/Digital Light Source; **560** Holger Winkler/zefa/CORBIS; **563** James Cavallini/Photo Researchers; **564** Holger Winkler/zefa/CORBIS; **565** (t to b,2,3,5)Macmillan/McGraw-Hill, (c,r,4)Hutchings Photography/Digital Light Source; **566** Ron Chapple/Getty Images; **567** Hutchings Photography/Digital Light Source; **568** Image Source/Jupiterimages; **571** Image Source/Jupiterimages; **572** Hutchings Photography/Digital Light Source; **575** (all)Hutchings Photography/Digital Light Source; **576** Nati Harnik/AP Images; **577,579** Hutchings Photography/Digital Light Source; **583** (t)Andrew J. Martinez/Photo Researchers, (b)Steve Hamblin/Alamy, (bkgd)Jeffrey L. Rotman/CORBIS; **584** Dr P. Marazzi/Photo Researchers, Inc.; **585** Hutchings Photography/Digital Light Source; **586** C Squared Studios/Getty Images; **589** (l,r)C Squared Studios/Getty Images; **590** (tl,r)Hutchings Photography/Digital Light Source, (cl)Macmillan/McGraw-Hill, (bl)Science Source/Photo Researchers; **591** (all)Science Source/Photo Researchers; **592** (t)Holger Winkler/zefa/CORBIS, (c)Image Source/Jupiterimages, (b)C Squared Studios/Getty Images; **595** Bo Tornvig/Getty Images; **598–599** Steve Gschmeissner/Photo Researchers; **600** (inset)Scott Camazine/Photo Researchers, (bkgd)Science Source/Photo Researchers; **601** (t)Hutchings Photography/Digital Light Source, (b)Eurelios/Photo Researchers; **602** (l) Wellcome Library, London, (r)Michael Abbey/Photo Researchers; **603** The McGraw-Hill Companies; **604** Dr. Dennis Kunkel/Visuals Unlimited/Getty Images; **605** Hutchings Photography/Digital Light Source; **607** (tl)Tomi/PhotoLink/Getty Images, (tr)Zephyr/Photo Researchers, (bl,br)St Bartholomew's Hospital/Photo Researchers; **608** (t)Michael Abbey/Photo Researchers, (b)St Bartholomew's Hospital/Photo Researchers; **609** (tcl) Macmillan/McGraw-Hill, (others)Hutchings Photography/Digital Light Source; **610** Science Pictures Ltd/Photo Researchers; **611** (tr)Hutchings Photography/Digital Light Source; **612** Hutchings Photography/Digital Light Source; **613** (tl)SPL Photo Researchers, (tr)Dr. Kessel & Dr. Kardon/Tissues & Organs/Visuals Unlimited/Getty Image, (c)RubberBall Productions, (cr)Eye of Science/Photo Researchers, (bl)Eye of Science/Photo Researchers, (br)Purestock/Getty Images; **614,618** PHOTOTAKE Inc./Alamy; **619** (t) Image Source/Getty Images, (b)George Musil/Visuals Unlimited/Getty Images, (bkgd)Stockbyte/SuperStock; **620** OJO Images/SuperStock; **621** (tr) Hutchings Photography/Digital Light Source; **622** (t)C Squared Studios/Getty Images, (bl)C Squared Studios/Getty Images, (br)D. Hurst/Alamy; **623** (t)George Diebold/Getty Images, (b)Hutchings Photography/Digital Light Source; **624** JUPITERIMAGES/Thinkstock/Alamy; **625** (tl)George Diebold/Getty Images, (cl)Creatas/PunchStock, (r)D. Hurst/Alamy, (bl)C Squared Studios/Getty Images; **626** (tr,br)Hutchings Photography/Digital Light Source, (t to b)Oleg Shpak/Alamy, (2–5,9)Hutchings Photography/Digital Light Source, (6)Macmillan/McGraw-Hill; **628** (t)Science Source/Photo Researchers, (b)OJO Images/SuperStock; **631** Steve Gschmeissner/Photo Researchers, Inc.; **634–635** Aflo Foto Agency/Photolibrary; **636** Dr. John Mazziotta et al./Photo Researchers; **637** Hutchings Photography/Digital Light Source; **638** David Stoecklein/CORBIS; **640** Doug Pensinger/Getty Images; **643** Hutchings Photography/Digital Light Source; **644** Doug Pensinger/Getty Images; **645** Hutchings Photography/Digital Light Source; **646** Steven Puetzer/Getty Images; **647** Hutchings Photography/Digital Light Source; **651** (l,r)CORBIS/Alamy; **653** Hutchings Photography/Digital Light Source; **654** Steven Puetzer/Getty Images; **655** (l)David J. Green-technology/Alamy, (r)Digital Vision/PunchStock; **656** ISM/Phototake; **657** Hutchings Photography/Digital Light Source; **659** Mark Andersen/Getty Images; **662** Hutchings Photography/Digital Light Source; **663** Hutchings Photography/Digital Light Source; **664** (t)Doug Pensinger/Getty Images, (c) Corbis Premium RF/Alamy, (b)ISM/Phototake; **667** Aflo Foto Agency/Photolibrary; **670–671** Yorgos Nikas/Getty Images; **672** P.M. Motta & J. Van Blerkom/Photo Researchers; **673** (tr)Hutchings Photography/Digital Light Source; **676** Eye of Science/Photo Researchers; **677** Yorgos Nikas/Getty Images; **679** (r)Hutchings Photography/Digital Light Source; **680** Eye of Science/Photo Researchers; **681** (t)Y. Nikas/Biol Reprod/Getty Images, (c) Keystone/Getty Images, (b)Mauro Fermariello/Photo Researchers; **682** Steve Allen/Getty Images; **683** (t)Vincenzo Lombardo/Getty Images, (b) Yellow Dog Productions/Getty Images; **685** (t)Visual&Written SL/Alamy, (b) Tony Freeman/PhotoEdit; **686** (tl)Dr G. Moscoso/Photo Researchers, (tc,tr) Anatomical Travelogue/Photo Researchers, (bl)Biophoto Associates/Photo Researchers, (br)Neil Bromhall/Photolibrary; **690** (t)Digital Vision, (b) Hutchings Photography/Digital Light Source; **691** (t,b)Biophoto Associates/Photo Researchers, (c)Anatomical Travelogue/Photo Researchers; **692** (t,b,bc)The McGraw-Hill Companies, (tc)Hutchings Photography/Digital Light Source; **693** (all)Marjorie L. Foreman; **694** (t)Eye of Science/Photo Researchers, (b)Yorgos Nikas/Getty Images; **697** Yorgos Nikas/Getty Images; **SR-00–SR-01**(bkgd)Gallo Images-Neil Overy/Getty Images; **SR-02** Hutchings Photography/Digital Light Source; **SR-06** Michell D. Bridwell/PhotoEdit; **SR-07** (t)The McGraw-Hill Companies, (b)Dominic Oldershaw; **SR-08** StudiOhio; **SR-09** Timothy Fuller; **SR-10** Aaron Haupt; **SR-42** (c)

Credits

NIBSC/Photo Researchers, Inc., (r)Science VU/Drs. D.T. John & T.B. Cole/Visuals Unlimited, Inc.; Stephen Durr; **SR-43** (t)Mark Steinmetz, (r)Andrew Syred/Science Photo Library/Photo Researchers, (br)Rich Brommer; **SR-44** (l)Lynn Keddie/Photolibrary, (tr)G.R. Roberts; David Fleetham/Visuals Unlimited/Getty Images; **SR-45** Gallo Images/CORBIS; **SR-46** Matt Meadows.